Plug-and-play control of interconnected systems

Dissertation zur Erlangung des Grades eines
Doktor-Ingenieurs
der Fakultät für Elektrotechnik und Informationstechnik
an der Ruhr-Universität Bochum

Sven Bodenburg
geboren in Braunschweig

Bochum, 2017

1. Gutachter: Prof. Dr.-Ing. Jan Lunze

2. Gutachter: Prof. Dr. Giancarlo Ferrari Trecate

Eingereicht am: 15. Mai 2017

Tag der mündlichen Prüfung: 01. Dezember 2017

Bibliografische Information der Deutschen Nationalbibliothek

Die Deutsche Nationalbibliothek verzeichnet diese Publikation in der Deutschen Nationalbibliografie; detaillierte bibliografische Daten sind im Internet über http://dnb.d-nb.de abrufbar.

ISBN 978-3-8325-4602-1

Logos Verlag Berlin GmbH
Comeniushof, Gubener Str. 47,
10243 Berlin
Tel.: +49 (0)30 42 85 10 90
Fax: +49 (0)30 42 85 10 92
INTERNET: http://www.logos-verlag.de

Acknowledgements

This thesis is the result of five years of research at the Institute of Automation and Computer Control of Prof. Jan Lunze at Ruhr-Universität Bochum, Germany.

I would like to thank Prof. Jan Lunze for the possibility to be his PhD-student even before I completed my master degree. My deepest gratitude for his confidence and support in both research and teaching all over the time. I would also like to thank Prof. Giancarlo Ferrari Trecate to review this thesis. His pioneering in the field of plug-and-play control is highly inspiring and has taken significant influence on my research.

Special thanks go to my fellow PhD-students Ozan Demir, Fabian Just, Andrey Mosebach, Yannick Nke, Tobias Noeßelt, Sebastian Pröll, Kai Schenk, Melanie Schuh, René Schuh, Alexander Schwab, Michael Schwung, Christian Stöcker, Daniel Vey, Philipp Welz, Christian Wölfel and Markus Zgorzelski. Thank you for valuable discussions, proofreading my thesis and the support in finding the final answer what plug-and-play control actually is. It has always been an enjoyable time with you at the institute. Beside my colleges, I would like to thank my student assistants Phil Geisler, David Gomez Cardeno, Alexander Knaub, Viktor Kraus, Simon Niemann, Fabian Peters, Simon Röchner, Christopher Schlichting, Alexander Schwab, Erman Seydioglu and Florian Werner. Thank you for the support, the discussions and the feedback. I have learned a lot. Thanks go also to the backbone of the institute Kerstin Funke, Susanne Malow, Andrea Marschall, Rolf Pura and Udo Wieser for the invaluable technical and administrative support.

I would like to express my gratitude to my family, who has always been interested in my work and supports me through my life.

Finally, I would to thank the most important persons in my life, my great love Julia and our son Bela. Thank you for giving me courage and strength and for showing me the really important things in life.

Dortmund, December 2017 Sven Bodenburg

Contents

Abstract

In the current status of the networked control of interconnected systems, the digital communication network is primarily used for the exchange of measured values amongst the control stations. Plug-and-play control extends the usage of this communication network towards the communication of models and algorithms with the aim to automatically design control stations at runtime. Therefore, plug-and-play control provides for every subsystem a design agent that initially knows only the model of its subsystem. To accomplish the design of a control station by a design agent first, a *suitable model* of the subsystem under the influence of other subsystems through the physical couplings has to be set up. Second, *local design conditions* have to be found that guarantee the adherence of the global control goal. Without a global coordination, the design agents itself have to request relevant information over the communication network from each other. If the designed control station is finally "plugged" into the control equipment, the overall closed-loop system will "play" as desired.

The aim of this thesis is to provide methods so as to enable the design agent to automatically accomplish the controller design at runtime. Therefore, the present work proposes three approaches which focus on the accuracy of the model that is used for the controller design with respect to the achievable overall closed-loop performance. An *exact model* enables the design without considerable limitations on the achievable closed-loop performance, while the design with a *local model* can only be enabled by strict design conditions. An *approximate model* reduces the amount of required models while preserving an adequate overall closed-loop performance.

If an exact model is needed for the design, the models from those subsystems that are strongly connected to the considered subsystems are required. This thesis presents a distributed search algorithm that assuredly finds all these subsystems, where its execution is distributed among the design agents. Whenever only the local model is accessible to the design agent, the design has to be performable without a model of the physical interaction. Two approaches are proposed to still enable the controller design. The first approach is based on a limited amplification of the coupling signals by each controlled subsystem. The second approach aims at restricting the operating region of each controlled subsystem. As a result, locally verifiable design conditions are derived that guarantee stability and a certain performance of the overall closed-loop system. To find a suitable approximate model, a decision rule is proposed so that the design agent is able to decide which interconnected subsystem models are relevant for the approximation and which can be considered as conscious model uncertainty. It is shown that a suitable approximation solely comprises neighbouring subsystem models and, moreover, that the conservatism of the local design conditions from the previous approach can significantly be reduced if an approximate model is used.

The main result of this thesis is a novel concept for the self-organised design of control stations by means of design agents. Plug-and-play control can be applied to various application scenarios such as fault-tolerant control or the integration of new subsystems as successfully demonstrated in this thesis. The proposed methods are tested and evaluated through simulations and experiments on an interconnected thermofluid process and on a multizone furnace.

1 Introduction to plug-and-play control

1.1 Plug-and-play control

The increasing digital networking of technical processes enables control stations to exchange information among each other with the aim to fulfil an overall control aim collectively. Figure 1.1 shows the typical structure of networked control systems, where the interconnected system consists of N physically coupled subsystems S_i and the network controller is realised by N local control stations C_i that are interconnected through the communication network. In the current status, the digital communication network is primarily used for the exchange of measured values (solid arrows in Fig. 1.1) among the control stations so as to guarantee a desired performance of the overall closed-loop system.

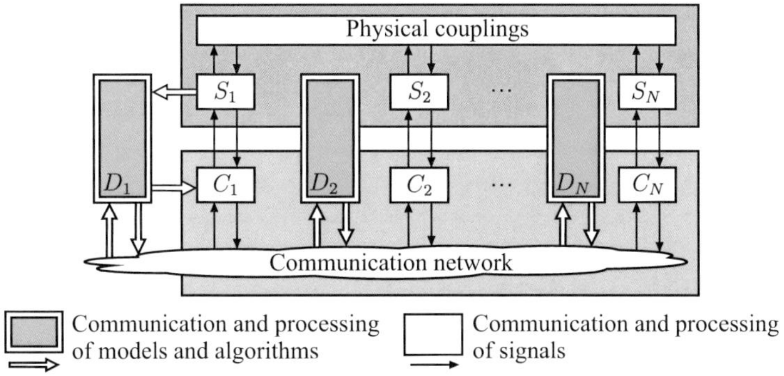

Figure 1.1: Structure of plug-and-play control of interconnected systems.

Plug-and-play control extends the usage of this communication network towards the exchange of models and algorithms (double arrows in Fig. 1.1). This information is required whenever the control station C_i has to be adapted to a new situation such as the occurrence of a fault in one of the subsystems or the integration of new sensors, actuators or whole subsystems. The meaning of plug-and-play control is the following:

> Plug-and-play control is a concept to automatically design and implement control station C_i at runtime.

In order to separate the execution phase of a control algorithm from its design phase, *design agents* D_i are introduced (highlighted by double framed blocks in Fig. 1.1), which are responsible for the design of control station C_i and its implement into the control equipment of subsystem S_i. To enable the design agent D_i to accomplish the controller design, two essential aspects has to be considered beforehand which are

1. the set-up of a model that describes the dynamics of the subsystem S_i under the influence of the physical interaction to other subsystems and

2. the formulation of local design conditions which ensure global closed-loop stability and guarantee a suitable input-output behaviour of the overall closed-loop system.

This design process is characterised by the fact that no global coordinator exists that provides the required information for the design agent so that the design agents itself have to request relevant information over the communication network from each other. Accordingly, for each subsystem S_i, plug-and-play control provides a design agent D_i that has available the model of its subsystem. To accomplish the design of control station C_i, design agent D_i has to decide which information is relevant and has to request this information form other design agents. Eventually, the design agent "plugs" the control algorithm into the control hardware so that the overall closed-loop system assuredly "plays" as desired.

The aim of this thesis is to provide methods to assign relevant model information, to formulate local design conditions and to design the control algorithm in order to enable design agent D_i to automatically design control station C_i.

1.2 Motivation and application scenarios

1.2.1 Motivation example: Power networks

This example illustrates why a global coordinator cannot be used to design the control algorithms of an interconnected system and emphasises the indispensability of local design agents.

Consider the power transmission grid in Germany as illustrated in Fig. 1.2. The power network consists of thousands components ranging from, of course, the producers and the consumers as well as transmission lines and converter stations which are spread over a vast geographical area. These components are interconnected through the transmission grids and, thus, represent a physically interconnected system. The control objectives of such a power network are amongst others the balancing of the generated and demanded power while sharing the electrical load, the synchronisation of the voltage frequencies and the reduction of the voltage magnitudes [121, 150]. A power plant is usually controlled by a control station that accesses the local measurements to act on the local actuators. A digital communication network additionally enables the control station to share their local measurements. The complexity of such a power network as well as the network access of the control stations brings challenges in the controller design.

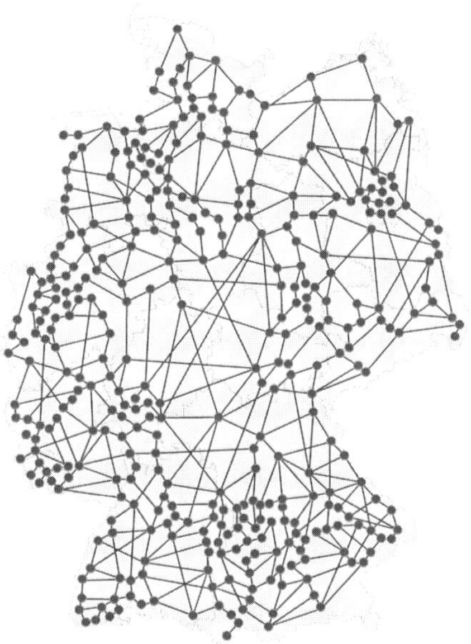

Figure 1.2: Scheme of Germany's power transmission grid: Nodes symbolise power generating stations and vertices illustrate the interconnection.

Now, particular attention on the processing of the controller design is paid. Especially for power networks, the design cannot be coordinated by a central entity for several reasons. First, the daily varying load forces the power network to change the operating points continuously. Such a set-point change leads to changes of the dynamics of the power network. Moreover, nowadays the power is no longer produced from the conventional bulk power systems but more and more from small-scale producers such as microgrids, like solar systems, wind stations or pumped-storage power plants, which can operate in grid-connected-mode (as a part of the power network) or in island-mode (separated from the power network) [71]. This starting-up and switching-off of producers changes the dynamics of the power network. Accordingly, the permanent *availability of an up-to-date model* of the overall power network at one place, if it is even possible, requires a enormous persistent communication network load to gather all the information and a high computational effort to set up the overall power network model. In addition, the accomplishment of the design itself based on such a complex model necessitates *high computational power*. Second, many companies offer their power based on the price and public demand. These companies are *not willing to share* their strategies and, of course, prohibit the central storage of their data in order to maintain their tactical advantages to other companies. Third, the connection to a multi-purpose network bears the risk of *cyber attacks* [59, 172]. The central storage of all information enables attackers to manipulate or steal all information at once. Moreover, a central entity represents a *single-point-of-failure* in the sense that a brake down of the central coordinator cause in the loss of all information and services.

Due to these physical and human made constraints, it is preferable to tore the models locally and to carry out the design locally. This example motivates the concept of local design agents, being introduced in this thesis. On demand, the design agents interact through the communication network to set up the required model with necessary (or permissible) accuracy to accomplish the controller design.

1.2.2 Application scenarios used in this thesis

Plug-and-play control is a concept that can be applied in any situation in which a control algorithm shall be designed or redesigned. This section describes the application of plug-and-play control to fault-tolerant control and the integration of a new subsystem. Throughout this thesis it is repeatedly referred to these application scenarios.

Fault-tolerant control. Dynamical systems are vulnerable to faults in actuators, sensors or the process itself. The effect of a fault may lead to misbehaviours of the controlled system which can end up into damages of the system and its environment. Fault-tolerant control aims at inhibiting the influence of a fault and maintaining the nominal control specifications [52, 53]. To make a system tolerant to faults, two steps are required: First, fault diagnosis to detect and identify the fault and, second, controller redesign to adapt the controller to the faulty situation so that the nominal specifications remain satisfied.

Both steps, the design of a diagnostic unit as well as the redesign of the control station are of particular interest in *plug-and-play diagnosis* (Chapter 8) and *plug-and-play reconfiguration* (Chapters 4, 5 and 7), respectively. In this thesis, the main focus is on the necessary accuracy of the model that is used for the (re)design in order to guarantee the adherence of overall closed-loop objectives. For fault diagnoses, the objectives are guaranteed fault detection and guaranteed fault identification, whereas for the controller reconfiguration it is focussed on the achievable performance of the reconfigured closed-loop system. With the model and the design objectives at hand, existing design methods or reconfiguration methods can be applied, like pole-placement or H_∞-design.

Integration of a new subsystem. Physically interconnected systems are of persistent changes in terms of subsystems that are added or removed. For instance in power networks single power plants or microgrids are started-up or switched-off depending on the public demand [71, 145]. A flexible supermarket refrigerator system or a multizone furnace is considered in [5, 130], where whole segments of the system are start-up if they are needed. Moreover, in batch processes the interconnection between tanks as well as the number of tanks change with the product to be produced as focused on in [2]. Furthermore, single vehicles of a platoon consistently leave or enter the platoon [133].

In general, the connection of new subsystem to the an existing interconnected system entails new physical links to the existing subsystems and vice versa. The new subsystem is successfully integrated when the overall closed-loop system still satisfies its global control task. The integration can be processed in two steps. First, from the view of the new subsystem, a controller initially has to be designed taking the physical couplings from other subsystems into account. Second, from the view of existing subsystems the influence from the physical interactions changes. These changes necessitate a redesign of the existing control stations.

Both phases are considered in *plug-and-play integration* (Chapter 6), where it is not primary focused on the specific (re)design method but on the model that is required to accomplish the design so as to guarantee a certain overall closed-loop performance. Although the behaviour of the subsystem is influenced by all other subsystems through the physical couplings, it is often reasonable to ignore some of the interacting subsystem dynamics for the controller design. Thus, the main issue is to decide which dynamics can be ignored and which dynamics are relevant with respect to the achievable performance of the overall closed-loop system.

1.2.3 Further application scenarios

In the previous section, the scenarios of fault-tolerant control and the integration of a new subsystem which represent the application scenarios considered in this thesis are explained. The reasons to initiate a design are much more versatile. This section briefly outlines some of these reasons.

Ordinary controller design. The *ordinary controller design* is the most obvious reason to initiate a design and is, therefore, research focus over decades [40, 80, 136] and is still recent [38, 69, 75, 104]. The design methods range from the independent design of the control stations over the design in a predestined order to the simultaneous design of all control stations. It is not the aim of plug-and-play control to compete with these methods from literature. Nevertheless, the flexibility to exchange models and algorithms among the design agents introduces new aspects in the controller design phase such as to forgo the adherence of a rigid information structure.

Changing control objectives. An online redesign of a local control algorithm can also be initiated if the control objectives change at runtime. Reconsider the platoon of vehicles. To reduce the communication effort, only a velocity control is active if the distance is within an acceptable bound. If a disturbance affects the platoon an additional distance control is activated to attenuate the disturbance and, thus, avoid collisions [66, 158]. A similar situation constitutes the merging of two car lanes [133]. Another example focusses on the cooperative reconfiguration of the overall closed-loop specifications after fault occurrence. If a fault affects a subsystem, it is often not desirable or even possible to retrieve the global control aim by reconfiguring the control station of the faulty subsystem only. In this case the overall control task is re-distributed among the faulty and the non-faulty subsystems so as to enable the overall system to satisfy its global goal in spite of the fault [32, 124, 152]. The re-distribution is enabled by assigning local tasks which, if collectively fulfilled, guarantee the satisfaction of the global task. The decomposition of global specifications into local control goals is one of the primal concerns of plug-and-play control.

1.3 Problem formulation and fundamental questions

The main focus of plug-and-play control is to organise the design of a control station C_i in the sense that design agent D_i has to know a model that represents the dynamics of its subsystem S_i under consideration of the physical interaction with other subsystems and has to have available local design requirements that regard the global control aim. This issue is formulated as plug-and-play control problem (Problem 1.2). Thereafter, this section points out approaches to the plug-and-play control problem followed by leading questions of this thesis.

1.3.1 The plug-and-play control problem

Let the preferred control solution be completely decentralised. With the focus on the application scenarios, the situation is considered in which control station C_i has to be designed and the control stations C_j, $(j = 1, ..., N,\ j \neq i)$ already exist. This design is accomplished by design agent D_i, which, therefore, has to solve the well-known design problem:

Problem 1.1 (Controller design problem)

Given:	• *Model of the system*
	• *Design conditions*
Find:	*Control station* C_i

From the local view of subsystem S_1, the system model that is required for the design of C_1 has to represent the behaviour of S_1 interacting with all other controlled subsystems through the physical couplings as highlighted by the yellow box in Fig. 1.3b. Due to this physical interaction, the design of control station C_1 has effect on the dynamics of all other controlled subsystems and, thus, on the behaviour of the overall closed-loop system. Hence, with the model that represents this physical interaction at hand of D_1, the overall closed-loop behaviour can be considered within the design of control station C_1. Accordingly, if the model of the plant and design conditions for the global control aim are available to design agent D_1, it can solve Problem 1.1 by well-established design methods from literature.

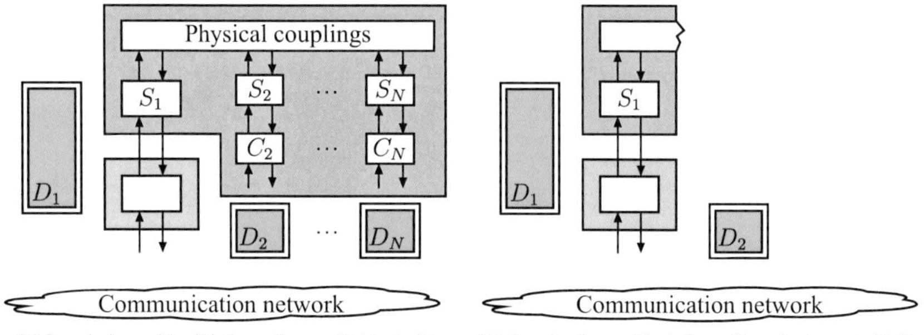

(a) Local view with all information available to D_1

(b) Local view with information that are initially available to D_1

Figure 1.3: Local view of the global system from subsystem S_1. The control stations C_i, $i = 2, ..., N$ exist and the control station C_1 has to be designed by D_1.

Plug-and-play control considers the situation in which the desired global control aim is known, but the design agents D_i have only available the model of its subsystem S_i, have sparse information about the physical couplings and are aware about the existence of some further design agents. That means that the pictorial representation of the interconnected system from the perspective of design agent D_1 with its available information results to Fig. 1.3b. Due to this situation, the main issues of plug-and-play control are reflected by the following two fundamental questions:

- What are local conditions for the design of control station C_i that guarantee the adherence of the global control aim?
- How detailed shall the behaviour of subsystem S_i under the influence of the physical interaction be modelled and how can this model be set up by D_i?

A solution to these two issues, enables design agent D_1 to accomplish the design of control station C_1 (i.e., to solve Problem 1.1). In summary, the plug-and-play control problem reads as follows:

Problem 1.2 (Plug-and-play control problem)

Given:
- *Global control aim*
- *Design agents D_i, $(\forall i \in \mathcal{N})$ which know*
  - *the subsystem model S_i,*
  - *some information about the physical couplings and*
  - *that further design agents exist*

Find:
- *Local design conditions that guarantee the adherence of the global control aim and*
- *a decision rule that enables D_1 to set up a suitable model of subsystem S_1 under the influence of the physical interaction*

to enable design agent D_1 to solve Problem 1.1

1.3.2 Approaches and fundamental questions

The achievable performance of the overall closed-loop system depends on the accuracy of the model that is used for the design of the control algorithms. Intuitively, with the availability of the overall system model, necessary and sufficient design conditions can be formalised. In contrast to this, if only neighbouring models are taken into account, the overall system dynamics become uncertain with the consequence that the design conditions are merely sufficient for the overall closed-loop objectives. Accordingly, the solution of the plug-and-play control problem concerns a trade-off between the accuracy of the model and the acceptable conservatism of the design conditions as portrayed in Fig. 1.4.

As mentioned before, design agent D_1 initially knows only the model of the associated subsystem S_1 and is only aware about the local couplings from other subsystems visualised by the block diagram on the left-hand side of Fig. 1.4. Nevertheless, the design of C_1 has to aim at the satisfaction of the global control aim. Starting from this situation, three approaches to the plug-and-play control problem are considered.

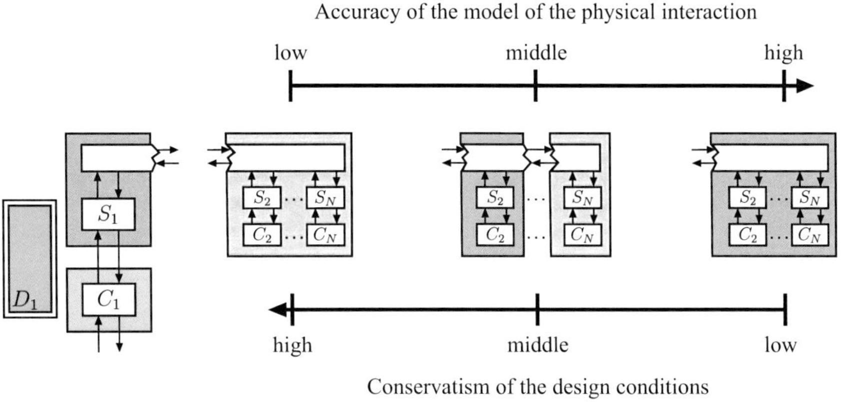

Figure 1.4: Ways of solution to the plug-and-play control problem: Model accuracy versus conservatism of design conditions. Known and unknown model information for the design of C_1 are illustrated by the yellow box and the grey box, respectively. Starting with the situation in which D_1 only knows the local models (left-hand side of the figure), the controller design has to accomplished by means of an exact model of the physical interaction (right-hand side of the bar), an approximate model (centre of the bar) or without a model of the physical interaction (left-hand side of the bar).

The first approach considers the set up of a model that represents the behaviour of the physical interaction exactly. Such an *exact model* is required, if C_1 has to be designed without limiting the achievable performance of the overall closed-loop system (right-hand side of the bar in Fig. 1.4). Thus, this approach is addressed to the construction of this model by means of gathering local model information of other design agents and the combination of those. From this situation arises the following question:

- How do the design agents need to interact in order to set up a model that represents the dynamics of the physical interaction?

The second approach aims at enabling the controller design using the initially available model information, i.e., without a model of the physical interaction (left-hand side of the bar in Fig. 1.4). Hence, no additional models has to be gathered by design agent D_1. To handle the concomitant model uncertainty, strict design requirements have to be formulated to still guarantee a desired closed-loop performance. The key questions of this approach are the following:

- How to enable the controller design without a model of the physical interaction?

- What are local design conditions to guarantee input-output stability and a certain input-output performance of the overall closed-loop system?
- Which degree of conservatism has to be accepted within the local design conditions?

Naturally, some controlled subsystems have a greater influence on subsystem S_1 than others. Hence, it is reasonable to model only those dynamics, while the dynamics of the controlled subsystems with a minor influence are considered as conscious model uncertainty. The third approach to the plug-and-play control problem encompasses a trade-off between the accuracy of the model used for the design and the achievable overall closed-loop performance (centre of the bar in Fig. 1.4). The leading questions focussing on this issue are the following:

- How can D_1 decide which models are relevant for the design of control station C_1 and which models are negligible?
- Which relaxation of the restrictive local design conditions from the second approach is entailed by a more precise model?

1.4 Literature on plug-and-play control

In computer science, the keyword "plug-and-play" describes the ability of a computer to automatically configure its hardware and devices [131]. In accordance with this definition, plug-and-play as used in this thesis is the ability of a networked control system to automatically design and implement the control algorithms (configure) to control the subsystems (devices) in a way that they behave as desired.

As the main methodological issue of plug-and-play control focusses on the required amount of model information for controller design, Section 1.4.1 outlines existing controller design methods that forgo a complete model of the overall system. For the realisation of plug-and-play control, it should be possible to replace the existing control algorithm that is processed in the control equipment at runtime. Thus, in Section 1.4.4, solutions are proposed which enable the usage of plug-and-play control to industrial processes. In the theory of networked control system and the control of manufacturing systems concepts exist which are called "plug-and-play". In Sections 1.4.2 and 1.4.3 these concepts are outlined.

1.4.1 Controller design with limited model information

In the control of physically interconnected systems or large-scale systems, the available information for control (i.e., signal information) as well as for the design (i.e., model information) is of prime importance since decades [40, 75, 80, 109, 117]. Making decisions with limited information is well studied in decision theory (see for example the references [89, 90]). In this context information refers to the knowledge about the decision rules and signals of other members. Amongst others the publication [80] draws the parallel between the value of information in decision theory and in the control of interconnected systems. The publications [40, 117] pay particular attention on making decisions in terms of the control signal to be generated or the control algorithm to be designed under constrained access of measurements or models, respectively. Accordingly, the control task is performed in a decentralised manner by N control stations which only have access to the local measurements of the respective subsystem. Similarly, the design task is decentralised too in the sense that a complete model of the overall system is never used in any step of the design process. Later on in networked control systems, these hard constraints are softened in the sense that the control stations are able to exchange signals in order distribute a global control task on local control stations. Of course, a heavy exchange of real time signals is not preferable due to limitations of the band with of the communication network as well as the trust of the received measurements [42, 118, 187]. Similarly, all model information become accessible through the digital communication network for the local design. Nevertheless, due to complexity reasons and privacy issues, the focus of this thesis is to distinguish between relevant model information and those which can purposely be ignored for the controller design.

Some textbooks are devoted to the reduction of the complexity within the analyses and the controller design for large-scale systems [40, 67, 109, 117, 164, 184, 185]. Therefore, the overall system is decomposed into subsystems (not necessarily with a physical interpretation). The control stations are then separately be designed based on the respective subsystem model. In a final step, the fulfilment of the overall closed-loop system specifications is tested by means of sufficient conditions. This method is referred to as composite-system method [35, 117]. Furthermore, this method provides the additional aggregation of the subsystem models in order to reduce the design effort. Accordingly, each subsystem is represented by a Lyapunov function or a comparison system in contrast to a complete model [41, 116] (here, the subsystems are physical entities). In relation to these aggregation techniques, the publications [48, 142, 151] have proposed model reduction techniques that reduce the order of each subsystem model while preserving the interconnection structure and having the overall system dynamics in mind. A different approach to the analysis of the overall system has been proposed in [109], where the overall system dynamics from the local perspective are approximated by the model of the subsystem together with a course model of the physical interaction. This course model is obtained by a heuristic procedure based on model reduction techniques. From the analysis of the overall system by means of simplified models, controller design techniques have been derived for instance in [66, 117, 164]. However, since no complete model is used for the design the achievable performance of the overall closed-loop system is poor. To improve the performance while reducing the complexity of the design, the control stations are designed in a predestined order [95, 117, 136]. Therefore, those controlled subsystems are taken into account which have been designed in the previous step. Although, the first control station can be designed with the corresponding subsystem model, the design of control station C_N requires all model information.

The above mentioned design techniques have in common that for the controller design at least a piece of information from all subsystems is required. For specific interconnection structures or control strategies, the controller can be designed using the subsystem model only. A string of vehicles has been investigated in [39, 117], where the controlled vehicle platoon has to satisfy a string stability condition. For this specific interconnection structure and global performance requirement, local design conditions are derived which only require the model S_i of the corresponding vehicle. In [157] a self-organising controller has been proposed which consists of N control stations that interact depending on the situation. Whenever the local state exceeds a local threshold, a communication to other control stations is established to drive the state back to the origin. Due to this mechanism, the overall closed-loop system is asymptotically stable if the local controlled subsystems are asymptotically stable. Therefore, only the local subsystem model is required for the design. In contrast to these special cases, the publications [75–77, 105] have considered the controller design for a system with arbitrary interconnection structure (which are unknown during the design phase). It has been shown that no design method exists that simultaneously generates static state-feedback gains $\boldsymbol{K}_i$, $(i = 1, ..., N)$ so that the closed-loop system is stable. That means that there always exists an interconnection structure that destabilises the controlled subsystems. Only for fully actuated discrete time subsystems the design of a deadbeat controller will result in a stable closed-loop system with the best possible performance. Hence, to accomplish the controller design for arbitrarily interconnected subsystems, more information than only the local subsystem model is required.

Design based on local and neighbouring model information. Various authors have proposed local design conditions which can be checked with the local subsystem model and some information from the direct neighbours. Essentially, these design conditions ground on *small-gain conditions* in the sense that each controlled subsystem has to have a limited amplification from the coupling input to the coupling output [104, 130, 145, 188].

In [104] a local design condition has been derived by the decomposition of a global closed-loop stability condition which is based on vector Lyapunov functions. This decomposition is processed by means of comparison systems. The local design that has been proposed in [188] ground also on control Lyapunov functions. The resulting control station stabilises the subsystem and suppresses the coupling influence so that the overall closed-loop is stable. In [130] each controlled subsystems should satisfy their local specification while being robustly stable to the specifications of all other controlled subsystems. The required information to process the respective designs encompass, amongst the local subsystem model, the (control) Lyapunov functions from the direct neighbours or the specifications of all other controlled subsystems, respectively.

In [145] a *plug-and-play design* of control stations has been considered which is characterised by the fact that the accessible information is constrained to the local subsystem model and the model of the subsystems which have a direct coupling influence. Based on an upper bound on the free motion of the controlled subsystem, a small-gain condition has been derived in [65] that has been refined in [145]. Basically, the static reinforcement of each controlled subsystems should be small with respect to the physical couplings so that the overall closed-loop is asymptotically stable.

It has to be emphasised that [104, 130, 145, 188] have presented *independent* and *scalable* controller designs in the sense that control station C_i can independently be designed from other control stations and the design requires only model information from neighbouring subsystems. Chapters 5 and 6 also proposes *independent* and *scalable* controller designs.

No design conditions but an interaction measure has been proposed by [86]. This interaction measure evaluates the admissible coupling strength between the subsystems until the overall system becomes unstable. One of the proposed stability tests grounds on small gains of each subsystem. As a result, the overall system stability can be checked individually by each subsystem using only the local model information and information about the physical couplings from the direct neighbouring subsystems. It is worth noting that the results from [86] have been used in [94, 163] to derive an independent design of decentralised controllers. Basically, the input-output amplification of each controlled subsystem should to be smaller than a given bound. This bound, however, is calculated based on the model of the overall system to be controlled.

Optimisation under limited model information. The design of control stations with respect to minimise a global cost function is of particular interest by many researchers. This paragraph reviews results, which do not use a complete model of the overall system.

In [68, 69] an iterative design method for a distributed controller has been presented that aims at minimising a global cost function in a distributed manner. Therefore, the optimisation problem is decomposed into N dependent optimisation problems. At each iteration, the local problem is solved taking the solutions from the direct neighbours into account. Similarly, a distributed reconfiguration method of an actuator network after the break down of one actuator has been proposed in [173, 174]. The reconfiguration goal constitutes the exact reconstruction of the nominal closed-loop behaviour using the remaining actuators. Therefore, an optimisation problem is formalised and solved by the remaining intelligent actuators in a distributed manner by means of a gradient method.

Furthermore, the design of a suboptimal distributed controller for a platooning system has been considered in [31]. Therefore, the locally optimal state feedback controller have to be designed in a predestined order taking only the model of the controlled subsystem into account that has been designed in the previous step.

A decentralised and distributed model predictive control scheme has been presented in [148, 149, 189]. In addition to asymptotic stability of the overall closed-loop system constraints in terms of the subsystem states and the control signal are considered. The local model predictive control stations make use of the local subsystem model only and are designed using additional models of the direct neighbouring subsystems. The proposed design is based on the small-gain condition from [145]. To improve the precision of the state prediction, the references [73, 118, 179] have proposed distributed model predictive control concepts, where the model predictive control stations use amongst the local subsystem model also the models of the subsystems with a direct coupling influence.

1.4.2 Plug-and-play in networked control systems

In the literature of networked control systems, the term plug-and-play is mainly referred to the ability of a controller to automatically react to changes in the process in terms of instrumentation and in terms of subsystems that are added or removed [168]. The following paragraphs survey existing frameworks to enable these properties.

Integration of new instrumentations. The research project *Plug and Play Process Control* (P^3C) focusses on concepts for process control, which enables the integration of new sensors and actuators that are plugged in at runtime [181]. The integration of a new device is processed in two steps. First, a model of the new hardware has to be identified on-line and, second, the controller has to be adapted based on the identified model to make use of the new device [47, 168]. Accordingly, on-line identification methods are proposed in [44, 101, 102] to obtain a model of the new actuator or new sensor and to evaluate its usefulness for control. To make use of new sensor signals, sensor fusion methods are presented in [45, 169]. Moreover, to handle additional sensors and actuators a reconfiguration method based on a Youla Kucera parametrisation of the controller and the system has been proposed in [176–178] in order to switch to the new controller in a smooth manner [135]. The results are applied to a livestock stable [177], to a hydraulic network [97], to a evaporator unit [96] and to a district heating system [102]

Integration of new subsystems. The integration of new subsystems is the main focus in [145]. As mentioned above, the plug-and-play design only depends on the local subsystem model and the subsystem model of its direct neighbours. Thus, if a new subsystem is connected, first, the corresponding control station is designed gathering the relevant subsystem models. In a second step, the control stations of those subsystems are redesigned for which the new subsystem is a neighbour (cf. Section 1.2.2).

In [188] a passive approach to integrate new subsystems has been presented. That is, the control stations are designed to guarantee global closed-loop stability in spite of changing interconnection structures and a changing number of subsystems. Accordingly, the overall closed-loop system can handle failures that occur in the interconnection among the subsystems.

In [46, 99, 129, 138] a hierarchical predictive controller has been proposed which is composed of a global coordinator and decentralised control stations which collectively aim at minimising a cost function. Due to the modular structure of the involved cost function a new subsystem can easily be integrated. In particular, the global coordinator provides the local control units with relevant information about the physical interaction. If a new subsystem is added, the global coordinator has to update its calculation so that the local units receive the information about the new coupling influence.

The issue of a variable number of subsystems has gathered considerable attention in the field of fault-tolerant control of networked control systems. The concept that has been presented in [99, 138] inherently provide fault tolerance for minor faults such as offsets in the measurements. In [146, 147] a fault-tolerant control framework has been proposed which consists of a distributed fault diagnosis scheme presented in [55] and a local controller redesign from [145] (see Section 1.4.1). If a fault has been detected, the faulty subsystem is subsequently unplugged to avoid the propagation of the fault to the other subsystems through the physical couplings. In order to still guarantee fault detection, the local diagnosis systems that have exploited the measurements of the unplugged subsystem has to updated their diagnosis algorithm. If the faulty subsystem is repaired it can be re-plugged-in. Therefore, the previously adapted diagnosis algorithm has to be modified once again and, furthermore, the neighbouring control stations has to be redesigned as mentioned above.

1.4.3 Plug-and-play in manufacturing systems

In the context of Industry 4.0, manufacturing systems such as production lines, assembly processes or robot cells shall become more flexible in terms of adding, replacing and removing components and plant modules or changing the production configuration [159]. This section surveys existing methods to enable this flexibility. All presented solutions aim at a modular program structure in contrast to a monolytic one. As a consequence, only the affected program segments need to be adapted to achieve the above mentioned agility.

A modular programming paradigm has been proposed in [84, 137, 186]. The overall software program to control the plant is split into sub-programs, the so called *software agents* which are assigned to the sub-processes of the plant. The agents exchange signals and algorithms among each other to fulfil the overall program specification [83]. The application to industrial cases has been studies in [128].

Similar is the idea of *plug-and-procedure* [33, 34], where the application program is separated into a hierarchy of modular software components (called holons) which have autonomy and cooperation ability. A holon for example is attached to the device and are installed onto the application program if the device is added to the process. In [134] plug-and-produce has been applied to robot cells.

In [91, 92, 113] a knowledge-based control sequence generation has been presented that is based on atomic services. An atomic service represents a sensor, an actuator, a control algorithm or a communication interface by means of a semantic model. In order to describe the behaviour of a drilling machine or a conveyor belt for example, atomic services are assembled. If one of the above mentioned changes occur, the semantic model of the overall process is reassembled to enable the reconfiguration of the control sequence diagram. This procedure is called *dynamic orchestration*. In [113] a new field device is integrated to a manufacturing system by means of dynamic orchestration.

Moreover, the process control solution ACPLT allows read and write access of the application program at runtime and, thus, enables the adaptation of this program [28].

1.4.4 Implementation of plug-and-play control

For the application of plug-and-play control it is essential to manipulate the control algorithms at runtime. Due to the modularity of the application program as surveyed in Section 1.4.3 the mentioned paradigms can be used to replace existing control algorithms by new algorithm. Furthermore, the dSPACE rapid prototyping system allows the adaptation of controller parameters at runtime [156]. Similarly, FOUNDATION Fieldbus enables the manipulation of the parameters of a PID controller at runtime by means of control in the field [162].

1.4.5 Classification of this thesis

Plug-and-play control focuses on the organisation of the design of control stations for physical interconnected systems by means of design agents D_i. This organisation comprises the modelling of the subsystem dynamics under the influence of the physical interaction and the design of the control station regarding the adherence of a global control aim. The situation is considered in which each subsystem is equipped with a design agent that store the model of its subsystem and no global coordination exists that provides the design agents with the required information. Hence, to accomplish the design of control station C_i, design agent D_i has to contact other design agents and has to procure models and algorithms over the communication network. The present work answers the following questions:

- Which models are required for the controller design and which models can be neglected?
- How can the relevance of a model be evaluated by design agent D_i?

As a result, plug-and-play control is characterised by the fact that the required amount of model information is chosen during the design phase by the respective design agent. Therefore, the design is naturally accompanied by the exchange of model information among the design agents over the communication network. The exchange of models and trajectories has also be considered explicitly in [68, 133, 180]. Moreover, various concepts exist in which it is a priori defined which model information is required for the controller design. [31, 68, 75, 94, 95, 104, 117, 130, 145, 173, 188].

Throughout this thesis, the subsystems are considered to have linear dynamics. The global control aim is boundedness, input-output stability and a certain input-output performance of the overall closed-loop system which has to be fulfilled by a complete decentralised control structure. The communication network is assumed to be ideal, such that no delays, collisions or package losses occur.

1.5 Contributions of this thesis

This thesis proposes different solutions to the plug-and-play design problem (Problem 1.2) in accordance with the three approaches presented in Section 1.3.2. These approaches differ with respect to the accuracy of the model that represents the *physical interaction* of S_1 with other controlled subsystems (see Fig. 1.4).

Modelling the physical interaction exactly (Chapter 4). The model that represents the behaviour of the physical interaction is composed of the models of those controlled subsystems which are strongly connected to subsystem S_1. A distributed algorithm (Algorithm 4.1) is presented that assuredly finds all these controlled subsystems (Theorem 4.1). This algorithm essentially grounds on a distributed realisation of a depth-first search algorithm. Since, the models of the subsystems which are not strongly connected to subsystem S_1 are ignored, the computational workload for the design is reduced. The algorithm is distributed among the design agents so that its execution involves the interaction of the design agents. This interaction is analysed in Section 4.3 by means of a *communication graph*. It is shown that this communication graph usually yields a subgraph of the bidirectional representation of the interconnection graph. Moreover, it is shown that Algorithm 4.1 has linear complexity.

The proposed algorithm is used to achieve fault-tolerance by the reconfiguration of the control station C_1 after fault occurrence in subsystem S_1. The plug-and-play reconfiguration that is processed by design agent D_1 is summarised by Algorithm 4.2.

Leave the physical interaction unknown (Chapter 6 and Chapter 5). Two approaches are proposed which enable the controller design without a model of the physical interaction. Both approaches ground on the idea to limit the influence of each controlled subsystem to the physical couplings. This limitation is achieved

1. by restricting the operating set (Section 5.2) or
2. by restricting the amplification from the coupling input to the coupling output (Section 6.2)

of all N controlled subsystems. As a result, the physical interaction becomes bounded either in its operating set or in its amplification as highlighted in Fig. 1.5.

It is shown that both, the restricted operating set as well as upper bound of the physical interaction can be determined without gathering models from other design agents. Based on this result, sufficient local stability conditions (Proposition 5.1 and Theorem 6.1) and sufficient local performance conditions (Proposition 5.2 and Theorem 6.2) are derived to ensure boundedness, input-output stability or a certain performance of the overall closed-loop system. Controller design methods that are based on invariant sets (Theorem 5.1 and Corollary 5.1) and on the H_∞-norm (Theorem 6.3) are proposed.

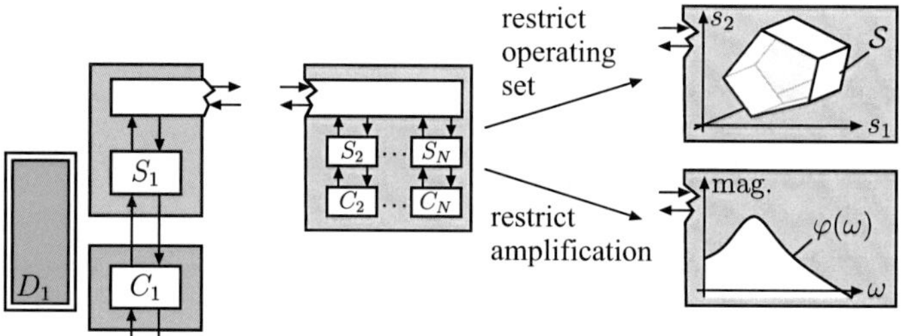

Figure 1.5: Results of the two approaches to remain the dynamics of the physical interaction (grey box) unknown: The dynamics of the physical interaction are bounded either in their operating set (signal $\boldsymbol{s}(t) = (s_1(t), s_2(t), s_3(t))^\top$ is located inside the polytope $\mathcal{S}$) or in their amplification (the amplitude plot is bounded from above by the function $\varphi(\omega)$).

Both approaches are evaluated in simulations and experiments on a thermofluid process (Sections 5.5 and 6.5) by means of plug-and-play reconfiguration (Algorithm 5.3) and plug-and-play integration (Algorithm 6.1). The result of this evaluations indicates that both approaches are applicable to real world processes despite the conservatism of the local design requirements.

Approximate the physical interaction (Chapter 7 and Chapter 8). Naturally, the dynamics of the physical interaction are mainly characterised by the dynamics of some controlled subsystems, whereas the dynamics of other controlled subsystems are irrelevant for the controller design. Hence, these irrelevant dynamics can reasonably be treated as acceptable uncertainty of the model to be set up. Results to this issue are with respect to the design of a control algorithm as well as to the design of a diagnosis system.

- **Controller design (Chapter 7):** A local decision rule is derived to separate all controlled subsystems into *strongly coupled* and *weakly coupled* to subsystem S_1 (Theorem 7.1). The strongly coupled controlled subsystems constitute the approximation, while the weakly coupled controlled subsystems are considered as model uncertainty. It is shown that the physical interaction is appropriately approximated by the consideration of a small number of controlled subsystems (Section 7.2.3). Due to the approximation of the physical interaction, it can be shown that the conservatism of the local stability and performance conditions from Chapter 6 can be reduced (Corollaries 7.2 and 7.4). The approach is applied to a multizone furnace to make the process tolerant to actuator failures (Algorithm 7.3). The simulation emphasises the distinct reduction of the conservatism of the local stability condition if just the neighbouring zone is taken into account for the controller reconfiguration.

- **Design of a diagnosis system (Chapter 8):** The accuracy of the model of the physical interaction is rated by means of the *detectability* and *isolability* of considered faults (Algorithm 8.1). The uncertainty of the generated residual must be as small as necessary to guarantee fault detection (Theorem 8.1) and fault isolation (Theorem 8.2). Moreover, it is shown that the uncertainty of the residual can be reduced by using a PI-based residual generator (Corollary 8.1). Plug-and-play diagnosis is evaluated on a multizone furnace. The result of this example highlights that the accuracy of the interaction model depends upon the severity of the considered faults.

1.6 Structure of this thesis

The main results of this thesis are distinct methods for plug-and-play control which are presented in Chapters 4–8. The results differ in the accuracy of the model of the physical interaction which is used for controller design and, consequently, in the involved conservatism of the design claims (see Fig. 1.4). Each of these chapters can be read independently of one another.

Chapter 2 introduces the notation, relevant definitions and mathematical basics that are used in this these. Well-known results from robust control, fault-tolerant control and the theory of convex sets are recapitulated as they serve as the basis for the proposed approaches for plug-and-play control. Finally, a themofluid process and a multizone furnace are outlined which are used to evaluate the approaches by means of simulations and experiments.

Chapter 3 states tools to handle model information as this is essential in plug-and-play control. The operator con is introduced to compactly describe the connection of models. The notation of frequently used models are introduced, followed by the tasks of the design agents and the visualisation of their interactions by means of a communication graph.

Chapter 4 presents a distributed algorithm to enable design agent D_1 to set up a model that exactly describes the dynamics of the physical interaction. The algorithms finds the subsystems that are strongly connected to subsystem S_1 and procures the respective models without using a global coordinator. This chapter analyses the interaction among the design agents during the execution of the algorithm by means of the application to fault-tolerant control.

Chapter 5 proposes local design conditions so that the control stations can be designed without a model of the physical interaction. The underlying idea to derive these conditions is to force each subsystem state to remain in a desired operating region by design. Furthermore, two controller design methods are presented. The solution of the first method is a static state-feedback, whereas the second method designs a tube-based model predictive control. This concept is used to make a thermofluid process tolerant to faults

Chapter 6 presents local controller design conditions that only need the locally available model but aim at input-output stability and the adherence of a certain input-output performance of the overall closed-loop system. These local conditions ground on the idea to attenuate the effect of the coupling input on the coupling output of each controlled subsystem. An H_∞-design is proposed which is eventually tested by experiments on a thermofluid process for the integration of a newly added tank.

Chapter 7 presents a measure of the coupling strength among the subsystems with the aim to classify the subsystems into those which have a strong coupling to the considered subsystem and into those whose coupling is weak. It is shown that the dynamics of the subsystems which are chosen to be strongly coupled essentially characterise the behaviour of the physical interaction. Moreover, the conservatism of the local design conditions presented in Chapter 6 is significantly reduced due to the more precise model of the physical interaction.

Chapter 8 deals with the design of a model-based diagnosis system which only evaluated the signals of the subsystem to be monitored by exploiting an approximate model of the physical interaction. It is shown that only an approximation is sufficient to guarantee fault detection and fault isolation. Based on a detectability condition and an isolability condition it can locally be decided upon the required accuracy of the model.

Chapter 9 summarises and concludes the thesis and gives an outlook about possible research directions.

2 Preliminaries

This chapter starts with the introduction of the notation and relevant definitions that will be used throughout this thesis. Afterwards, basic notations and definitions in graph theory are stated. Thereafter, the models of the subsystems, their physical interconnection and their control stations are given, followed by basics on robust control and fault-tolerant control. Moreover, an interconnected tank process and a multizone furnace are presented which serve as demonstration examples for plug-and-play control.

2.1 Notation and definitions

2.1.1 General notation

Throughout this thesis, scalars are denoted by italic letters (d), vectors by bold italic letters ($\boldsymbol{b}$) and matrices by upper-case bold italic letters ($\boldsymbol{A}$). Models are represented by capital italic letters (S) and subscripts distinguish distinct parts (S_i). Transfer functions or impulse response functions of a model bear the same letter. Accordingly, matrix-valued functions and vector-valued functions are denoted by upper-case bold italic letters ($\boldsymbol{S}(s)$, $\boldsymbol{C}(t)$), scalar-valued functions are denoted by upper-case italic letters ($S(s)$, $C(t)$) and subscripts denote the I/O relation ($\boldsymbol{S}_{\mathrm{yu}}(s)$, $\boldsymbol{C}_{\mathrm{uw}}(t)$)

$\mathbb{N}$, $\mathbb{R}$, $\mathbb{C}$ denote the set of natural numbers, real numbers and complex numbers, respectively. $\mathbb{N}_+$ denotes the set of positive natural numbers and $\mathbb{N}_{+0} := \mathbb{N}_+ \cup \{0\}$. Further sets are denoted by calligraphic letters ($\mathcal{N}$), where $|\mathcal{N}|$ denotes the cardinality of the set $\mathcal{N}$. The operator $\leftarrow$ denotes the inclusion of an element into a set ($\mathcal{N} \leftarrow j$) or the choice of an element from a set ($j \leftarrow \mathcal{N}$), respectively.

$\mathbf{I}$ and $\mathbf{O}$ denote the identity matrix and the zeros matrix, respectively. $\mathbf{1}$ and $\mathbf{0}$ represent vectors, where each element is 1 and 0, respectively. The dimension of these matrices and vectors is always clear from the context.

$\dim(\boldsymbol{x})$ denotes the dimension of a vector $\boldsymbol{x}$ and $\mathrm{rank}(\boldsymbol{A})$ represents the rank of a matrix $\boldsymbol{A}$. The transpose of a vector $\boldsymbol{x}$ or a matrix $\boldsymbol{A}$ is denoted by $\boldsymbol{x}^\top$ or $\boldsymbol{A}^\top$, respectively. $\boldsymbol{A}^{\mathrm{H}}$ refers to the conjugate transpose of the matrix $\boldsymbol{A}$. A symmetric matrix (Hermitian matrix) satisfy the relation $\boldsymbol{A} = \boldsymbol{A}^\top$ ($\boldsymbol{A} = \boldsymbol{A}^\mathrm{H}$).

The i-th eigenvalue of a square matrix $\boldsymbol{A} \in \mathbb{R}^{n\times n}$ is denoted by $\lambda_i(\boldsymbol{A})$. A matrix is called *Hurwitz* if all its eigenvalues have negative real part. The notation $\boldsymbol{A} \prec 0$ ($\boldsymbol{A} \preceq 0$) for a Hermitian matrix means that the matrix is negative (semi-) definite, meaning that all its eigenvalues have strictly negative (non-positive) real part. Correspondingly, the symbol $\succ$ ($\succeq$) denotes positive (semi-) definiteness of Hermitian matrices, i.e, all its eigenvalues have strictly positive (non-negative) real part. For two vectors $\boldsymbol{v}$, $\boldsymbol{w} \in \mathbb{R}^n$ the relation $\boldsymbol{v} > \boldsymbol{w}$ ($\boldsymbol{v} \geq \boldsymbol{w}$) holds element-wise, i.e., $v_i > w_i$ ($v_i \geq w_i$) for all $i = 1, ..., n$, where v_i and w_i are the i-th element of the vectors $\boldsymbol{v}$ and $\boldsymbol{w}$, respectively. Accordingly, for the two matrices $\boldsymbol{V}$, $\boldsymbol{W} \in \mathbb{R}^{n\times m}$ which are composed of the elements v_{ij} and w_{ij} for $i = 1, ..., n$ and $j = 1, ..., m$ the relation $\boldsymbol{V} > \boldsymbol{W}$ ($\boldsymbol{V} \geq \boldsymbol{W}$) holds element-wise too, i.e., $v_{ij} > w_{ij}$ ($v_{ij} \geq w_{ij}$) for all i, j.

A symmetric block matrix is abbreviated using the symbol $\star$, e.g.,

$$\begin{pmatrix} \boldsymbol{A} & \boldsymbol{B} \\ \star & \boldsymbol{C} \end{pmatrix} \quad \text{means} \quad \begin{pmatrix} \boldsymbol{A} & \boldsymbol{B} \\ \boldsymbol{B}^\top & \boldsymbol{C} \end{pmatrix}.$$

Consider the matrices $\boldsymbol{A}_1, ..., \boldsymbol{A}_N$ and the set $\mathcal{S} \subseteq \{1, ..., N\}$. The notation $\mathrm{diag}(\boldsymbol{A}_i)_{i\in\mathcal{S}}$ is used to denote the block diagonal matrix with the elements $\boldsymbol{A}_i$, ($\forall i \in \mathcal{S}$) on its diagonal, e.g.,

$$\mathrm{diag}(\boldsymbol{A}_i)_{i\in\{2,4,7\}} = \begin{pmatrix} \boldsymbol{A}_2 & & \\ & \boldsymbol{A}_4 & \\ & & \boldsymbol{A}_7 \end{pmatrix}.$$

Let $\boldsymbol{B}_{ij} \in \mathbb{R}^{n\times m}$ for $i = 1, ..., r$ and $j = 1, ..., c$ and $\mathcal{R} \subseteq \{1, ..., r\}$, $\mathcal{C} \subseteq \{1, ..., c\}$. The notation $\boldsymbol{B}_{\mathcal{R},\mathcal{C}}$ represents a block matrix which consists of the entries $\boldsymbol{B}_{ij}$, ($\forall i \in \mathcal{R}$, $\forall j \in \mathcal{C}$), for example

$$\boldsymbol{B}_{\{1,3\},\{2,4\}} = \begin{pmatrix} \boldsymbol{B}_{12} & \boldsymbol{B}_{14} \\ \boldsymbol{B}_{32} & \boldsymbol{B}_{34} \end{pmatrix}, \quad \boldsymbol{B}_{\{1\},\{2,4\}} = \begin{pmatrix} \boldsymbol{B}_{12} & \boldsymbol{B}_{14} \end{pmatrix}, \quad \boldsymbol{B}_{\{1,3\},\{4\}} = \begin{pmatrix} \boldsymbol{B}_{14} \\ \boldsymbol{B}_{34} \end{pmatrix}.$$

$\boldsymbol{B}_\mathcal{N}$ is the concisely notation for $\boldsymbol{B}_{\mathcal{N},\mathcal{N}}$.

Consider the time-dependent matrix $\boldsymbol{G}(t)$ and vector $\boldsymbol{u}(t)$. The convolution-operator is denoted by the asterisk $*$, i.e,

$$\boldsymbol{G}(t) * \boldsymbol{u}(t) = \int_0^t \boldsymbol{G}(t-\tau)\boldsymbol{u}(\tau)\mathrm{d}\tau.$$

Consider a transfer function matrix $\boldsymbol{G}(s)$, where s denotes the complex frequency. The corresponding Fourier transformation and amplitude plot is represented by $\boldsymbol{G}(\mathrm{j}\omega)$ and $|\boldsymbol{G}(\mathrm{j}\omega)|$, respectively.

For a scalar d, $|d|$ denotes the absolute value. For a vector $\boldsymbol{x} \in \mathbb{R}^n$ or a matrix $\boldsymbol{A} \in \mathbb{R}^{n\times m}$ the operator $|\cdot|$ holds element-wise, i.e.,

$$|\boldsymbol{x}| = \begin{pmatrix} |x_1| \\ \vdots \\ |x_n| \end{pmatrix}, \qquad |\boldsymbol{A}| = \begin{pmatrix} |a_{11}| & \cdots & |a_{1n}| \\ \vdots & \ddots & \vdots \\ |a_{n1}| & \cdots & |a_{nn}| \end{pmatrix}.$$

$\|\boldsymbol{x}\|$ and $\|\boldsymbol{A}\|$ represent an arbitrary vector norm and the corresponding induced matrix norm according to

$$\|\boldsymbol{x}\|_p := \left(\sum_{i=1}^{n} |x_i|^p\right)^{\frac{1}{p}}, \qquad \|\boldsymbol{A}\|_p := \max_{\boldsymbol{x}\neq \boldsymbol{0}} \frac{\|\boldsymbol{A}\boldsymbol{x}\|_p}{\|\boldsymbol{x}\|_p}$$

with the real scalar $p \geq 1$. The spectral norm of a matrix is

$$\|\boldsymbol{A}\|_2 = \sqrt{\lambda_{\max}\left(\boldsymbol{A}^{\mathrm{H}}\boldsymbol{A}\right)} = \sigma_{\max}\left(\boldsymbol{A}\right),$$

where $\sigma_{\max}$ and $\lambda_{\max}$ denote the maximal spectral radius and the maximal eigenvalue, respectively. The absolute value and the norm of the time-dependent matrix $\boldsymbol{G}(t)$ or the transfer function matrix $\boldsymbol{G}(s)$ are applied for every time t or every complex frequency s, respectively.

2.1.2 Definitions on stability

In this thesis, asymptotic stability and input-output (I/O) stability are considered which are introduced in this section.

Definition 2.1 (Input-output (I/O) stability [116]) *The system described by*

$$\boldsymbol{y}(t) = \boldsymbol{G}(t) * \boldsymbol{u}(t)$$

or by

$$\boldsymbol{y}(s) = \boldsymbol{G}(s)\,\boldsymbol{u}(s)$$

is said to be I/O stable *if*

$$\int_0^\infty \|\boldsymbol{G}(t)\|\mathrm{d}t < \infty \tag{2.1}$$

or

$$\|\boldsymbol{G}(\mathrm{j}\omega)\| < \infty, \quad \forall\omega \in \mathbb{R}, \tag{2.2}$$

respectively. The impulse response function $\boldsymbol{G}(t)$ *and the transfer function* $\boldsymbol{G}(s)$ *satisfying* (2.1) *and* (2.2)*, respectively, are called* I/O stable.

Definition 2.2 (Asymptotic stability [116]) *The autonomous system*

$$\dot{\boldsymbol{x}}(t) = \boldsymbol{f}(\boldsymbol{x}(t)), \quad \boldsymbol{x}(0) = \boldsymbol{x}_0$$

is said to be asymptotically stable *with respect to the equilibrium point* $\boldsymbol{x} = \boldsymbol{0}$ *if for each* $\epsilon > 0$ *there exists* $\delta > 0$ *such that* $\|\boldsymbol{x}_0\| \leq \delta$ *implies* $\|\boldsymbol{x}(t)\| \leq \varepsilon$, $(\forall t \geq 0)$ *and the condition* $\lim_{t\to\infty} \|\boldsymbol{x}(t)\| = 0$ *holds.*

Theorem 2.1 (Asymptotic stability [116]) *The system*

$$\dot{\boldsymbol{x}}(t) = \boldsymbol{A}\boldsymbol{x}(t) + \boldsymbol{B}\boldsymbol{u}(t), \quad \boldsymbol{x}(0) = \boldsymbol{x}_0$$

is asymptotically stable if and only if the system matrix $\boldsymbol{A}$ *is Hurwitz.*

2.1.3 Convex sets

A convex set has the property that for two arbitrary point in this set, every point on the straight line segment that joints these two points is also contained in this set. The mathematical definition is the following:

Definition 2.3 (Convex set [57]) *A set* $\mathcal{A} \subseteq \mathbb{R}^n$ *is called* convex *if*

$$\alpha\boldsymbol{x}_1 + (1-\alpha)\boldsymbol{x}_2 \in \mathcal{A}$$

for any $\boldsymbol{x}_1$, $\boldsymbol{x}_2 \in \mathcal{A}$ *and any* $\alpha \in [0, 1]$.

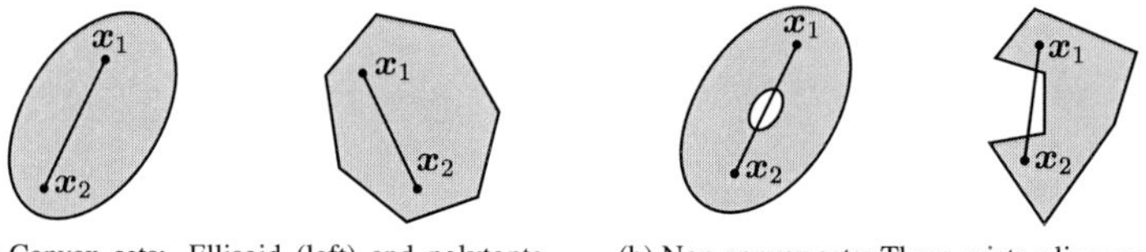

(a) Convex sets: Ellisoid (left) and polytopte (right). The line segment that joints two arbitrary points lies completely in the set.

(b) Non-convex sets: There exists a line segment that joints two points that does not lies completely in the grey set.

Figure 2.1: Examples of convex sets and non-convex sets.

Figure 2.1 illustrates examples of convex and non-convex sets. In this thesis ellipsoids and polytopes as shown in Fig. 2.1a are of special interest.

Definition 2.4 (Ellipsoidal set [57]) *An ellipsoidal set is given by*

$$\mathcal{A} = \left\{ \boldsymbol{x} \in \mathbb{R}^n : \boldsymbol{x}^\top \boldsymbol{P} \boldsymbol{x} \leq 0 \right\}$$

with $\boldsymbol{P} \in \mathbb{R}^{n \times n}$ *is symmetric and positive definite.*

Definition 2.5 (Polytopic set [193]) *A polytopic set is given by*

$$\mathcal{A} = \{ \boldsymbol{x} \in \mathbb{R}^n : \boldsymbol{P} \boldsymbol{x} \leq \boldsymbol{b} \}$$

with $\boldsymbol{P} \in \mathbb{R}^{m \times n}$ *and* $\boldsymbol{b} \in \mathbb{R}^m$.

Note that Definition 2.5 defines a so called $\mathcal{H}$-polytope that corresponds to an intersection of a finite number of closed halfspaces [193]. Without loss of generality, it is assumed that all polytopic sets considered in this thesis contain the origin in their interior.

In the following the definitions for the addition and subtraction of sets are stated.

Definition 2.6 (Minkowski sum [100]) *The Minkowski sum of two sets* $\mathcal{A}$, $\mathcal{B} \subseteq \mathbb{R}^n$ *is defined by*

$$\mathcal{A} \oplus \mathcal{B} := \{ \boldsymbol{a} + \boldsymbol{b} : \boldsymbol{a} \in \mathcal{A}, \boldsymbol{b} \in \mathcal{B} \} .$$

Definition 2.7 (Pontryagin difference [100]) *The Pontryagin difference of two sets* $\mathcal{A}$, $\mathcal{B} \subseteq \mathbb{R}^n$ *is defined by*

$$\mathcal{A} \ominus \mathcal{B} := \{ \boldsymbol{c} \in \mathbb{R}^n : \boldsymbol{c} + \boldsymbol{b} \in \mathcal{A}, \boldsymbol{b} \in \mathcal{B} \} .$$

The processing of the Minkowski sum and the Pontryagin difference is visualised in Fig. 2.2a and Fig. 2.2b, respectively.

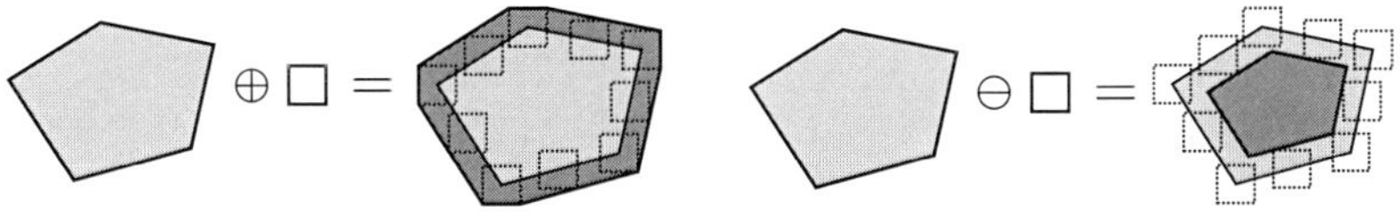

(a) Minkowski sum (b) Pontryagin difference

Figure 2.2: Minkowski sum and Pontryagin difference of two polytops. The dark grey polytope represents the resulting set (adapted from [167]).

Note that the Pontryagin difference is not the complement of the Minkowski sum. For two polytopes $\mathcal{A}$ and $\mathcal{B}$, only the relation $(\mathcal{A} \ominus \mathcal{B}) \oplus \mathcal{B} \subseteq \mathcal{A}$ holds. This fact will be of particular interest in Chapter 5.

An important means in the study of convex sets constitute the support function as well as the support halfspace and support set.

Definition 2.8 (Support function [155]) *For a given vector* $\boldsymbol{a} \in \mathbb{R}^n$*, the* support function $h_{\mathcal{A}}(\boldsymbol{a})\colon \mathbb{R}^n \to \mathbb{R}$ *for the nonempty, compact and convex set* $\mathcal{A} \subset \mathbb{R}^n$ *is defined by*

$$h_{\mathcal{A}}(\boldsymbol{a}) := \max_{\boldsymbol{x} \in \mathcal{A}} \boldsymbol{x}^\top \boldsymbol{a}.$$

The *support halfspace* is defined by

$$\mathcal{H}_{\mathcal{A}}(\boldsymbol{a}) := \left\{\boldsymbol{x} \in \mathcal{R}^n \colon \boldsymbol{x}^\top \boldsymbol{a} \le h_{\mathcal{A}}(\boldsymbol{a})\right\}, \quad \forall \boldsymbol{a} \in \mathbb{R},\ |\boldsymbol{a}| = 1$$

and the *support set* is defined by

$$\mathcal{F}_{\mathcal{A}}(\boldsymbol{a}) := \mathcal{H}_{\mathcal{A}}(\boldsymbol{a}) \cap \mathcal{A}, \quad \forall \boldsymbol{a} \in \mathbb{R},\ |\boldsymbol{a}| = 1.$$

The support function $h_{\mathcal{A}}(\boldsymbol{a})$ represents the distance between the origin and support halfspace $\mathcal{H}_{\mathcal{A}}(\boldsymbol{a})$ as illustrated in Fig. 2.3. For ellipsoidal sets, the support set represents a point on the boundary of $\mathcal{A}$ (cf. Fig. 2.3a), whereas for polytopes, the support set represents either an edge (e.g., $\boldsymbol{a}_1$ in Fig. 2.3b) or a vertex (e.g., $\boldsymbol{a}_2$ in Fig. 2.3b).

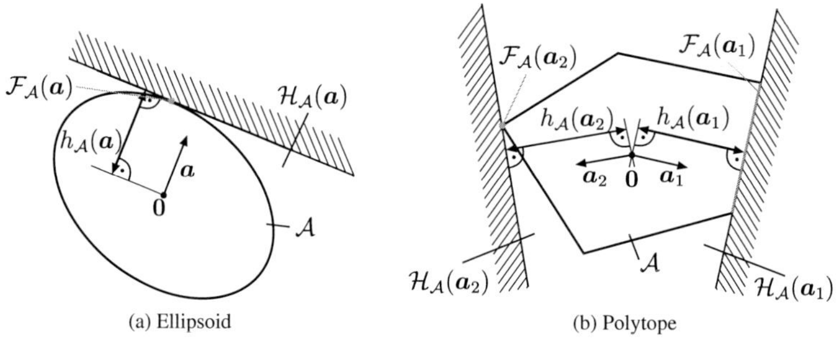

Figure 2.3: Interpretation of the support function $h_{\mathcal{A}}(\boldsymbol{a})$, the support halfspace $\mathcal{Y}_{\mathcal{A}}(\boldsymbol{a})$ and the support set $\mathcal{F}_{\mathcal{A}}(\boldsymbol{a})$ (grey).

2.1.4 Linear matrix inequalities

Many problems in control theory can be expressed by linear matrix inequalities [58, 154].

Definition 2.9 (Linear matrix inequality [58]) *Let $\boldsymbol{F}_i \in \mathbb{R}^{m\times m}$ for $i = 0, 1, ..., m$ be symmetric matrices. The inequality*

$$\boldsymbol{F}(\boldsymbol{x}) = \boldsymbol{F}_0 + \sum_{i=1}^{m} x_i \boldsymbol{F}_i \prec 0 \tag{2.3}$$

is a linear matrix inequality (LMI) and $\boldsymbol{x} = (x_1, ..., x_m)^\top \in \mathbb{R}^m$ is the decision variable.

Note that in many applications, the decision variable is a matrix $\boldsymbol{X} \in \mathbb{R}^{q\times r}$ (i.e., $\boldsymbol{F}(\boldsymbol{X}) \prec 0$) rather than a vector. Those LMIs can be viewed as a special case of (2.3) by defining an arbitrary basis $\boldsymbol{E}_1, ..., \boldsymbol{E}_n$ of $\mathbb{R}^{q\times r}$ and expanding $\boldsymbol{X}$ as $\boldsymbol{X} = \sum_{i=1}^{n} x_i \boldsymbol{E}_i$.

A linearisation of nonlinear matrix inequalities can often be achieved by means of a congruence transformation.

Definition 2.10 (Congruence transformation [154]) *Let $\boldsymbol{A}$ be a Hermitian matrix and $\boldsymbol{T}$ be a square non-singular matrix. The transformation*

$$\boldsymbol{T}^{\mathrm{H}}\boldsymbol{A}\boldsymbol{T}$$

is called a congruence transformation *of $\boldsymbol{A}$. Such a transformation does not change the number of positive and negative eigenvalues of $\boldsymbol{A}$.*

As a consequence of Definition 2.10, congruence transformations are equivalence conditions on LMIs in the sense that $\boldsymbol{F}(\boldsymbol{X}) \prec 0$ if and only if $\boldsymbol{T}^{\mathrm{H}}\boldsymbol{F}(\boldsymbol{X})\boldsymbol{T} \prec 0$. Especially the following Lemma provides a powerful result for the linearisation of nonlinear matrix inequalities.

Lemma 2.1 (Schur Complement [58]) *Let the matrices $\boldsymbol{V}$, $\boldsymbol{W}$ be symmetric and non-singular. The following statements are equivalent*

1. $\begin{pmatrix} \boldsymbol{V} & \boldsymbol{S} \\ \boldsymbol{S}^\top & \boldsymbol{W} \end{pmatrix} \prec 0$
2. $\boldsymbol{V} \prec 0$ *and* $\boldsymbol{W} - \boldsymbol{S}^\top\boldsymbol{V}^{-1}\boldsymbol{S} \prec 0$
3. $\boldsymbol{W} \prec 0$ *and* $\boldsymbol{V} - \boldsymbol{S}\boldsymbol{W}^{-1}\boldsymbol{S}^\top \prec 0$.

The equivalence relations of Lemma 2.1 also hold for negative semi-definiteness as well as for positive (semi-) definiteness.

LMIs are efficiently solvable by numerically stable methods such as the interior point method [57] which has been implemented in software like YALMIP, Sedumi and SDPT3 [112, 170, 175] for MATLAB.

2.1.5 Set invariance in control

This section briefly recapitulates basics on set invariance in control.

Consider the nonlinear system

$$\dot{\boldsymbol{x}}(t) = \boldsymbol{f}\big(\boldsymbol{x}(t), \boldsymbol{s}(t)\big), \quad \boldsymbol{x}(0) = \boldsymbol{x}_0, \tag{2.4}$$

where $\boldsymbol{x} \in \mathbb{R}^n$ and $\boldsymbol{s} \in \mathbb{R}^m$ denote the system state and the disturbance signal, respectively. A set $\mathcal{I} \subset \mathbb{R}^n$ is called *positively invariant* for the autonomous system (2.4) with $\boldsymbol{s}(t) = \boldsymbol{0}$ if whenever the system state is contained in this set for some time, the state will never leave this set in the future. This basically means that for all states $\boldsymbol{x}(t)$ on the border of the set $\mathcal{I}$, the derivative $\dot{\boldsymbol{x}}(t)$ has to point inside the set as illustrated in Fig. 2.4.

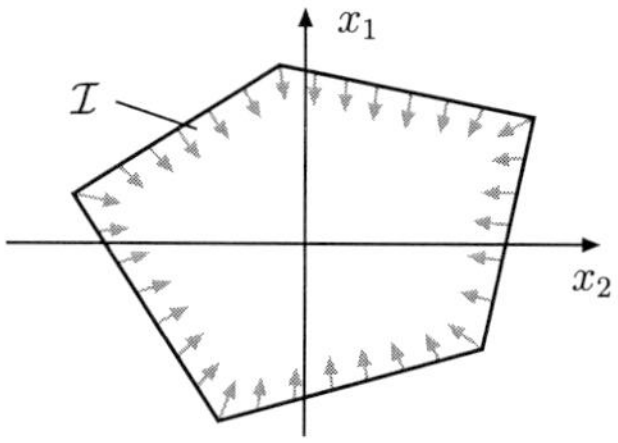

Figure 2.4: Positively invariant set $\mathcal{I}$ for an autonomous system with $\boldsymbol{x} = (x_1\ x_2)^\top$: Grey arrows represent $\dot{\boldsymbol{x}}$.

The attribute *positively* explicitly refers to the fact that the state may be outside the set for past times. Accordingly, the set $\mathcal{I}$ is called *invariant* if the state will remain within the set for future and past times. In case the system (2.4) is disturbed (i.e., $\boldsymbol{s}(t) \neq \boldsymbol{0}$), the set $\mathcal{I}$ is called *robustly positively invariant* if it remains positively invariant for a bounded disturbance.

Definition 2.11 (Robustly positively invariant set [50]) *The set $\mathcal{I} \subset \mathbb{R}^n$ is called robustly positively invariant (RPI) for the system* (2.4) *if for all $\boldsymbol{x}_0 \in \mathcal{I}$ and $\boldsymbol{s} \in \mathcal{S} \subset \mathbb{R}^m$ the solution $\boldsymbol{x}(t) \in \mathcal{I}$ for all $t > 0$.*

For a linear discrete-time system

$$\boldsymbol{x}(k+1) = \boldsymbol{A}\boldsymbol{x}(k) + \boldsymbol{E}\boldsymbol{s}(k), \quad \boldsymbol{x}(0) = \boldsymbol{x}_0,$$

the set $\mathcal{I}$ is robustly positively invariant if and only if the condition

$$\boldsymbol{A} \cdot \mathcal{I} \oplus \boldsymbol{E} \cdot \mathcal{S} \subseteq \mathcal{I} \tag{2.5}$$

is fulfilled.

2.1.6 Graph theory

A directed graph is denoted by $\mathcal{G} = (\mathcal{V}, \mathcal{E})$, where the set $\mathcal{V}$ represents the set of vertices and the set $\mathcal{E} \subseteq \mathcal{V} \times \mathcal{V}$ denotes the set of edges. An edge from vertex i to vertex j is denoted by (i, j).

A path that starts at vertex i and ends in vertex j traversing several vertices is denoted by $\mathrm{P}(i,j) = ((k_0, k_1), ..., (k_{l-1}, k_l))$, where $k_0 = i, k_l = j$ and $k_m \neq k_n$, $(m, n = 1, ..., l,\ m \neq n)$. With $\mathrm{P}(i, \mathcal{K}, j)$ a path is denoted including edges to and from vertices listed in the set $\mathcal{K}$.

On basis of the given notation the definition of strong connection between two vertices is formulated.

Definition 2.12 (Strong connectivity [117]) *The vertex i is called strongly connected to vertex j if in the graph $\mathcal{G}$ there exists a path $\mathrm{P}(i,j)$ and $\mathrm{P}(j,i)$.*

The inversion of a graph $\mathcal{G}$ is denoted by $\mathcal{G}^\top = \left(\mathcal{V}, \mathcal{E}^\top\right)$, where $\mathcal{E}^\top = \{(j,i) \colon (i,j) \in \mathcal{E}\}$. A weighted graph is denoted by $\mathcal{G} = (\mathcal{V}, \mathcal{E}, w)$, where w represents the weight function that assigns each edge a specific value. The reachability graph with respect to the vertex i of a given graph $\mathcal{G}$ is a directed graph denoted by $\mathcal{G}_{\mathrm{R}i}(\mathcal{G}) = (\mathcal{V}_{\mathrm{R}i}, \mathcal{E}_{\mathrm{R}i})$. The set of vertices and edges is defined by $\mathcal{V}_{\mathrm{R}i} = \{j \colon \mathrm{P}(i,j) \text{ in } \mathcal{G}\}$ and $\mathcal{E}_{\mathrm{R}i} = \{(j,k) \colon j, k \in \mathcal{V}_{\mathrm{R}i},\ (j,k) \in \mathcal{E}\}$, respectively.

2.2 Models

2.2.1 Interconnected system

The interconnected systems is composed of N physically coupled subsystems as highlighted by the yellow box in Fig. 2.5. The subsystem $i \in \mathcal{N} := \{1, ..., N\}$ is represented by the linear state-space model

$$S_i : \begin{cases} \dot{\boldsymbol{x}}_i(t) = \boldsymbol{A}_i \boldsymbol{x}_i(t) + \boldsymbol{B}_i \boldsymbol{u}_i(t) + \boldsymbol{E}_i \boldsymbol{s}_i(t), \quad \boldsymbol{x}_i(0) = \boldsymbol{x}_{i,0} \\ \boldsymbol{y}_i(t) = \boldsymbol{C}_i \boldsymbol{x}_i(t) \\ \boldsymbol{z}_i(t) = \boldsymbol{C}_{\mathrm{z}i} \boldsymbol{x}_i(t), \end{cases} \tag{2.6}$$

where $\boldsymbol{x}_i \in \mathbb{R}^{n_i}$, $\boldsymbol{u}_i \in \mathbb{R}^{m_i}$, $\boldsymbol{y}_i \in \mathbb{R}^{r_i}$, $\boldsymbol{s}_i \in \mathbb{R}^{m_{si}}$ and $\boldsymbol{z}_i \in \mathbb{R}^{r_{zi}}$ denote the state, the control input, the measurement output, the coupling input and the coupling output, respectively. Subsystem S_i is interconnected with the remaining subsystems according to the relation

$$K_i: \quad \boldsymbol{s}_i(t) = \sum_{j \in \mathcal{P}_i} \boldsymbol{L}_{ij} \boldsymbol{z}_j(t), \tag{2.7}$$

where the matrix $\boldsymbol{L}_{ij} \in \mathbb{R}^{m_{si} \times r_{zi}}$ represents the coupling from subsystem S_j to subsystem S_i and the set $\mathcal{P}_i$ denotes the *predecessors* of subsystem S_i and is defined as follows:

$$\mathcal{P}_i := \{j \colon \boldsymbol{L}_{ij} \neq \boldsymbol{O}\}.$$

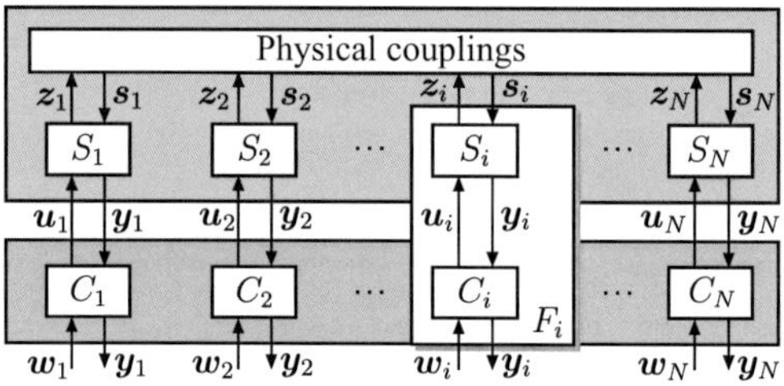

Figure 2.5: Structure of the decentralised controlled interconnected system.

Concerning the models (2.6) and (2.7), the following assumptions are made:

A 2.1 *The subsystem dynamics and the physical couplings are exactly represented by the model* (2.6) *and* (2.7).

A 2.2 *The pair* $(\boldsymbol{A}_i, \boldsymbol{B}_i)$ *is stabilisable and the pair* $(\boldsymbol{A}_i, \boldsymbol{C}_i)$ *is detectable for each* $i \in \mathcal{N}$.

A 2.3 $\boldsymbol{L}_{ii} = \boldsymbol{O}$ *for all* $i \in \mathcal{N}$, *i.e., the coupling input* $\boldsymbol{s}_i(t)$ *does not directly depend on the coupling output* $\boldsymbol{z}_i(t)$.

The latter assumption is without loss of generality since all internal dynamics can be modelled in the system matrix $\boldsymbol{A}_i$ for all $i \in \mathcal{N}$.

Locally interconnected system. A special interconnection structure is represented by a local interconnection among the subsystems, where subsystem S_i only influences the subsystems S_{i-1} and S_{i+1} and, conversely, is only influenced by subsystems S_{i-1} and S_{i+1}. This chain like interconnection is modelled by

$$K_1: \quad \boldsymbol{s}_1(t) = \boldsymbol{L}_{12}\boldsymbol{z}_2(t) \tag{2.8a}$$

$$K_i: \quad \boldsymbol{s}_i(t) = \boldsymbol{L}_{ii-1}\boldsymbol{z}_{i-1}(t) + \boldsymbol{L}_{ii+1}\boldsymbol{z}_{i+1}(t), \quad i = 2, .., N-1 \tag{2.8b}$$

$$K_N: \quad \boldsymbol{s}_N(t) = \boldsymbol{L}_{NN-1}\boldsymbol{z}_{N-1}(t) \tag{2.8c}$$

as visualised by the interconnection graph $\mathcal{G}_\mathrm{K}$ shown in Fig. 3.7a on page 60.

Overall system model. The model (2.6), (2.7) is called the *interconnection oriented model* of the overall system [117]. Alternatively, the overall plant can be represented by the *unstructured model*

$$S: \begin{cases} \dot{\boldsymbol{x}}(t) = \boldsymbol{A}\boldsymbol{x}(t) + \boldsymbol{B}\boldsymbol{u}(t), \quad \boldsymbol{x}(0) = \boldsymbol{x}_0 \\ \boldsymbol{y}(t) = \boldsymbol{C}\boldsymbol{x}(t) \end{cases} \tag{2.9}$$

with the accumulated signals

$$\boldsymbol{x}(t) = \begin{pmatrix} \boldsymbol{x}_1(t) \\ \vdots \\ \boldsymbol{x}_N(t) \end{pmatrix}, \quad \boldsymbol{u}(t) = \begin{pmatrix} \boldsymbol{u}_1(t) \\ \vdots \\ \boldsymbol{u}_N(t) \end{pmatrix}, \quad \boldsymbol{y}(t) = \begin{pmatrix} \boldsymbol{y}_1(t) \\ \vdots \\ \boldsymbol{y}_N(t) \end{pmatrix},$$

the combined matrices

$$\begin{aligned} \boldsymbol{A} &= \mathrm{diag}(\boldsymbol{A}_i)_{i\in\mathcal{N}} + \mathrm{diag}(\boldsymbol{E}_i)_{i\in\mathcal{N}} \cdot \boldsymbol{L} \cdot \mathrm{diag}(\boldsymbol{C}_{\mathrm{z}i})_{i\in\mathcal{N}} \\ \boldsymbol{B} &= \mathrm{diag}(\boldsymbol{B}_i)_{i\in\mathcal{N}} \\ \boldsymbol{C} &= \mathrm{diag}(\boldsymbol{C}_i)_{i\in\mathcal{N}} \end{aligned}$$

and the interconnection matrix

$$\boldsymbol{L} = \begin{pmatrix} \mathrm{O} & \boldsymbol{L}_{12} & \cdots & \boldsymbol{L}_{1N} \\ \boldsymbol{L}_{21} & \mathrm{O} & \cdots & \boldsymbol{L}_{2N} \\ \vdots & \vdots & \ddots & \vdots \\ \boldsymbol{L}_{N1} & \boldsymbol{L}_{N2} & \cdots & \mathrm{O} \end{pmatrix}. \tag{2.10}$$

Concerning the overall system S, the following assumption is made:

A 2.4 *The overall system S does not contain any decentralised fixed modes*

This assumption guarantees that the overall system (2.9) can be stabilised by a decentralised controller.

2.2.2 Controlled interconnected system

Throughout this thesis the interconnected system is to be controlled by a decentralised controller (blue box in Fig. 2.5). The control station $i \in \mathcal{N}$ is represented by the linear state-space model

$$C_i : \begin{cases} \dot{\boldsymbol{x}}_{\mathrm{C}i}(t) = \boldsymbol{A}_{\mathrm{C}i}\boldsymbol{x}_{\mathrm{C}i}(t) + \boldsymbol{B}_{\mathrm{C}i}\left(\boldsymbol{w}_i(t) - \boldsymbol{y}_i(t)\right), & \boldsymbol{x}_{\mathrm{C}i}(0) = \boldsymbol{x}_{\mathrm{C}i,0} \\ \boldsymbol{u}_i(t) = \boldsymbol{C}_{\mathrm{C}i}\boldsymbol{x}_{\mathrm{C}i}(t) + \boldsymbol{D}_{\mathrm{C}i}\left(\boldsymbol{w}_i(t) - \boldsymbol{y}_i(t)\right), \end{cases} \tag{2.11}$$

where $\boldsymbol{x}_{\mathrm{C}i} \in \mathbb{R}^{n_{\mathrm{C}i}}$ and $\boldsymbol{w}_i \in \mathbb{R}^{r_i}$ denote the state of the control station and the reference input.

The subsystem, which is controlled by the control station is called *controlled subsystem* and is represented by the linear state-space model

$$F_i : \begin{cases} \dot{\boldsymbol{x}}_{\mathrm{F}i}(t) = \boldsymbol{A}_{\mathrm{F}i}\boldsymbol{x}_{\mathrm{F}i}(t) + \boldsymbol{B}_{\mathrm{F}i}\boldsymbol{w}_i(t) + \boldsymbol{E}_{\mathrm{F}i}\boldsymbol{s}_i(t), & \boldsymbol{x}_{\mathrm{F}i}(0) = \boldsymbol{x}_{\mathrm{F}i,0} \\ \boldsymbol{y}_i(t) = \boldsymbol{C}_{\mathrm{F}i}\boldsymbol{x}_{\mathrm{F}i}(t) \\ \boldsymbol{z}_i(t) = \boldsymbol{C}_{\mathrm{Fz}i}\boldsymbol{x}_{\mathrm{F}i}(t), \end{cases} \tag{2.12}$$

where $\boldsymbol{x}_{\mathrm{F}i}(t) = \begin{pmatrix} \boldsymbol{x}_i^\top(t) & \boldsymbol{x}_{\mathrm{C}i}^\top(t) \end{pmatrix}^\top$.

2.2.3 Communication network

In this thesis, the communication network is assumed to be ideal in the following sense:

A 2.5 *Information is transmitted over the communication network without time-delays, package losses or falsifications.*

Indeed, this assumption is impossible to satisfy. However, compared to the dynamics of the controlled plant the transmission of information happens much faster which justifies the assumption A 2.5.

2.3 Robust control techniques

In the field of robust control, the control stations are designed based on an uncertain or an incomplete model [116, 192]. This section introduces comparison systems that are used to represent unknown-but-bounded model uncertainties. Afterwards, the H_∞-norm is introduced as a basis for the design of robust control stations.

2.3.1 Systems with unknown-but-bounded model uncertainty

Consider the model uncertainty to be in feedback structure with the system as depicted in Fig. 2.6a. Let the system be represented by the frequency domain model

$$S : \begin{cases} \boldsymbol{y}(s) = \hat{\boldsymbol{S}}(s)\,\boldsymbol{u}(s) + \boldsymbol{S}_{\mathrm{yq}}(s)\,\boldsymbol{q}(s) \\ \boldsymbol{p}(s) = \boldsymbol{S}_{\mathrm{pu}}(s)\,\boldsymbol{u}(s) + \boldsymbol{S}_{\mathrm{pq}}(s)\,\boldsymbol{q}(s), \end{cases} \tag{2.13}$$

where the additional signals $\boldsymbol{q}(s) \in \mathbb{C}^{m_{\mathrm{q}}}$ and $\boldsymbol{p}(s) \in \mathbb{C}^{r_{\mathrm{p}}}$ denote the error input signal and the error output signal, respectively. The model uncertainty is represented by the *uncertainty model*

$$E : \quad \boldsymbol{q}(s) = \boldsymbol{E}(s)\,\boldsymbol{p}(s), \tag{2.14}$$

where the transfer function matrix $\boldsymbol{E}(s)$ is assumed to be unknown. The model

$$\hat{S} : \quad \hat{\boldsymbol{y}}(s) = \hat{\boldsymbol{S}}(s)\,\boldsymbol{u}(s), \tag{2.15}$$

is referred to as the *approximate model* whereas the difference

$$\boldsymbol{y}_{\Delta}(s) = \boldsymbol{y}(s) - \hat{\boldsymbol{y}}(s)$$

between the approximated output $\hat{\boldsymbol{y}}(s)$ and the original output $\boldsymbol{y}(s)$ is modelled by the *error model*, which is represented by

$$S_{\Delta} : \quad \boldsymbol{y}_{\Delta}(s) = \boldsymbol{S}_{\Delta}(s)\,\boldsymbol{u}(s) \tag{2.16}$$

with

$$\boldsymbol{S}_{\Delta}(s) = \boldsymbol{S}_{\mathrm{yq}}(s)\boldsymbol{E}(s)\big(\mathbf{I} - \boldsymbol{S}_{\mathrm{pq}}(s)\boldsymbol{E}(s)\big)^{-1}\boldsymbol{S}_{\mathrm{pu}}(s)$$

provided that the inverse exists. The approximate model (2.15) and the error model (2.16) are connected in parallel as visualised in Fig. 2.6b.

(a) Feedback connection of the system S with the model uncertainty E

(b) Parallel connection of the approximate model $\hat{S}$ and the error model S_{Δ}

Figure 2.6: Structure of the uncertain system (adapted from [116]).

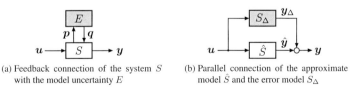

Comparison systems. The uncertainty is described by means of an upper bound on the frequency response $\boldsymbol{E}(\mathrm{j}\omega)$. This upper bound is modelled by a so called *comparison system*

$$\bar{E}: \quad \bar{q}(\omega) = \bar{E}(\omega)\,\|\boldsymbol{p}(\mathrm{j}\omega)\|, \tag{2.17}$$

where $\bar{E}(\omega) \in \mathbb{R}_+$.

Definition 2.13 (Norm-related comparison system [116]) *The system* (2.17) *is a comparison system of the uncertainty model* (2.14) *if*

$$\bar{E}(\omega) \geq \|\boldsymbol{E}(\mathrm{j}\omega)\|, \quad \forall \omega \in \mathbb{R}$$

holds. Then $\bar{q}(\omega) \geq \|\boldsymbol{q}(\mathrm{j}\omega)\|$ *holds for all* $\omega \in \mathbb{R}$ *and for arbitrary bounded functions* $\mathbf{p}(\mathrm{j}\omega)$.

As a consequence,

$$\bar{S}_\Delta: \quad \bar{y}_\Delta(\omega) = \bar{S}_\Delta(\omega)\,\|\boldsymbol{u}(\mathrm{j}\omega)\|$$

with

$$\bar{S}_\Delta(\omega) := \|\boldsymbol{S}_{\mathrm{yq}}(\mathrm{j}\omega)\|\,\bar{E}(\omega)\left(1 - \|\boldsymbol{S}_{\mathrm{pq}}(\mathrm{j}\omega)\|\,\bar{E}(\omega)\right)^{-1}\|\boldsymbol{S}_{\mathrm{pu}}(\mathrm{j}\omega)\|$$

represents a comparison system of the error model (2.16) (i.e., $\bar{y}_\Delta(\omega) \geq \|\boldsymbol{y}_\Delta(\mathrm{j}\omega)\|,\ \forall \omega \in \mathbb{R}$) if the relation

$$\|\boldsymbol{S}_{\mathrm{pq}}(\mathrm{j}\omega)\|\,\bar{E}(\omega) < 1, \quad \forall \omega \in \mathbb{R} \tag{2.18}$$

holds.

In direct analogy, a comparison system for the time-domain uncertainty model

$$E: \quad \boldsymbol{q}(t) = \boldsymbol{E}(t) * \boldsymbol{p}(t), \tag{2.19}$$

is represented by

$$\bar{E}: \quad \bar{\boldsymbol{q}}(t) = \bar{\boldsymbol{E}}(t) * |\boldsymbol{p}(t)|, \tag{2.20}$$

where the comparison applies element-wise.

Definition 2.14 (Element-by-element-related comparison system [116, 117]) *The system* (2.20) *is a comparison system of the uncertainty model* (2.19) *if*

$$\bar{\boldsymbol{E}}(t) \geq |\boldsymbol{E}(t)|, \quad \forall t \geq 0$$

holds. Then $\bar{\boldsymbol{q}}(t) \geq |\boldsymbol{q}(t)|$ *holds for all* $t \geq 0$ *and for arbitrary bounded functions* $\mathbf{p}(t)$.

Accordingly, the comparison system of the error model is represented by

$$\bar{S}_{\Delta}: \quad \bar{\boldsymbol{y}}(t) = \bar{\boldsymbol{S}}_{\Delta}(t) * |\boldsymbol{u}(t)|$$

with

$$\bar{\boldsymbol{S}}_{\Delta}(t) := |\boldsymbol{S}_{\mathrm{yq}}(t)| * \bar{\boldsymbol{\Psi}}(t) * |\boldsymbol{S}_{\mathrm{pu}}(t)|$$

and

$$\bar{\boldsymbol{\Psi}}(t) = \bar{\boldsymbol{E}}(t) + \bar{\boldsymbol{E}}(t) * |\boldsymbol{S}_{\mathrm{pq}}(t)| * \bar{\boldsymbol{\Psi}}(t) \tag{2.21}$$

providing that $\bar{\boldsymbol{\Psi}}(t) \geq \mathbf{O}$ for all $t \geq 0$ holds.

2.3.2 Basics on H_∞-norm

A popular controller design method for uncertain systems is based on the H_∞-norm [72, 192]. This section recapitulates the basics on H_∞-norm in control.

Consider the unstructured system model (2.9) in I/O frequency domain representation

$$S: \quad \boldsymbol{y}(s) = \boldsymbol{S}(s)\,\boldsymbol{u}(s)$$

with

$$\boldsymbol{S}(s) = \boldsymbol{C}\big(s\mathbf{I} - \boldsymbol{A}\big)^{-1}\boldsymbol{B}. \tag{2.22}$$

The H_∞-norm of the transfer function (2.22) means

$$\|\boldsymbol{S}(s)\|_{\mathrm{H}_\infty} = \sup_{\omega\in\mathbb{R}} \sigma_{\max}\left(\boldsymbol{S}(\mathrm{j}\omega)\right). \tag{2.23}$$

That is, $\|\boldsymbol{S}(s)\|_{\mathrm{H}_\infty} > \|\boldsymbol{S}(\mathrm{j}\omega)\|_2$ for all $\omega \in \mathbb{R}$. For a scalar transfer function, the H_∞-norm yields the maximal magnitude of the amplitude plot

$$\|S(s)\|_{\mathrm{H}_\infty} = \sup_{\omega\in\mathbb{R}} |S(\mathrm{j}\omega)|.$$

The H_∞-norm of a transfer function can also be represented by the corresponding state-space model (2.9). The following alternative characterisation of the H_∞-norm will used in this thesis.

Lemma 2.2 (Characterisation of the H_∞-norm [58, 154]) *For a given scalar $\gamma \in \mathbb{R}_+$ and the transfer function* (2.22) *of the system* (2.9) *the following statements are equivalent:*

1. *There exists a symmetric matrix $\boldsymbol{P} \succ 0$ such that*

$$\begin{pmatrix} \boldsymbol{A}^\top \boldsymbol{P} + \boldsymbol{P}\boldsymbol{A} & \boldsymbol{P}\boldsymbol{B} & \boldsymbol{C}^\top \\ \star & -\gamma \mathbf{I} & \mathbf{O} \\ \star & \star & -\gamma \mathbf{I} \end{pmatrix} \prec 0.$$

2. *$\boldsymbol{A}$ is Hurwitz and $\|\boldsymbol{S}(s)\|_{\mathrm{H}_\infty} < \gamma$.*

To analyse whether the H_∞-norm of the transfer function $\boldsymbol{S}(s)$ conforms a specific shape, an additional weight function $\boldsymbol{W}(s)$ has to be introduced. Therefore, the system to be analysed is extended by the transfer function matrix $\boldsymbol{W}(s)$ as shown in Fig. 2.7. Hence, if the H_∞-norm from the input $\boldsymbol{u}(s)$ to the performance output $\boldsymbol{v}(s)$ is limited according to

$$\|\boldsymbol{S}(s)\boldsymbol{W}(s)\|_{\mathrm{H}_\infty} < \gamma, \tag{2.24}$$

then, for scalar transfer functions, the amplitude plot of the system satisfies the relation

$$|S(\mathrm{j}\omega)| < \gamma \cdot \left(|W(\mathrm{j}\omega)|\right)^{-1}.$$

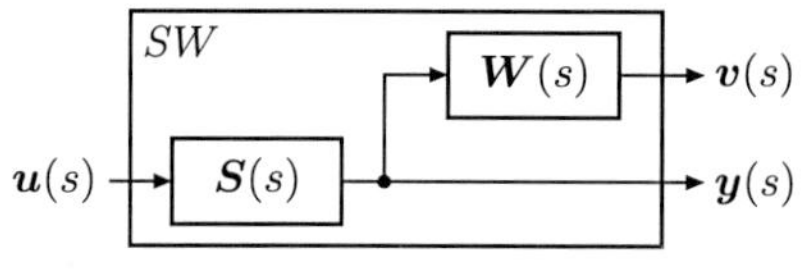

Figure 2.7: Extended system SW.

The weight function is assumed to be representable as a state-space model

$$W: \begin{cases} \dot{\boldsymbol{x}}_{\mathrm{W}}(t) = \boldsymbol{A}_{\mathrm{W}}\boldsymbol{x}_{\mathrm{W}}(t) + \boldsymbol{B}_{\mathrm{W}}\boldsymbol{y}(t), & \boldsymbol{x}_{\mathrm{W}}(0) = \boldsymbol{0} \\ \boldsymbol{v}(t) = \boldsymbol{C}_{\mathrm{W}}\boldsymbol{x}_{\mathrm{W}}(t) + \boldsymbol{D}_{\mathrm{W}}\boldsymbol{y}(t), & \end{cases} \tag{2.25}$$

where $\boldsymbol{W}(s) = \boldsymbol{D}_{\mathrm{W}} + \boldsymbol{C}_{\mathrm{W}}(s\mathbf{I} - \boldsymbol{A}_{\mathrm{W}})^{-1}\boldsymbol{B}_{\mathrm{W}}$. Accordingly, to check whether the relation (2.24) holds, Lemma 2.2 can be used for the extended system

$$SW: \begin{cases} \begin{pmatrix} \dot{\boldsymbol{x}}(t) \\ \dot{\boldsymbol{x}}_{\mathrm{W}}(t) \end{pmatrix} = \begin{pmatrix} \boldsymbol{A} & \mathrm{O} \\ \boldsymbol{B}_{\mathrm{W}}\boldsymbol{C} & \boldsymbol{A}_{\mathrm{W}} \end{pmatrix} \begin{pmatrix} \boldsymbol{x}(t) \\ \boldsymbol{x}_{\mathrm{W}}(t) \end{pmatrix} + \begin{pmatrix} \boldsymbol{B} \\ \mathrm{O} \end{pmatrix} \boldsymbol{u}(t), \quad \begin{pmatrix} \boldsymbol{x}(0) \\ \boldsymbol{x}_{\mathrm{W}}(0) \end{pmatrix} = \begin{pmatrix} 0 \\ 0 \end{pmatrix} \\ \boldsymbol{v}(t) = \begin{pmatrix} \boldsymbol{D}_{\mathrm{W}}\boldsymbol{C} & \boldsymbol{C}_{\mathrm{W}} \end{pmatrix} \begin{pmatrix} \boldsymbol{x}(t) \\ \boldsymbol{x}_{\mathrm{W}}(t) \end{pmatrix} \\ \boldsymbol{y}(t) = \begin{pmatrix} \boldsymbol{C} & \mathrm{O} \end{pmatrix} \begin{pmatrix} \boldsymbol{x}(t) \\ \boldsymbol{x}_{\mathrm{W}}(t) \end{pmatrix} \end{cases}$$

which is the serial connection of the models (2.9) and (2.25).

2.4 Fault-tolerant control methods

Fault-tolerant control (FTC) aims at making technical systems tolerant to faults in the sense that the system shall fulfil its function in the presence of faults [53]. This section starts with the introduction of the considered faults and its modelling. Thereafter, reconfiguration and diagnosis techniques are recapitulated.

2.4.1 Model of the faulty subsystem

A fault is denoted by $f \in \mathcal{F}$, where $\mathcal{F} := \{f_1, ..., f_q\}$ is the known *set of considered faults*. It is assumed that the fault affects a single subsystems, which is, without loss of generality subsystem S_1. Three different kinds of faults are considered, namely actuator faults, sensor faults and process faults. An actuator fault or a sensor fault lead to a malfunctions of the actuator or corruptions on the measurements, respectively. A process fault changes the internal dynamics of the subsystem. Moreover, a failure of an actuator or a sensor makes it useless.

The class of the considered faults and its occurrence is restricted as follows:

A 2.6 *All considered faults $f \in \mathcal{F}$ are persistent and only a single fault occurs. The fault occurs in subsystem S_1 at the unknown time $t_{\mathrm{f}} \geq 0$ when all subsystems are in their operating point, i.e., $\boldsymbol{x}_i(t_{\mathrm{f}}) = \mathbf{0}$, $(\forall i \in \mathcal{N})$.*

The faulty subsystem S_1 is represented by the state-space model

$$S_{1f}: \begin{cases} \dot{\boldsymbol{x}}_{1f}(t) &= \boldsymbol{A}_{1f}\boldsymbol{x}_{1f}(t) + \boldsymbol{B}_{1f}\boldsymbol{u}_{1f}(t) + \boldsymbol{E}_1\boldsymbol{s}_{1f}(t), \quad \boldsymbol{x}_{1f}(t_{\mathrm{f}}) = \mathbf{0} \\ \boldsymbol{y}_{1f}(t) &= \boldsymbol{C}_{1f}\boldsymbol{x}_{1f}(t) \\ \boldsymbol{z}_{1f}(t) &= \boldsymbol{C}_{\mathrm{z}1}\boldsymbol{x}_{1f}(t). \end{cases} \tag{2.26}$$

A process fault is modelled by changing the corresponding element $\boldsymbol{A}_{1,ij}$ in $\boldsymbol{A}_1$, henceforth denoted by $\boldsymbol{A}_{1f}$. Similarly, a sensor or an actuator fault are modelled by changing the corresponding row $\boldsymbol{c}_{1,j}^{\top}$ in $\boldsymbol{C}_1$ (denoted by $\boldsymbol{C}_{1f}$) or by changing the corresponding column $\boldsymbol{b}_{1,j}$ in $\boldsymbol{B}_1$ (denoted by $\boldsymbol{B}_{1f}$). An actuator failure is represented by setting the corresponding column $\boldsymbol{b}_{1,j}$ in $\boldsymbol{B}_1$ to zero.

2.4.2 Controller reconfiguration by a virtual actuator

This section recapitulates the controller reconfiguration after actuator failures by means of a virtual actuator [53, 144, 166].

Let the overall system (2.9) be controlled by a linear controller that is represented in state space by

$$C: \begin{cases} \dot{\boldsymbol{x}}_{\mathrm{C}}(t) = \boldsymbol{A}_{\mathrm{C}}\boldsymbol{x}_{\mathrm{C}}(t) + \boldsymbol{B}_{\mathrm{C}}\left(\boldsymbol{w}(t) - \boldsymbol{y}(t)\right), \quad \boldsymbol{x}_{\mathrm{C}}(0) = \boldsymbol{0} \\ \boldsymbol{u}(t) = \boldsymbol{C}_{\mathrm{C}}\boldsymbol{x}_{\mathrm{C}}(t) + \boldsymbol{D}_{\mathrm{C}}\left(\boldsymbol{w}(t) - \boldsymbol{y}(t)\right) \end{cases}$$

in order that the model of the nominal closed-loop is represented by

$$F: \begin{cases} \dot{\boldsymbol{x}}_{\mathrm{F}}(t) = \boldsymbol{A}_{\mathrm{F}}\boldsymbol{x}_{\mathrm{F}}(t) + \boldsymbol{B}_{\mathrm{F}}\boldsymbol{w}(t), \quad \boldsymbol{x}_{\mathrm{F}}(0) = \boldsymbol{0} \\ \boldsymbol{y}(t) = \boldsymbol{C}_{\mathrm{F}}\boldsymbol{x}_{\mathrm{F}}(t). \end{cases}$$

If an actuator fails, the faulty system is modelled by

$$S_f: \begin{cases} \dot{\boldsymbol{x}}_f(t) &= \boldsymbol{A}\boldsymbol{x}_f(t) + \boldsymbol{B}_f\boldsymbol{u}_f(t), \quad \boldsymbol{x}_f(t_{\mathrm{f}}) = \boldsymbol{0} \\ \boldsymbol{y}_f(t) &= \boldsymbol{C}\boldsymbol{x}_f(t) \end{cases}$$

To recover asymptotic stability of the closed-loop system, a *virtual actuator*

$$VA: \begin{cases} \dot{\boldsymbol{x}}_\delta(t) = \boldsymbol{A}_\delta\boldsymbol{x}_\delta(t) + \boldsymbol{B}\mathring{\boldsymbol{u}}(t), \quad \boldsymbol{x}_\delta(t_{\mathrm{f}}) = \boldsymbol{0} \\ \mathring{\boldsymbol{y}}(t) = \boldsymbol{y}_f(t) + \boldsymbol{C}\boldsymbol{x}_\delta(t) \\ \boldsymbol{u}_f(t) = \boldsymbol{M}\boldsymbol{x}_\delta(t) \end{cases} \tag{2.27}$$

with $\boldsymbol{A}_\delta = \boldsymbol{A} - \boldsymbol{B}_f\boldsymbol{M}$ is placed in between the faulty system and the nominal controller as shown in Fig. 2.8. The reconfigured signals are denoted by $\mathring{\boldsymbol{y}}$ and $\mathring{\boldsymbol{u}}$. Accordingly, the virtual actuator VA together with the nominal controller C represents the reconfigured controller $\mathring{C}$. The main idea of the virtual actuator is to make use of alternative actuators to mimic the failed actuator in order to stabilise the faulty system. Therefore, the virtual actuator aims at driving the difference state $\boldsymbol{x}_\delta(t) = \boldsymbol{x}(t) - \boldsymbol{x}_f(t)$ to zero such that the faulty system behaves like the nominal system and, thus, the effect of the fault is hidden from the nominal controller.

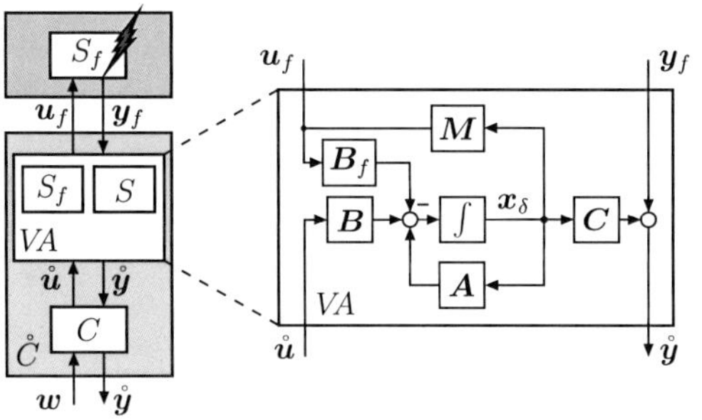

Figure 2.8: Structure of the reconfigured closed-loop system and the virtual actuator. Reconfiguration of controller C using a virtual actuator VA. The result is the reconfigured controller $\mathring{C}$. For the design and the execution of the virtual actuator the models S and S_f are required.

The analysis of the reconfigured closed-loop system $\mathring{F}$ that is modelled by

$$\mathring{F}: \begin{cases} \frac{\mathrm{d}}{\mathrm{d}t}\begin{pmatrix} \tilde{\boldsymbol{x}}_{\mathrm{F}}(t) \\ \boldsymbol{x}_\delta(t) \end{pmatrix} = \begin{pmatrix} \boldsymbol{A}_{\mathrm{F}} & \mathrm{O} \\ \begin{pmatrix} -\boldsymbol{B}\boldsymbol{D}_{\mathrm{C}}\boldsymbol{C} & \boldsymbol{B}\boldsymbol{C}_{\mathrm{C}} \end{pmatrix} & \boldsymbol{A}_\delta \end{pmatrix} \begin{pmatrix} \tilde{\boldsymbol{x}}_{\mathrm{F}}(t) \\ \boldsymbol{x}_\delta(t) \end{pmatrix} + \begin{pmatrix} \boldsymbol{B}_{\mathrm{F}} \\ \boldsymbol{B}\boldsymbol{D}_{\mathrm{C}} \end{pmatrix} \boldsymbol{w}(t) \\ \mathring{\boldsymbol{y}}(t) = \begin{pmatrix} \mathrm{O} & -\boldsymbol{C}_{\mathrm{F}} \end{pmatrix} \begin{pmatrix} \tilde{\boldsymbol{x}}_{\mathrm{F}}(t) \\ \boldsymbol{x}_\delta(t) \end{pmatrix}, \end{cases}$$

where $\tilde{\boldsymbol{x}}_{\mathrm{F}}(t) = \begin{pmatrix} \boldsymbol{x}_f^\top(t) - \boldsymbol{x}_\delta^\top(t) & \boldsymbol{x}_{\mathrm{C}}^\top(t) \end{pmatrix}^\top$ indicates that asymptotic stability of the overall closed-loop system is recovered if and only if $\boldsymbol{A}_\delta$ is Hurwitz providing that the nominal closed-loop system has been asymptotically stable before.

Theorem 2.2 (Stabilising reconfiguration using a virtual actuator [166]) *The virtual actuator* (2.27) *stabilises the faulty closed-loop system if and only if the feedback gain $\boldsymbol{M}$ is designed such the matrix $\boldsymbol{A}_\delta = \boldsymbol{A} - \boldsymbol{B}_f\boldsymbol{M}$ is Hurwitz. There exists a feedback gain $\boldsymbol{M}$ if and only if the pair $(\boldsymbol{A}, \boldsymbol{B}_f)$ is stabilisable.*

Note that in direct analogy to the reconfiguration after actuator failures, the reference [166] has presented a *virtual sensor* as a reconfiguration solution after sensor failures. All reconfiguration solutions that are proposed in this thesis which are based on a virtual actuator can easily be adapted to sensor failures by means of a virtual sensor.

2.4.3 Model-based fault diagnosis

This subsection recapitulates basics on model-based diagnosis [53, 60, 70].

Consider a dynamical system modelled by (2.9) that is subject to some faults $f \in \mathcal{F}$. A model-based diagnosis system monitors the system by means of its signals $\boldsymbol{u}(t)$ and $\boldsymbol{y}(t)$. The basic principle is to test whether or not the measurements $\boldsymbol{u}(t)$ and $\boldsymbol{y}(t)$ are consistent with the system behaviour. Therefore, the diagnosis system exploits a model S of the fault-free system as well as models S_f, $(\forall f \in \mathcal{F})$ of the faulty system. Regarding the model-based diagnosis, it is distinguished between the following diagnosis steps:

- *Fault detection*: Decide whether or not a fault has occurred. A fault is called detected if the measurements $\boldsymbol{u}(t)$ and $\boldsymbol{y}(t)$ are inconsistent with the behaviour of the faultless system S.
- *Fault isolation and identification*: Find the faulty component and identify the impact of the fault. A fault $f \in \mathcal{F}$ is called isolated if the measurements $\boldsymbol{u}(t)$ and $\boldsymbol{y}(t)$ are consistent with the behaviour of the faulty system S_f and inconsistent with the behaviours of the healthy system S and all other systems S_k, $(\forall k \in \mathcal{F}, k \neq f)$.

The issue under which conditions a certain fault can be detected and isolated concerns the *fault detectability* and *fault isolability*, respectively, which is considered in Chapter 8.

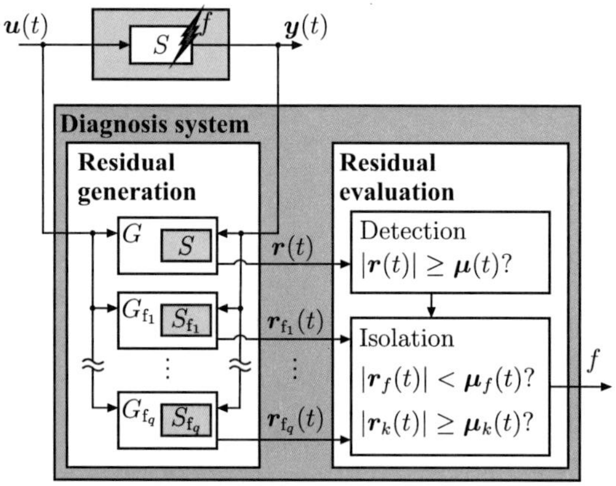

Figure 2.9: Structure of the diagnosis system: Residual generation consists of a detection residual generator G and q isolation residual generators G_k, $(k \in \mathcal{F})$. Residual evaluation detects and isolates the fault by means of thresholds.

Figure 2.9 shows the structure of a model-based diagnosis system. The structure comprises two stages of residual generation and residual evaluation. The purpose of residual generation is to generate a *residual* (fault indication signal) that indicates whether or not a fault has occurred. In the fault-free case, this residual is zero and differs from zero if a fault has occurred. The algorithm that generates the residual is called *residual generator*. The residual evaluation analysis the generated residual. Based on a decision rule it is concluded if any fault has occurred (fault detection) and which of the considered faults has occurred (fault isolation).

For fault detection, a *detection residual generator* is used that is represented by the I/O-oriented model

$$G:\ \boldsymbol{r}(t) = \boldsymbol{G}_{\mathrm{ru}}(t) * \boldsymbol{u}(t) + \boldsymbol{G}_{\mathrm{ry}}(t) * \boldsymbol{y}(t),$$

where $\boldsymbol{r}(t)$ denotes the detection residual (Fig. 2.9). This residual generator exploits the model S of the fault-free system. The fault is detected if for a given detection threshold $\boldsymbol{\mu}(t)$

$$|\boldsymbol{r}(t)| \geq \boldsymbol{\mu}(t), \quad \exists t \geq t_{\mathrm{f}},$$

where t_{D} denotes the time instant when the fault is detected. At $t = t_{\mathrm{D}}$ the detection process ends and the isolation process is started.

To isolate a fault $f \in \mathcal{F} = \{\mathrm{f}_1, ..., \mathrm{f}_q\}$, q additional *isolation residual generators*

$$G_k:\ \boldsymbol{r}_k(t) = \boldsymbol{G}_{\mathrm{ru}k}(t) * \boldsymbol{u}(t) + \boldsymbol{G}_{\mathrm{ry}k}(t) * \boldsymbol{y}(t),$$

for $k = \mathrm{f}_1, ..., \mathrm{f}_q$ are introduced, where $\boldsymbol{r}_k(t)$ denotes the isolation residual for the fault $k \in \mathcal{F}$ (Fig. 2.9). These residual generators make use of the respective model S_f of the faulty system. Initially, all faults $f \in \mathcal{F}$ are fault candidates. The fault candidate k is excluded from the set $\mathcal{F}$ of fault candidates if for a given isolation threshold, the relation

$$|\boldsymbol{r}_k(t)| \geq \boldsymbol{\mu}_k(t), \quad \exists t \geq t_{\mathrm{D}} \tag{2.28}$$

holds. Accordingly, the fault f is isolated if (2.28) holds for all $k \in \mathcal{F} \subset \{f\}$ and

$$|\boldsymbol{r}_f(t)| < \boldsymbol{\mu}_f(t), \quad \forall t \geq t_{\mathrm{D}}$$

holds. The moment the fault f is isolated is denoted by t_{I}.

It has to be emphasised that according to assumption A 2.6, the residuals $\boldsymbol{r}(t)$ and $\boldsymbol{r}_i(t)$ can only differ from zero if the faulty system is stimulated, (i.e., $\boldsymbol{u}(t) \neq \mathbf{0}$). The time instant when the system is stimulated is denoted by t_{S}.

Fig. 2.10 illustrates the evaluation of the residuals for the situation in which the fault $f = \mathrm{f}_1$ has occurred at time instant t_{f}. The system is stimulated at $t_{\mathrm{S}} > t_{\mathrm{f}}$. After the fault has been detected at t_{D}, the fault detection ends and the isolation process is started. The fault f_1 is isolated the moment the residual $|\boldsymbol{r}_{\mathrm{f}_2}|$ has exceeded its threshold $\boldsymbol{\mu}_{\mathrm{f}_2}$. At $t = t_{\mathrm{I}}$ the diagnosis process terminates with the result that $f = \mathrm{f}_1$ has occurred.

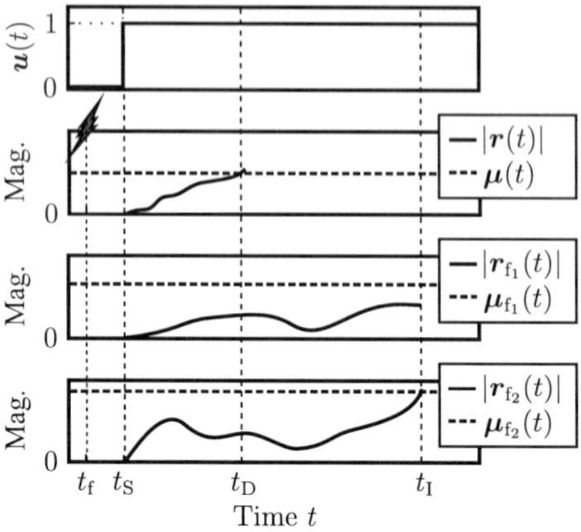

Figure 2.10: Evaluation of the detection residual $\boldsymbol{r}(t)$ and the identification residuals $\boldsymbol{r}_f(t)$ for two possible faults $f \in \mathcal{F} = \{\mathrm{f}_1, \mathrm{f}_2\}$.

2.5 Demonstration examples

The proposed methods for plug-and-play control are evaluated by simulations and experiments on two distinct processes. As the first demonstration example serves a thermofluid process realised at the Institute of Automation and Computer Control at Ruhr-Universität Bochum. The second example constitute a multizone furnace to grow GaAs crystals with the highest possible purity.

Figure 2.11: Pilot plant: Highlighted tanks are considered process.

2.5.1 Interconnected thermofluid process

The pilot plant (Fig. 2.11) consists of four cylindrical storage tanks, three reactors and a buffer tank connected over pipes. Over 70 sensors and 80 actuators provide the measurements and control actions, respectively. The tanks that are used for the demonstration process are highlighted in Fig. 2.11. The thermofluid process is used for simulations and experiments in Chapters 5 and 6.

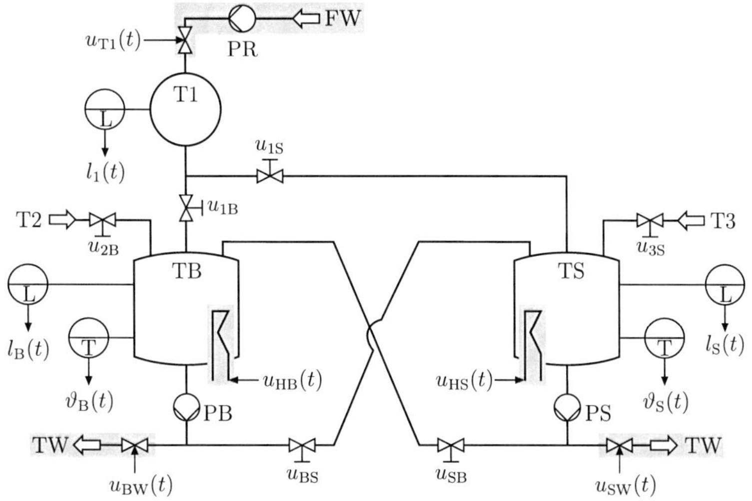

Figure 2.12: Setup of the interconnected thermofluid process: The actuator used for control are highlighted grey.

Process description. The setup of the thermofluid process is shown in Fig. 2.12. The process consists of the storage tanks T1 and the two reactors TB and TS used to realise continuous flow processes. The tank T1 is fed by the fresh water supply (FW) through the pump PR, where the inflow is controlled by means of the valve angle $u_{T1}(t)$. The liquid in T1 is passed on into the reactors TB and TS where the ratio is set by the valve angles u_{1B} and u_{1S}. In addition to the inflow from tank T1, the reactor TB is supplied with a constant inflow from the storage tank T2. Via pump PB the outflow of TB is pumped to the reactor TS while the remaining outflow is conducted to the buffer tank TW for further processing. The volume flow to TW is controlled by the valve angle $u_{BW}(t)$. The temperature $\vartheta_B(t)$ of the liquid in TB is adjustable by the heating rods using the input $u_{HB}(t)$. In addition, reactor TS is constantly fed by the liquid of tank T3. Equivalently to reactor TB, the outflow is split where one part is conveyed via the pump PS to reactor TB and the second part that is dosed through the valve angle $u_{\mathrm{SW}}(t)$ is passed to the buffer tank TW. The water in tank TS can be heated up by the heating rods that are driven by the signal $u_{\mathrm{HS}}(t)$.

The tanks are equipped with sensors that continuously measure the level $l_1(t)$, $l_{\mathrm{B}}(t)$, $l_{\mathrm{S}}(t)$ and the temperature $\vartheta_{\mathrm{B}}(t)$, $\vartheta_{\mathrm{S}}(t)$ of the content. The signals $u_{\mathrm{T1}}(t)$, $u_{\mathrm{BW}}(t)$, $u_{\mathrm{SW}}(t)$, $u_{\mathrm{HB}}(t)$ and $u_{\mathrm{HS}}(t)$ constitute the control signals while $u_{1\mathrm{S}}$, $u_{1\mathrm{B}}$, $u_{2\mathrm{B}}$, $u_{3\mathrm{S}}$, u_{BS} and u_{SB} are constants.

The three tanks are physically coupled by the flow from T1 to TS and TB as well as by the flow from TS to TB and vice versa. The coupling strength is adjustable by means of the valve angles $u_{1\mathrm{S}}$, $u_{1\mathrm{B}}$, u_{BS} and u_{SB}. The control of $u_{\mathrm{BW}}(t)$ and $u_{\mathrm{SW}}(t)$ is realised in a way that it does not influence the volume passed on to reactor TS and to reactor TB, respectively.

The nonlinear model of the process as well as the linearised models are given in Appendix A.1.

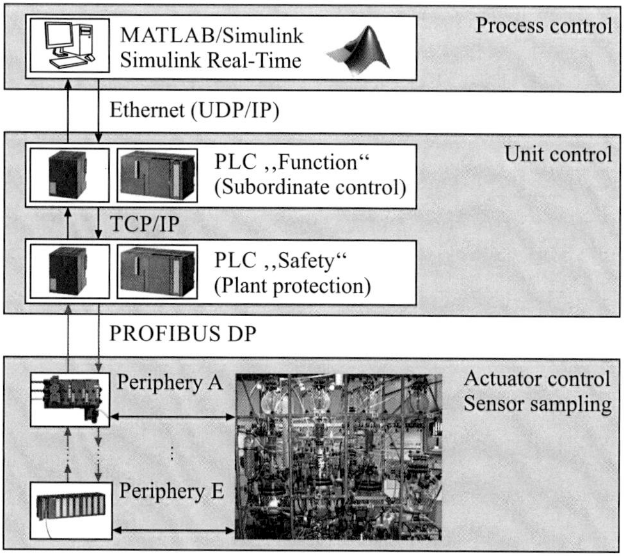

Figure 2.13: Automation concept for the pilot plant.

Hardware description. The automation concept for the pilot plant consists of three levels (Fig. 2.13). On the top level, the control algorithms are implemented in MATLAB/Simulink using Simulink Real-time with the sampling time $T = 0.2\,s$. The MATLAB workstation has an Intel Core i7-2600 Quad processor with 3.4 GHz each, 4 GB RAM and operates with Windows 7 64-bit. The MATLAB workstation and the programmable logic controllers (PLC) are connected over a 100 Mbits/s Ethernet network using UDP/IP. The PLCs communicate over PROFIBUS DP with the peripheral units, which are connected to the actuators and sensors.

As T is approximately 300 times faster than the time constants of the thermofluid process, the control can be considered to be continuous and the communication network can be assumed to be ideal in accordance with assumption A 2.5.

2.5.2 Multizone furnace

The multizone furnace (Fig. 2.14) consists of a pipe with a large number of heating zones and an ampulla that contains the GaAs mixture. Each zone is equipped with a separating sensing and actuating unit in order that a decentralised control solution is preferably used. To produce a GaAs crystal with high purity, first, the zones shall be heated up to melt the GaAs mixture. The gradient-freeze methods requires to cool down the liquid in a coordinated way. Hence, beginning from the left zone the zones are cooled down one after the other such that the pollution of the mixture will float to the end of the crystal. The heating zones are physically coupled by the heat transfer between neighbouring zones. Though the coupling strength depends upon the temperature difference between the subsystems. During the simultaneous heating up, the temperature difference between the zones is small such that the physical interactions are weak. In contrast to this, during the cooling-down phase, the temperature difference between neighbouring zones increases and, thus, the system interactions have a strong effect on the dynamics of the heating zone.

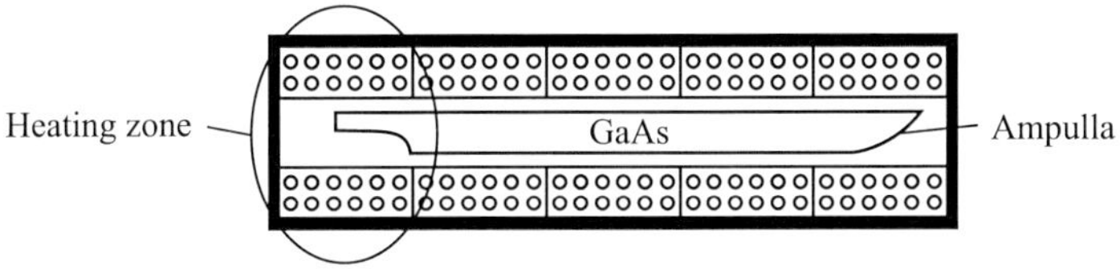

Figure 2.14: Multizone furnace.

To fulfil the specifications of the gradient-freeze method, it is reasonable to take the neighbouring dynamics into account for the design of the control stations. Hence, the multizone furnace serves as demonstration example to evaluate the methods from Chapters 7 and 8 by means of simulations.

The model of a multizone furnace has been derived in [29] and is presented in Appendix A.2.

3 Handling of model information

The first part of this chapter focusses on models. Frequently used models are introduced and the operator con *is presented to process the connection of models. The second part focusses on the design agents. The abilities of the design agents are listed, followed by the introduction of the communication that visualises the interactions among the design agents during the design process.*

3.1 Connection of models

This section introduces the operator con for the connection of models. Essentially, the operator connects the input signals of the models with equivalently named output signals. In the following, it is distinguished between the connection of models and the connection of models with comparison systems.

Connection of models. The connection of models is performed by

$$H = \mathrm{con}\left(\mathcal{H}, \boldsymbol{a}, \boldsymbol{o}\right),$$

where $\mathcal{H}$ denotes a set of models to be connected. $\boldsymbol{a}$ and $\boldsymbol{o}$ denote the input signal vector and the output signal vector of the resulting model H, respectively. In the following, the connection of models in I/O representation and state-space representation is proposed in detail.

Consider the models to be connected are represented in frequency domain. First, the model equations are accumulated to yield

$$\boldsymbol{o}(s) = \boldsymbol{H}_{\mathrm{oa}}(s)\,\boldsymbol{a}(s) + \boldsymbol{H}_{\mathrm{od}}(s)\,\boldsymbol{d}(s) \tag{3.1}$$

$$\boldsymbol{d}(s) = \boldsymbol{H}_{\mathrm{da}}(s)\,\boldsymbol{a}(s) + \boldsymbol{H}_{\mathrm{dd}}(s)\,\boldsymbol{d}(s), \tag{3.2}$$

where the signal vector $\boldsymbol{d}(s)$ accumulates all input signals of the models in $\mathcal{H}$ which are not part of the input signal vector $\boldsymbol{a}(s)$. Second, the accumulated equations (3.1) and (3.2) are connected to the desired model

$$H:\ \boldsymbol{o}(s) = \Big(\boldsymbol{H}_{\mathrm{oa}}(s) + \boldsymbol{H}_{\mathrm{od}}(s)\big(\mathbf{I} - \boldsymbol{H}_{\mathrm{dd}}(s)\big)^{-1}\boldsymbol{H}_{\mathrm{da}}(s)\Big)\,\boldsymbol{a}(s). \tag{3.3}$$

Models which are represented in time domain by I/O-relations can be connected in the exact same way, where the resulting model H reads

$$H: \boldsymbol{o}(t) = \boldsymbol{H}_{\mathrm{oa}}(t) + \boldsymbol{H}_{\mathrm{od}}(t) * \boldsymbol{\Psi}(t) * \boldsymbol{H}_{\mathrm{da}}(t) * \boldsymbol{a}(t) \tag{3.4}$$

with $\boldsymbol{\Psi}(t) = \delta(t)\mathbf{I} + \boldsymbol{H}_{\mathrm{dd}}(t) * \boldsymbol{\Psi}(t)$.

Finally, consider the situation in which state-space models need to be connected. The accumulation of the state-space equations of all models in $\mathcal{H}$ results to

$$\begin{aligned}
\dot{\boldsymbol{x}}_{\mathrm{H}}(t) &= \boldsymbol{A}\boldsymbol{x}_{\mathrm{H}}(t) + \boldsymbol{B}\boldsymbol{a}(t) + \boldsymbol{E}\boldsymbol{d}(t)\\
\boldsymbol{d}(t) &= \boldsymbol{C}_{\mathrm{d}}\boldsymbol{x}_{\mathrm{H}}(t) + \boldsymbol{D}_{\mathrm{d}}\boldsymbol{a}(t) + \boldsymbol{F}_{\mathrm{d}}\boldsymbol{d}(t)\\
\boldsymbol{o}(t) &= \boldsymbol{C}_{\mathrm{o}}\boldsymbol{x}_{\mathrm{H}}(t) + \boldsymbol{D}_{\mathrm{o}}\boldsymbol{a}(t) + \boldsymbol{F}_{\mathrm{o}}\boldsymbol{d}(t),
\end{aligned}$$

where $\boldsymbol{x}_{\mathrm{H}}(t)$ is the accumulated state vector. The connected model H is represented by

$$H: \begin{cases} \dot{\boldsymbol{x}}_{\mathrm{H}}(t) = \left(\boldsymbol{A} + \boldsymbol{E}(\mathbf{I} - \boldsymbol{F}_{\mathrm{d}})^{-1}\boldsymbol{C}_{\mathrm{d}}\right)\boldsymbol{x}_{\mathrm{H}}(t) + \left(\boldsymbol{B} + \boldsymbol{E}(\mathbf{I} - \boldsymbol{F}_{\mathrm{d}})^{-1}\boldsymbol{D}_{\mathrm{d}}\right)\boldsymbol{a}(t)\\ \boldsymbol{o}(t) = \left(\boldsymbol{C}_{\mathrm{o}} + \boldsymbol{D}_{\mathrm{o}}(\mathbf{I} - \boldsymbol{F}_{\mathrm{d}})^{-1}\boldsymbol{C}_{\mathrm{d}}\right)\boldsymbol{x}_{\mathrm{H}}(t) + \left(\boldsymbol{F}_{\mathrm{o}} + \boldsymbol{D}_{\mathrm{o}}(\mathbf{I} - \boldsymbol{F}_{\mathrm{d}})^{-1}\boldsymbol{D}_{\mathrm{d}}\right)\boldsymbol{a}(t). \end{cases} \tag{3.5}$$

Throughout this thesis the well-postness of the equations (3.3)–(3.5) are considered, i.e., the inversions in (3.3)–(3.5) exist.

The following example illustrates the application of con to close the loop.

Example 3.1 *Closing the loop by means of the operator* con

Consider that the model of subsystem S_1 is given in frequency domain representation

$$S_1: \begin{cases} y_1(s) = S_{\mathrm{yu1}}(s)u_1(s) + S_{\mathrm{ys1}}(s)s_1(s)\\ z_1(s) = S_{\mathrm{zu1}}(s)u_1(s) + S_{\mathrm{zs1}}(s)s_1(s) \end{cases} \tag{3.6}$$

as well as the the model of control station C_1

$$C_1: u_1(s) = C_1(s)(w_1(s) - y_1(s)),$$

shown as block diagrams in Fig. 3.1a.

Both models are connected to the model F_1 of the controlled subsystem, which represents the dynamics from the inputs $w_1(s)$ and $s_1(s)$ to the outputs $y_1(s)$ and $z_1(s)$ (Fig. 3.1b). The operator con is used as follows:

$$F_1 = \mathrm{con}\left(\{S_1, C_1\}, (w_1, s_1)^{\top}, (y_1, z_1)^{\top}\right).$$

In accordance with (3.1) and (3.2), the equations (3.6) and (3.6) are lumped together as follows:

$$\begin{pmatrix} y_1(s)\\ z_1(s)\end{pmatrix} = \begin{pmatrix} 0 & S_{\mathrm{ys1}}(s)\\ 0 & S_{\mathrm{zs1}}(s)\end{pmatrix}\begin{pmatrix} w_1(s)\\ s_1(s)\end{pmatrix} + \begin{pmatrix} 0 & S_{\mathrm{yu1}}(s)\\ 0 & S_{\mathrm{zu1}}(s)\end{pmatrix}\begin{pmatrix} y_1(s)\\ u_1(s)\end{pmatrix}$$

$$\begin{pmatrix} y_1(s) \\ u_1(s) \end{pmatrix} = \begin{pmatrix} 0 & S_{\text{ys1}}(s) \\ 0 & C_1(s) \end{pmatrix} \begin{pmatrix} w_1(s) \\ s_1(s) \end{pmatrix} + \begin{pmatrix} 0 & S_{\text{yu1}}(s) \\ -C_1(s) & 0 \end{pmatrix} \begin{pmatrix} y_1(s) \\ u_1(s) \end{pmatrix}.$$

The model of the controlled subsystem results to

$$F_1 : \begin{cases} y_1(s) = F_{\text{yw1}}(s) w_1(s) + F_{\text{ys1}}(s) s_1(s) \\ z_1(s) = F_{\text{zw1}}(s) w_1(s) + F_{\text{zs1}}(s) s_1(s) \end{cases}$$

with

$$\begin{aligned} F_{\text{yw1}}(s) &= S_{\text{yu1}}(s) C_1(s) \big(1 + C_1(s) S_{\text{yu1}}(s)\big)^{-1}, \\ F_{\text{ys1}}(s) &= S_{\text{ys1}}(s) - S_{\text{yu1}}(s) C_1(s) S_{\text{ys1}}(s) \big(1 + C_1(s) S_{\text{yu1}}(s)\big)^{-1}, \\ F_{\text{zw1}}(s) &= S_{\text{zu1}}(s) C_1(s) \big(1 + C_1(s) S_{\text{yu1}}(s)\big)^{-1}, \\ F_{\text{ys1}}(s) &= S_{\text{zs1}}(s) - S_{\text{zu1}}(s) C_{\text{uy1}}(s) S_{\text{ys1}}(s) \big(1 + C_1(s) S_{\text{yu1}}(s)\big)^{-1}. \end{aligned}$$

□

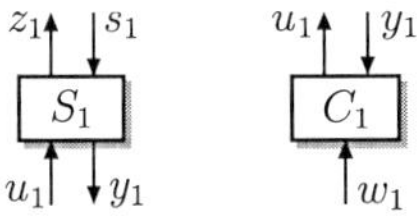

(a) Models S_1 and C_1 to be connected

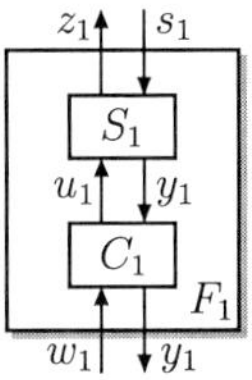

(b) Connected model F_1

Figure 3.1: Using the operator con to close the control loop.

Connection of models with comparison systems. In order to process the connection of comparison systems with models using the operator con, first, the norm and the absolute value of a model is introduced.

Definition 3.1 (Absolute value model and norm model) *Consider a model H that is representation in frequency domain by*

$$H : \boldsymbol{y}(s) = \boldsymbol{H}(s)\, \boldsymbol{u}(s).$$

or in time domain by

$$H : \boldsymbol{y}(t) = \boldsymbol{H}(t) * \boldsymbol{u}(t).$$

Then $|H|$ denotes the absolute value model

$$|H| : |\boldsymbol{y}(s)| \le |\boldsymbol{H}(s)| \cdot |\boldsymbol{u}(s)|$$

or

$$|H|: \ |\boldsymbol{y}(t)| \leq |\boldsymbol{H}(t)| * |\boldsymbol{u}(t)|,$$

respectively. Accordingly, $\|H\|$ *represents the* norm model

$$\|H\|: \ \|\boldsymbol{y}(s)\| \leq \|\boldsymbol{H}(s)\| \cdot \|\boldsymbol{u}(s)\|$$

or

$$\|H\|: \ \|\boldsymbol{y}(t)\| \leq \|\boldsymbol{H}(t)\| * \|\boldsymbol{u}(t)\|,$$

respectively. Absolute value models and norm models are comparison systems in accordance with Definitions 2.13 and 2.14.

The connection of absolute value models, norm models or comparison systems, which are collected in $\mathcal{H}$ is performed by

$$\bar{H} = \operatorname{con}\left(\mathcal{H}, |\boldsymbol{a}|, \bar{\boldsymbol{o}}\right).$$

The resulting model $\bar{H}$ is also a comparison system with the input signal vector $|\boldsymbol{a}| \geq \boldsymbol{0}$ ($\|\boldsymbol{a}\| \geq 0$) and the output signal vector $\bar{\boldsymbol{o}} \geq |\boldsymbol{o}| \geq \boldsymbol{0}$ ($\bar{o} \geq \|\boldsymbol{o}\| \geq 0$). The accumulation of the models and connection of those signals follows the same steps as presented in the previous paragraph.

The following example elucidates the connection of a comparison system with a model.

Example 3.2 *Connect a model with a comparison system using* con

The connection of the system model S described by (2.13) with the comparison system $\bar{E}$ described by (2.19) is considered. The result of the connection will be the comparison system $\bar{S}$. The operator con is applied as follows:

$$\bar{S} = \operatorname{con}\left(\{|S|, \bar{E}\}, |u|, \bar{y}\right),$$

where

$$|S|: \begin{cases} |y(t)| \leq |\hat{S}(t)| * |u(t)| + |S_{\mathrm{yq}}(t)| * |q(t)| \\ |p(t)| \leq |S_{\mathrm{pu}}(t)| * |u(t)| + |S_{\mathrm{pq}}(t)| * |q(t)| \end{cases} \tag{3.7}$$

is the scalar absolute value model of (2.13) and

$$\bar{E}: \ |q(t)| \leq \bar{q}(t) = \bar{E}(t) * |p(t)| \tag{3.8}$$

is the scalar comparison system of the model uncertainty.

In accordance with (3.1) and (3.2), the equations (3.7) and (3.8) are lumped together

$$|y(t)| \leq |\hat{S}(t)| * |u(t)| + \begin{pmatrix} |S_{\mathrm{yq}}(t)| & 0 \end{pmatrix} * \begin{pmatrix} |q(t)| \\ |p(t)| \end{pmatrix}$$

$$\begin{pmatrix} |q(t)| \\ |p(t)| \end{pmatrix} \leq \begin{pmatrix} 0 \\ |S_{\mathrm{pu}}(t)| \end{pmatrix} * |u(t)| + \begin{pmatrix} 0 & |S_{\mathrm{pq}}(t)| \\ \bar{E}(t) & 0 \end{pmatrix} * \begin{pmatrix} |q(t)| \\ |p(t)| \end{pmatrix}$$

and result in the comparison system

$$\bar{S}: \ |y(t)| \leq \bar{y}(t) = |\hat{S}(t)| * |u(t)| + |S_{\text{yq}}(t)| * \Psi(t) * |S_{\text{pu}}(t)| * |u(t)|$$

with $\Psi(t) = \bar{E}(t) + \bar{E}(t) * |S_{\text{pq}}(t)| * \Psi(t)$. □

Note that to connect comparison systems, it has to be guaranteed that the inversions exists (cf. eqns (2.18) and (2.21)). Then the resulting model is a comparison system too. Throughout this thesis, conditions are provided to guarantee the existence of the resulting comparison system.Ka

3.2 Models of plug-and-play control

This section introduces the notation and construction of models that are frequently used in this thesis. Therefore, the notion of local view and global view are stated beforehand.

3.2.1 Global view and local view

From the *global view*, the dynamics of the overall closed-loop system are analysed by the accumulated reference signal vector

$$\boldsymbol{w}(t) = \begin{pmatrix} \boldsymbol{w}_1^\top(t) & \boldsymbol{w}_2^\top(t) & \ldots & \boldsymbol{w}_N^\top(t) \end{pmatrix}^\top$$

and the accumulated output signal vector

$$\boldsymbol{y}(t) = \begin{pmatrix} \boldsymbol{y}_1^\top(t) & \boldsymbol{y}_2^\top(t) & \ldots & \boldsymbol{y}_N^\top(t) \end{pmatrix}^\top,$$

as illustrated in Fig. 3.2a.

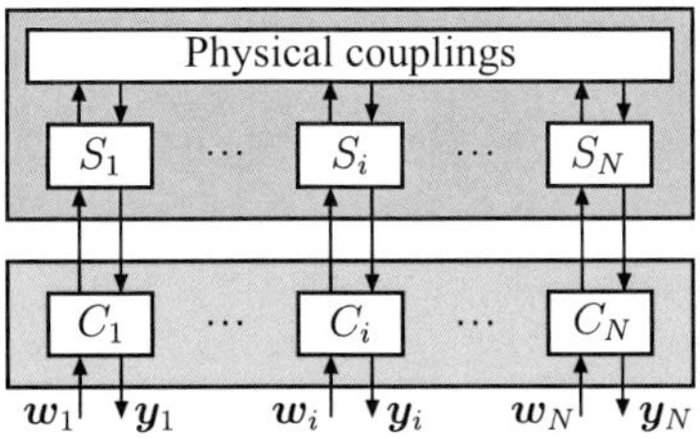

(a) Global view

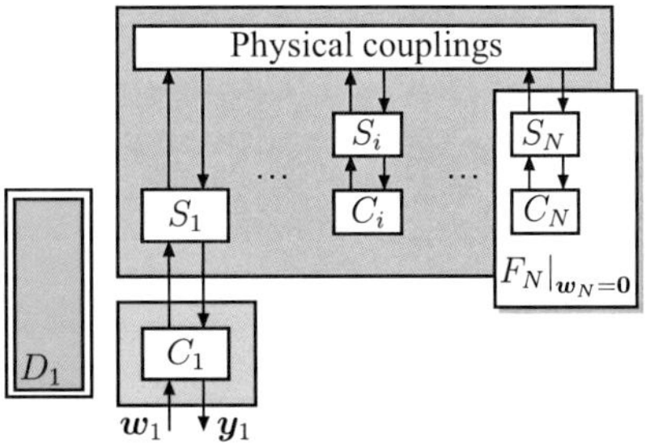

(b) Local view of subsystem S_1

Figure 3.2: Global view and local view of the overall closed-loop system.

From the *local view*, the overall closed-loop system is analysed by means of the local input $\boldsymbol{w}_k(t)$ and the local output $\boldsymbol{y}_k(t)$ as highlighted in Fig. 3.2b for the local view of subsystem S_1. Obviously, all other reference signals $\boldsymbol{w}_i(t)$, $(i \in \mathcal{N},\ i \neq k)$ can be understood as disturbance signals. To analyse the effect of the input $\boldsymbol{w}_k(t)$ on the output $\boldsymbol{y}_k(t)$ only, the following assumption is made:

A 3.1 *The $N-1$ controlled systems F_i, $(\forall i \in \mathcal{N},\ i \neq k)$ are in their operating points and are externally unforced, i.e., $\boldsymbol{x}_{\mathrm{F}i,0} = \mathbf{0}$ and $\boldsymbol{w}_i(t) = \mathbf{0}$.*

As a consequence, the models F_i of the $N-1$ controlled subsystems are represented by the reduced model

$$F_i|_{\boldsymbol{w}_i=\mathbf{0}} = \mathrm{con}\left(\{F_i\},\ \boldsymbol{s}_i,\ \boldsymbol{z}_i\right)$$

shown in Fig. 3.2b.

3.2.2 Models from the local view

From the local view of subsystem S_1, the effect of all other controlled subsystems F_i, $(i = 2, ..., N)$ on subsystem S_1 through the physical couplings is called *physical interaction*, which is represented by the model

$$P_1^{\mathcal{N}} = \mathrm{con}\Big(\{\, F_i|_{\boldsymbol{w}_i=\mathbf{0}},\ i \in \mathcal{N} \setminus \{1\}\} \cup \{K_i,\ i \in \mathcal{N}\},\ \boldsymbol{z}_1,\ \boldsymbol{s}_1\Big) \tag{3.9}$$

as highlighted in Fig. 3.3. The superscribe $\mathcal{N}$ is used to explicitly indicate which models from which design agent need to be connected to the desired model. Accordingly, to construct the model $P_1^{\mathcal{N}}$ the models that are initially stored by D_i, $(\forall i \in \mathcal{N})$ are required (cf. Section 3.3).

Moreover, the dynamics of subsystem S_1 under consideration of the physical interaction $P_1^{\mathcal{N}}$ are modelled by the *interacting subsystem model*

$$S_1^{\mathcal{N}} = \mathrm{con}\Big(\{S_1,\ P_1^{\mathcal{N}}\},\ \boldsymbol{u}_1,\ \boldsymbol{y}_1\Big).$$

Accordingly, the connection of the controlled subsystem F_1 with the physical interaction $P_1^{\mathcal{N}}$ yields the *controlled interacting subsystem* with input $\boldsymbol{w}_1$ and output $\boldsymbol{y}_1$. The corresponding model is denoted by $F_1^{\mathcal{N}}$ and results to

$$F_1^{\mathcal{N}} = \mathrm{con}\Big(\{F_1,\ P_1^{\mathcal{N}}\},\ \boldsymbol{w}_1,\ \boldsymbol{y}_1\Big).$$

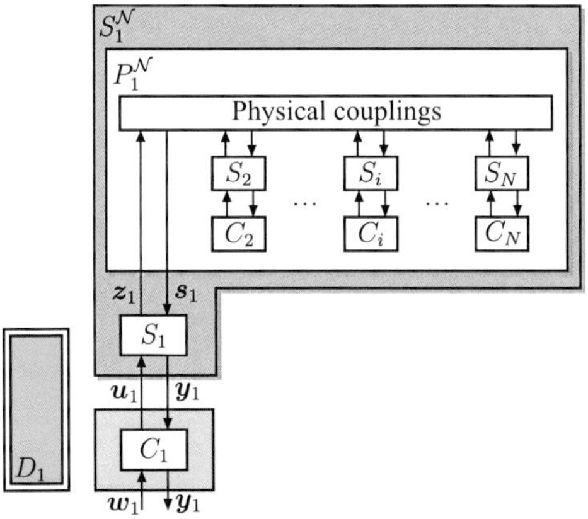

Figure 3.3: Structure of the the physical interaction $P_1^{\mathcal{N}}$ and interacting subsystem $S_1^{\mathcal{N}}$.

Subdivision of the physical interaction. The dynamics of the interacting subsystem $S_1^{\mathcal{N}}$ are essentially characterised by the dynamics of subsystem S_1 and of some neighbouring controlled subsystems F_i, $(\forall i \in \mathcal{N}_{\mathrm{S1}} := \{2, .., v-1\})$ whereas the dynamics of the other controlled subsystems F_i, $(\forall i \in \mathcal{N}_{\mathrm{W1}} := \{v, .., N\})$ have negligible influence. To distinguish between relevant and irrelevant model information for the controller design, the physical interaction $P_1^{\mathcal{N}}$ is subdivided into a *relevant part*

$$P_1^{\mathcal{N}_{\mathrm{S1}}} = \mathrm{con}\left(\left\{ F_i|_{\boldsymbol{w}_i=\boldsymbol{0}}, \, i \in \mathcal{N}_{\mathrm{S1}} \setminus \{1\}\right\} \cup \{K_i, \, i \in \mathcal{N}_{\mathrm{S1}}\}, \, \begin{pmatrix} \boldsymbol{z}_1 \\ \boldsymbol{q}_1 \end{pmatrix}, \, \begin{pmatrix} \boldsymbol{s}_1 \\ \boldsymbol{p}_1 \end{pmatrix}\right),$$

where $\boldsymbol{q}_1$ and $\boldsymbol{p}_1$ denote the error input signal and error output signal, respectively, and an *irrelevant part*

$$E_1^{\mathcal{N}_{\mathrm{W1}}} = \mathrm{con}\left(\left\{ F_i|_{\boldsymbol{w}_i=\boldsymbol{0}}, \, i \in \mathcal{N}_{\mathrm{W1}}\right\} \cup \{K_{v-1}, \, K_i, \, i \in \mathcal{N}_{\mathrm{W1}}\}, \, \boldsymbol{p}_1, \, \boldsymbol{q}_1\right),$$

as highlighted in Fig. 3.4. As a result of this separation, the physical interaction $P_1^{\mathcal{N}_1}$ is approximated by the *approximate model*

$$\hat{P}_1^{\mathcal{N}_{\mathrm{S1}}} = \mathrm{con}\left(\left\{ F_i|_{\boldsymbol{w}_i=\boldsymbol{0}}, \, i \in \mathcal{N}_{\mathrm{S1}} \setminus \{1\}\right\} \cup \{K_i, \, i \in \mathcal{N}_{\mathrm{S1}}\}, \, \boldsymbol{z}_1, \, \hat{\boldsymbol{s}}_1\right).$$

The error $s_{\Delta 1} = s_1 - \hat{s}_1$ between the actual coupling input s_1 and the approximated coupling input $\hat{s}_1$ is represented by the *error model*

$$P_{\Delta 1}^{\mathcal{N}_{S1}} = \mathrm{con}\left(\left\{P_1^{\mathcal{N}_{S1}},\, E_1^{\mathcal{N}_{W1}},\, \hat{P}_1^{\mathcal{N}_{S1}}\right\},\, z_1,\, s_{\Delta 1}\right).$$

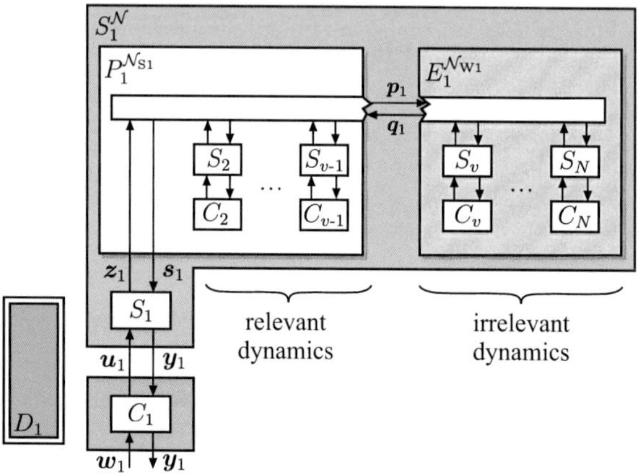

Figure 3.4: Subdivision of the physical interaction $P_1^{\mathcal{N}}$ into the relevant part $P_1^{\mathcal{N}_{S1}}$ and the irrelevant part $E_1^{\mathcal{N}_{W1}}$: The relevant part is used for controller design, whereas the irrelevant part is treated as model uncertainty.

Description of the irrelevant dynamics by a comparison system. For the purpose of plug-and-play control, the irrelevant dynamics are treated as unknown-but-bounded model uncertainty of the relevant part. Hence, the model uncertainty is described by a comparison system $\bar{E}_1^{\mathcal{N}_{W1}}$ according to (2.17) or (2.19). As a consequence, the error model $P_{\Delta 1}^{\mathcal{N}_{S1}}$ is only known as the comparison system

$$\bar{P}_{\Delta 1}^{\mathcal{N}_{S1}} = \mathrm{con}\left(\left\{\|P_1^{\mathcal{N}_{S1}}\|,\, \bar{E}_1^{\mathcal{N}_{W1}},\, \|\hat{P}_1^{\mathcal{N}_{S1}}\|\right\},\, \|z_1\|,\, \bar{s}_{\Delta 1}\right).$$

as well as the physical interaction $P_1^{\mathcal{N}}$ that is described by the comparison system

$$\bar{P}_1^{\mathcal{N}_{S1}} = \mathrm{con}\left(\left\{\|\hat{P}_1^{\mathcal{N}_{S1}}\|,\, \bar{P}_{\Delta 1}^{\mathcal{N}_{W1}}\right\},\, \|z_1\|,\, \bar{s}_1\right).$$

Note that absolute value model can also be used instead of norm models.

Models for controller design. The amount of model information used for the controller design differs with the methods proposed in this thesis.

In Chapter 4, the control station is designed based on a model

$$S_1^{\mathcal{N}_{\mathrm{S}1}} = \mathrm{con}\left(\left\{S_1,\, P_1^{\mathcal{N}_{\mathrm{S}1}},\, \boldsymbol{u}_1,\, \boldsymbol{y}_1\right\}\right) \tag{3.10}$$

that exactly represents the dynamics of the interacting subsystems $S_1^{\mathcal{N}}$. Accordingly, the set $\mathcal{N}_{\mathrm{S}1}$ collects the indices of all controlled subsystems that are *strongly connected* to subsystem S_1. Note that the model (3.10) is not an approximation of $S_1^{\mathcal{N}}$ and, thus, has no superscript $\hat{(\cdot)}$.

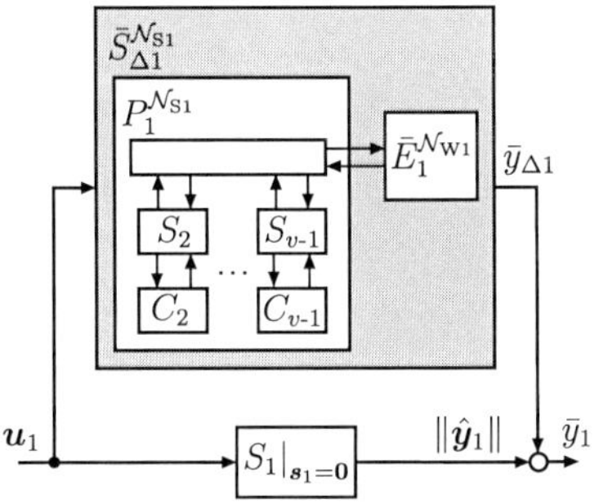

Figure 3.5: Structure of the uncertain model used for the design of control station C_1 considered in Chapters 5–7.

In Chapters 5–7, the control station is designed based on the isolates subsystem model

$$S_1|_{\boldsymbol{s}_1=\boldsymbol{0}} = \mathrm{con}\left(\{S_1\},\, \boldsymbol{u}_1,\, \hat{\boldsymbol{y}}_1\right).$$

The error $\boldsymbol{y}_{\Delta 1} = \boldsymbol{y}_1 - \hat{\boldsymbol{y}}_1$ is described by the comparison system

$$\bar{S}_{\Delta 1}^{\mathcal{N}_{\mathrm{S}1}} = \mathrm{con}\left(\left\{\|S_1\|,\, \bar{P}_1^{\mathcal{N}_{\mathrm{S}1}}\right\},\, \|\boldsymbol{u}_1\|,\, \bar{y}_{\Delta 1}\right)$$

as illustrated in Fig. 3.5. The aim of the design is to satisfy dynamical claims on the isolated controlled subsystem

$$F_1|_{\boldsymbol{s}_1=\boldsymbol{0}} = \mathrm{con}\left(\{F_1\},\, \boldsymbol{w}_1,\, \hat{\boldsymbol{y}}_1\right)$$

while attenuating the error. In Chapters 5 and 6 only the local model is used for the design, while all other models constitute the model uncertainty. That is, $\mathcal{N}_{\mathrm{S}1} = \{1\}$ and $\mathcal{N}_{\mathrm{W}1} = \{2, ..., N\}$. In contrast to this, in Chapter 7 an approximation of the physical interaction is used with the aim to approximate the error dynamics. Therefore, the set $\mathcal{N}_{\mathrm{S}1}$ collects the indices of all controlled subsystems that are *strongly coupled* to S_1.

In Chapter 8, the diagnosis system is designed based on an approximate model

$$\hat{S}_1^{\mathcal{N}_{S1}} = \text{con}\left(\left\{S_1,\ \hat{P}_1^{\mathcal{N}_{S1}}\right\},\ \boldsymbol{u}_1,\ \hat{\boldsymbol{y}}_1\right),$$

This model has to be as precise as necessary to detect and isolate a fault.

3.3 Design agents

As mentioned in Chapter 1, there exist N identical design agents D_i assigned to the N subsystems S_i. Each design agent D_i is able to perform the following listed tasks:

- Storing models, algorithms, comparison systems and parameters
- Establishing communication channels to other design agents
- Transmitting models, algorithms, comparison systems and parameters through the communication network to other design agents
- Connecting available model information (i.e., process con)
- Generating comparison systems from the available model information
- Executing algorithms for controller design (e.g., solve an optimisation problem) and for model procurement (e.g., Algorithm 4.1)
- Implementing algorithms into control hardware

The models, algorithms and comparison systems which are stored by design agent D_i are collected in a *model set* $\mathcal{M}_i$. For instance, the model set $\mathcal{M}_1 = \{S_1,\ K_1\}$ indicates that design agent D_1 knows the models S_1 and K_1.

Regarding the tasks of the design agents, the following assumptions are made:

A 3.2 *Initially, the design agents D_i, $(\forall i \in \mathcal{N})$ store the corresponding models S_i of the subsystem, K_i of the physical interaction and C_i of the control station (if existing), i.e., $\mathcal{M}_i = \{S_i,\ K_i,\ C_i\}$.*

A 3.3 *Design agent D_i is able to communicate with all other design agents D_j, $(j \in \mathcal{N})$ through the network.*

From the information technology point of view, a design agent can be viewed as a service-on-demand. Accordingly, whenever the service "design agent" is not needed, the computational power is available to execute other applications. In general the service-providing units are always equipped with enough computational power to execute the service as fast as possible. As a consequence, it is a reasonable assumption that a task of D_i is processed in negligible time compared to the dynamical behaviour of the process to be controlled. This fact leads to the following assumption:

A 3.4 *The execution of each task of a design agent does not consume time.*

3.4 Interconnection graph and communication graph

This section introduces the interconnection graph and the communication graph describing the physical interconnection among subsystems and the interactions between design agents, respectively.

3.4.1 Interconnection graph

The *interconnection graph* represents the overall system interconnection structure as a directed graph $\mathcal{G}_\mathrm{K} = (\mathcal{V}_\mathrm{K}, \mathcal{E}_\mathrm{K})$ with vertex set $\mathcal{V}_\mathrm{K} = \mathcal{N}$ and edge set $\mathcal{E}_\mathrm{K} = \{(j,i) \colon \boldsymbol{L}_{ij} \neq 0\} \subseteq \mathcal{V}_\mathrm{K} \times \mathcal{V}_\mathrm{K}$. Hence, a vertex represents a subsystem whereas an edge represents the physical coupling relation between two subsystems. An example of an interconnection graph that corresponds to six interconnected subsystems is illustrated in Fig. 3.6a.

(a) Interconnection graph $\mathcal{G}_\mathrm{K}$

(b) Local interconnection graph $\mathcal{G}_{\mathrm{K}1}$ constructed by D_1 with the model K_1 available

Figure 3.6: Example of an interconnection graph $\mathcal{G}_\mathrm{K}$ and a local interconnection graph $\mathcal{G}_{\mathrm{K}1}$.

As design agent D_i has available the model set $\mathcal{M}_i$, only a part of the interconnection graph can be constructed by D_i. This partial graph is called *local interconnection graph* and is described by the graph $\mathcal{G}_{\mathrm{K}i} = (\mathcal{V}_{\mathrm{K}i}, \mathcal{E}_{\mathrm{K}i})$ with vertex set $\mathcal{V}_{\mathrm{K}i} = \left\{k \colon k \in \bigcup_{j \in \mathcal{N}, \mathcal{P}_j \text{ available to } D_i} \mathcal{P}_j \cup \{i\}\right\}$ and edge set $\mathcal{E}_{\mathrm{K}i} = \left\{(j,k) \colon \boldsymbol{L}_{ij} \neq \mathbf{O},\ \boldsymbol{L}_{ij} \text{ available to } D_i\right\} \subseteq \mathcal{V}_{\mathrm{K}i} \times \mathcal{V}_{\mathrm{K}i}$. A local interconnection graph $\mathcal{G}_{\mathrm{K}1}$ of the interconnection graph shown in Fig. 3.6a illustrated in Fig. 3.6b. This

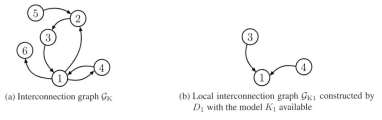

local interconnection is constructed by D_1 which stores the model set $\mathcal{M}_1 = \{S_1, K_1\}$. Essentially, the model K_1 provides design agent D_1 with the set of predecessors $\mathcal{P}_1$ and the coupling gains.

For locally interconnected systems, the corresponding interconnection graph is shown in Fig. 3.7. The local interconnection graph $\mathcal{G}_{\mathrm{K}2}$ that is constructed by D_2 with $\mathcal{M}_2 = \{S_2, K_2\}$ available is shown in Fig. 3.7b.

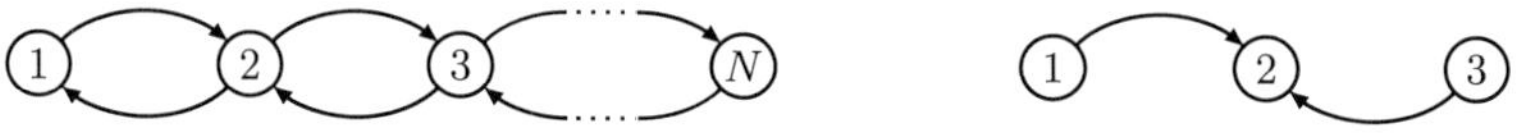

(a) Interconnection graph $\mathcal{G}_{\mathrm{K}}$ of a locally interconnected system

(b) Local interconnection graph $\mathcal{G}_{\mathrm{K}2}$ of a locally interconnected system constructed by D_2 with the model K_2 available

Figure 3.7: Example of an interconnection graph $\mathcal{G}_{\mathrm{K}}$ and a local interconnection graph $\mathcal{G}_{\mathrm{K}i}$ for locally interconnected subsystem.

3.4.2 Communication graph

Within the design process, design agent D_i has to procure relevant model information from other design design agents D_j, $(j \in \mathcal{N})$. Therefore, design agent D_i establishes a communication link to the target design agent. After the transmission of the requested model information, the communication link is removed. In accordance with assumption A 2.5, the whole transmission procedure does not consume any time and, thus, is gathered to the single time instant κ.

The exchange of information during the procurement phase is visualised by the *communication graph* $\mathcal{G}_{\mathrm{D}}(\kappa)$ which is a weighted directed graph $\mathcal{G}_{\mathrm{D}}(\kappa) = \big(\mathcal{V}_{\mathrm{D}}, \mathcal{E}_{\mathrm{D}}(\kappa), w_{\mathrm{D}}\big)$. The vertex set $\mathcal{V}_{\mathrm{D}} = \mathcal{N}$ represents the design agents, the edge set $\mathcal{E}_{\mathrm{D}}(\kappa) \subseteq \mathcal{V}_{\mathrm{D}} \times \mathcal{V}_{\mathrm{D}}$ is separated into the edge set $\mathcal{E}_{\mathrm{Q}}(\kappa) = \big\{(i,j)\colon D_i \text{ requests } D_j \text{ at event time } \kappa\big\}$ and the distinct edge set $\mathcal{E}_{\mathrm{T}}(\kappa) = \big\{(i,j)\colon D_i \text{ responses to } D_j \text{ at event time } \kappa\big\}$. The weight function $w_{\mathrm{D}}\colon \mathcal{E}_{\mathrm{D}}(\kappa) \to \mathcal{M}$, where $\mathcal{M} = \bigcup_{i\in\mathcal{N}} \mathcal{M}_i$ indicates which information is transmitted from design agent D_i to design agent D_j as a response of a request. If no information is transmitted (i.e., $w_{\mathrm{D}}((i,j)) = \emptyset$) the weight is omitted in the depiction of the graph. An example of a communication graph is shown in Fig. 3.8a.

The accumulation of the communication graphs $\mathcal{G}_{\mathrm{D}}(\kappa)$ for all events κ yields the *cumulated communication graph* $\mathcal{G}_{\mathrm{D}} = \big(\mathcal{V}_{\mathrm{D}}, \mathcal{E}_{\mathrm{D}}\big)$, where $\mathcal{E}_{\mathrm{D}} = \bigcup_{\kappa} \mathcal{E}_{\mathrm{Q}}(\kappa) \cup \mathcal{E}_{\mathrm{T}}(\kappa)$. In contrast to the communication graph, the cumulated communication graph does rather highlight the interaction of the design agents for all times κ than the particular transmitted model information. Figure 3.8b illustrated the cumulated communication graph of $\mathcal{G}_{\mathrm{D}}(\kappa)$ presented in Fig. 3.8a.

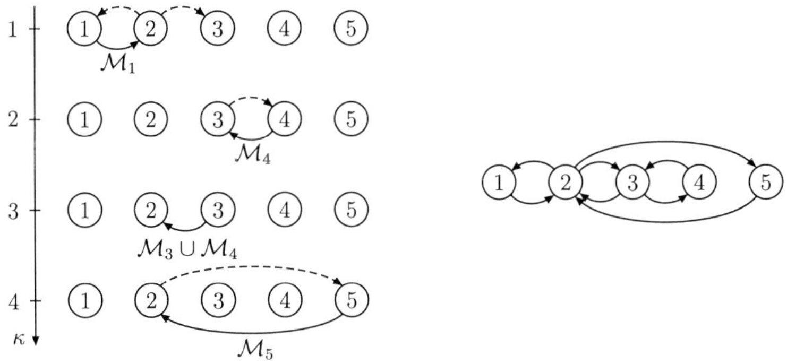

(a) Communication graph $\mathcal{G}_\mathrm{D}(\kappa)$: The dashed edges denote a request ($\mathcal{E}_\mathrm{Q}(\kappa)$) and the solid edges denote a response ($\mathcal{E}_\mathrm{T}(\kappa)$).

(b) Cumulated communication graph $\mathcal{G}_\mathrm{D}$.

Figure 3.8: Example of a communication graph $\mathcal{G}_\mathrm{D}(\kappa)$ and the corresponding cumulated communication graph $\mathcal{G}_\mathrm{D}$.

Note that in contrast to the interconnection graph, the communication graph does not represent a restriction on the possible communication, it rather illustrates the performed exchange of models among the design agents.

4 Plug-and-play control using the model of the physical interaction

This chapter focusses on the design of control station C_1 based on a model that exactly describes the dynamics of subsystem S_1 interacting with other subsystems through the physical couplings. The design shall be processed by design agent D_1, which a priori only knows the subsystem model S_1 and has sparse information about the physical couplings. The main result of this chapter is a distributed search algorithm that enables D_1 to set up the required model to accomplish the controller design. It is shown that this algorithm assuredly finds all subsystems that are strongly connected to subsystem S_1 and procures the respective models. This algorithm is distributed among the design agents so that the execution of the algorithm involves the cooperation of the design agents as analysed in this chapter. Moreover, the total correctness of the algorithm is proved. This chapter closes with a demonstration of plug-and-play control to guarantee fault-tolerance by reconfiguring control station C_1 using a virtual actuator.

4.1 Problem formulation

The situation is considered in which control station C_1 has to be designed based on the model of subsystem S_1 under the influence of the physical interaction $P_1^{\mathcal{N}}$. However, design agent D_1 initially stores only its subsystem model S_1 and has no information about the overall system interconnection structure. The focus of this chapter is to derive an algorithm that enables D_1 to find all subsystems which physically interact with subsystem S_1 so as to set up the model to be found. As a consequence, D_1 can design C_1 under consideration of the exact behaviour of subsystem S_1 together with its physical interaction. This design achieves a high overall closed-loop performance. Nevertheless, the design agents have to be equipped with enough computational power to handle complex, high dimensional models and to perform the controller design based on this model.

The model to be set up consists of subsystem S_1 itself together with all controlled subsystems that are *strongly connected* to the subsystem S_1 through the physical couplings as highlighted yellow in Fig. 4.1. The model to be found is denoted by $S_1^{\mathcal{N}_{S1}}$ and is constructed according to

$$S_1^{\mathcal{N}_{S1}} = \text{con}\left(\{S_1\} \cup \left\{ F_i|_{\boldsymbol{w}_i=\boldsymbol{0}}, i \in \mathcal{N}_{S1} \setminus \{1\}\right\} \cup \{K_i, i \in \mathcal{N}_{S1}\}, \boldsymbol{u}_1, \boldsymbol{y}_1\right), \tag{4.1}$$

where $\mathcal{N}_{S1}$ denotes the set which collects all indices of the subsystems S_i that are strongly connected to subsystem S_1 in accordance with Definition 2.12, i.e.,

$$\mathcal{N}_{S1} = \{i : \text{P}(1, i) \wedge \text{P}(i, 1) \text{ in } \mathcal{G}_K\}. \tag{4.2}$$

As the subsystems which are not strongly connected to subsystem S_1 are neither controllable nor observable through the I/O pair $(\boldsymbol{u}_1, \boldsymbol{y}_1)$, the model $S_1^{\mathcal{N}_{S1}}$ represents the I/O behaviour of subsystem S_1 together with the physical interaction exactly.

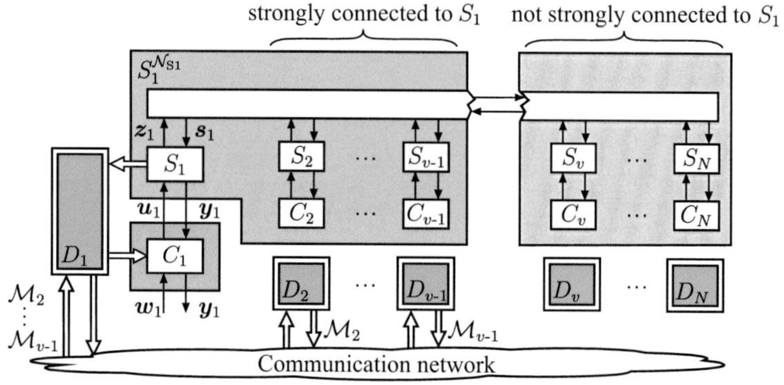

Figure 4.1: Structure of plug-and-play control using a model of the physical interaction: The physical interaction comprises the controlled subsystems which are strongly connected to S_1 (yellow box), whereas the other controlled subsystems (grey box) are irrelevant.

To solve the plug-and-play control problem (Problem 1.2), design agent D_1 must know all models $F_i|_{\boldsymbol{w}_i=\boldsymbol{0}}$, K_i, $(\forall i \in \mathcal{N}_{S1})$ to construct the model $S_1^{\mathcal{N}_{S1}}$. Based on the initially available information $\mathcal{M}_1 = \{S_1, K_1\}$, D_1 has to find the subsystems that are strongly connected to S_1 and has to gather the model set $\mathcal{M}_i = \left\{F_i|_{\boldsymbol{w}_i=\boldsymbol{0}}, K_i\right\}$ from D_i, $(\forall i \in \mathcal{N}_{S1})$ through the communication network as shown in Fig. 4.1 by double framed arrows. This modelling issue is considered in this chapter and is formalised as follows:

Problem 4.1 (Modelling the dynamics of the physical interaction)

Problem: *To enable design agent D_1 to accomplish the controller design (i.e., solve Problem 1.1), the model $S_1^{\mathcal{N}_{S1}}$ that is composed of S_1 and all its strongly connected controlled subsystems has to be known by D_1. However, D_1 initially knows the mode S_1 and has no information about the overall system interconnection structure.*

Given:
- *D_1 stores the model set $\mathcal{M}_1 = \{S_1, K_1\}$*
- *D_i, $(\forall i \in \mathcal{N} \setminus \{1\})$ store the corresponding model set $\mathcal{M}_i = \{F_i, K_i\}$*

Find: *Algorithm that automatically finds all subsystems that are strongly connected to S_1 and procures the respective model sets $\mathcal{M}_i$ from D_i.*

Approach: *Distribution of a graph search algorithm among the design agents so as to find the relevant subsystems collectively. $\mathcal{N}_{S1}$.*

A solution to Problem 4.1 states a solution to Problem 1.2. Hence, with the required model at hand, D_1 is able to solve the design problem (Problem 1.1).

Section 4.2 presents the search algorithm (Algorithm 4.1) to be implemented on each design agent. The communication among the design agents during the execution of the algorithm is analysed in Section 4.3. In Section 4.4 plug-and-play control is applied to reconfigure control station C_1 after an actuator has failed in subsystem S_1. The reconfiguration is accomplished by means of a virtual actuator, where the model $S_1^{\mathcal{N}_{S1}}$ to be set up by design agent D_1 is required to recover overall closed-loop asymptotic stability.

4.2 A distributed algorithm to model the physical interaction

This section presents a distributed search algorithm to find the subsystems which are strongly connected to subsystem S_1 and finally states the total correctness in Theorem 4.1.

Essentially, the proposed distributed algorithm aims at finding all vertices i of the interconnection graph $\mathcal{G}_K$ which are strongly connected to vertex 1 in accordance with Definition 2.12. The algorithm is distributed in the sense that it is implemented on each design agent such that the search is processed in a distributed manner. Since each design agent D_i knows its preceding design agents D_j by the set $\mathcal{P}_i$, the algorithm starts by D_1 and is continued by one of its preceding design agents and, later on, by its pre-predecessor etc. During the processing of the search, the model sets $\mathcal{M}_i$ are exchanged among the design agents so that D_1 finally is able to set up the desired model $S_1^{\mathcal{N}_{S1}}$.

Algorithm 4.1: Local search algorithm

Given: Model set $\mathcal{M}_1 = \{S_1, K_1\}$ with the set of predecessors $\mathcal{P}_1$ available to D_1 and
model set $\mathcal{M}_i = \{F_i, K_i\}$ with the set of predecessors $\mathcal{P}_i$ available to D_i,
$(\forall i \in \mathcal{N} \setminus \{1\})$

Initialise: $stat_1 = 1$, $stat_i = 0$, $(\forall i \in \mathcal{N} \setminus \{1\})$, $req_i = 0$, $(\forall i \in \mathcal{N})$,
$\mathcal{B}_i = \emptyset$, $(\forall i \in \mathcal{N})$, $\kappa = 0$

Processing on D_1:

   1. **run** res = getModel(1, 1)

Processing on D_i:

   2. **function** getModel(h,i)

   3.   **if** $req_i = 0$ **then** set $req_i = h$ **else** return $stat_i$

   4.   **while** $\mathcal{P}_i \setminus \mathcal{B}_i \neq \emptyset$

   5.   set $j \leftarrow \mathcal{P}_i \setminus \mathcal{B}_i$ and $\mathcal{B}_i \leftarrow j$

   6.   set $\kappa = \kappa + 1$

   7.   **run** res = getModel(i, j)   // Forward-search

   8.   **switch** res   // Backtracking

   9.   **case** res = $\mathcal{M}_j$

   10.   set $stat_i = 1$ and $\mathcal{M}_i = \mathcal{M}_i \cup \mathcal{M}_j$

   11.   **case** res = $stat_j$

   12.   **if** $stat_j = 1$ **then** set $stat_i = 1$

   13.   **end**

   14.   **end**

   15.   set $\kappa = \kappa + 1$

   16.   **end**

   17.   **if** $i = 1$ **then** return to D_{req_i} // Algorithm terminates

   18.   **if** $stat_i = 1$ **then** return $\mathcal{M}_i$ to D_{req_i} **else** return $stat_i$ to D_{req_i}

   19. **end**

Result: Design agent D_1 stores the model set $\mathcal{M}_1 = \{S_1, K_1\} \cup \{F_i, K_i, i \in \mathcal{N}_{S1} \setminus \{1\}\}$

The proposed Algorithm 4.1 consists mainly of the function getModel that is implemented on each design agent. The two arguments of the function getModel(h,i) denote the index of the requesting design agent D_h and the index of the requested agent D_i, respectively. Each

Table 4.2: Internal variables of the Algorithm 4.1

Variable	Description	Value
$stat_i \in \{0,1\}$	Connectivity status of S_i to S_1	$stat_i$=1: strongly connected
		$stat_i$=0: not strongly connected
$req_i \in \mathbb{N}_{+0}$	Index of the design agent requesting D_i	req_i=h: request from D_h
		req_i=0: not requested
$\mathcal{B}_i \subseteq \mathcal{P}_i$	Set of investigated nodes by D_i	

design agent D_i, ($i \in \mathcal{N}$) makes use of the local variables listed in Table 4.2. The local variable $stat_i$ states whether or not subsystem S_i is strongly connected to S_1. The local variable req_i stores the index of the design agent that has requested design agent D_i the first time. The local set $\mathcal{B}_i$ collects the indices of the design agents which has been requested by D_i. The counter κ counts the established and removed communication links for the request (Line 6) and the response (Line 15) and, thus, counts the iterations of the algorithm.

Initially, all subsystems S_i, ($\forall i \in \mathcal{N} \setminus \{1\}$) are considered to be not strongly connected to S_1 ($stat_i = 0$, ($\forall i \in \mathcal{N} \setminus \{1\}$) in Initialise). During the execution of the algorithm, this status will be changed if S_i is strongly connected to S_1 (Line 10 and 12).

As the subsystems that are strongly connected to S_1 have to be found, Algorithm 4.1 is started by D_1 (Line 1), that is, initially $i = h = 1$. In the following, the procedure of Algorithm 4.1 is explained by its execution on design agent D_i, where D_i has been requested by D_h (Line 2) and D_i will request D_j later on (Line 7). The first request of D_i by design agent D_h is stored ($req_i = h$ in Line 3) for the later response, whereas additional requests from other design agents are denied (in order to guarantee termination of the algorithm). In Line 5, the requested design agent D_i picks another not requested design agent D_j for a request from the set $\mathcal{P}_i \setminus \mathcal{B}_i$ ($j \leftarrow \mathcal{P}_i \setminus \mathcal{B}_i$), marks this agent ($\mathcal{B}_i \leftarrow j$) and performs the actual request in Line 7 (forward-search). The smallest index $j \in \mathcal{P}_i \setminus \mathcal{B}_i$ which has not already been chosen is selected. If there exist no further agents to be requested, D_i replies to D_h in Line 18 (backtracking). Design agent D_j can answer its request by transmitting its connectivity status ($stat_j$) in Line 11. If S_j is strongly connected to S_1 ($stat_j = 1$), S_i is for sure strongly connected to S_1 too, otherwise the connectivity status of D_i remains unchanged (Line 12). Alternatively, D_j can reply with its model information (Line 9) if D_i is the first design agent that has requested D_j. In this case D_i unites the model sets (Line 10). The distributed algorithm ends at design agent D_1 the moment all predecessors have been requested and have replied (Line 17).

The total correctness of Algorithm 4.1 is stated in the following theorem.

Theorem 4.1 (Total correctness of Algorithm 4.1) *Algorithm 4.1 finds all vertices of the interconnection graph $\mathcal{G}_K$ which are strongly connected to vertex 1 in accordance with Definition 2.12.*

Proof. See Appendix B.1. □

Figure 4.2 visualises the corresponding flow diagram of Algorithm 4.1 executed on design agent D_i. As mentioned above, D_h has requested D_i and D_i will request D_j so that D_i will response to D_h and D_j will response to D_i (double arrows in Fig. 4.2).

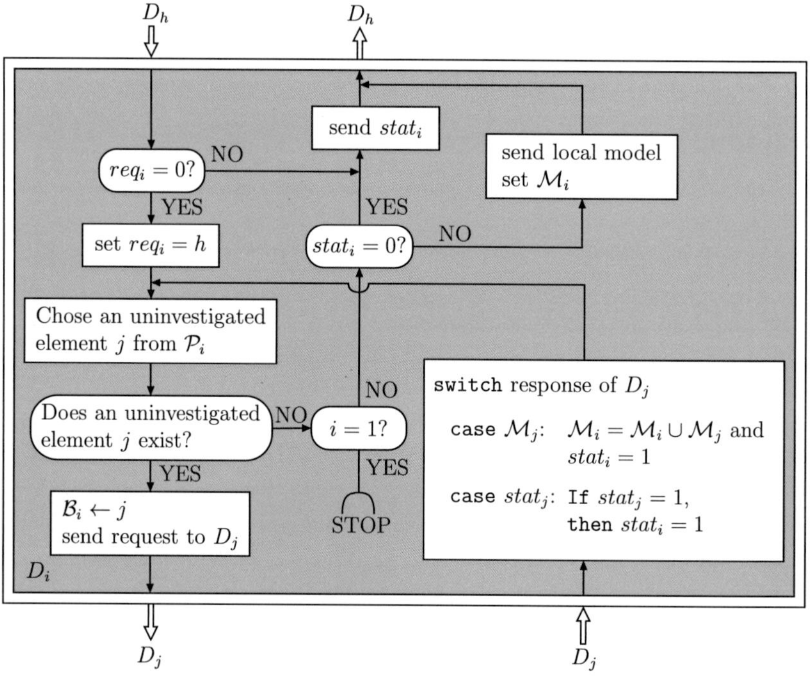

Figure 4.2: Flow diagram of Algorithm 4.1 executed at design agent D_i: D_h has requested D_i and D_i will request D_j.

Note that Algorithm 4.1 can easily be adapted to find all vertices which are strongly connected to the vertex k. Therefore, in Initialise and Lines 1 and 17, 1 has to be exchanged by k.

Remark 4.1 *The proposed Algorithm 4.1 applies the distributed depth-first-search algorithm* *presented in [61] and extends it by the evaluation of the connectivity of the vertices i to the start vertex 1. In [61] it has been proved that the distributed depth-first-search algorithm labels every vertex i which can be reached by vertex 1 (in Algorithm 4.1 all vertices i are labelled which reach vertex 1), terminates eventually in the start vertex and the traversal pattern is indeed depth-first.*

It should be emphasised that the proposed Algorithm 4.1 is performed serially. That is, the function `getModel` *is executed only at one design agent at a time. This yields a rudimentary distributed search algorithm. However, faster and more efficient algorithms exist which inherent a parallelisation of the depth-first-search algorithm that has been proposed in [61] (e.g., [36, 63, 123, 161]). All of them can be adapted to be performed for the procurement of model information for plug-and-play control.*

4.3 Analysis of the modelling algorithm

Algorithm 4.1 is applied to an interconnected tank system to analyse the processing of Algorithm 4.1and the involved interactions among the design agents.

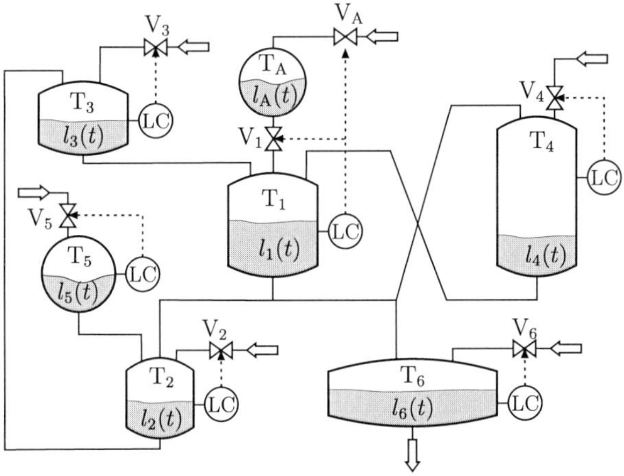

Figure 4.3: Setup of the interconnected tank process used for the analysis of Algorithm 4.1.

The interconnected tank process consists of six tanks T_i, $(i = 1, ..., 6)$ and an auxiliary tank T_A physically coupled through pipes as depicted in Fig. 4.3. Each tank represents a subsystem, where the auxiliary tank is allocated to tank T_1. Each tank is externally filled with a specific liquid to be mixed with the fillings of other tanks. The external inflow is controlled by existing level controllers LC adjusting the opening angle of the respective valves V_i, $(i = 1, ..., 6)$ and $\mathrm{V_A}$ with the aim to hold the levels $l_i(t)$, $(i = 1, ..., 6)$ close to their operating points.

The interconnection structure is represented by the interconnection matrix (2.10) with

$$\boldsymbol{L} = \begin{pmatrix} 0 & 0 & 1 & 1 & 0 & 0 \\ 1/3 & 0 & 0 & 0 & 1 & 0 \\ 0 & 1 & 0 & 0 & 0 & 0 \\ 1/3 & 0 & 0 & 0 & 0 & 0 \\ 0 & 0 & 0 & 0 & 0 & 0 \\ 1/3 & 0 & 0 & 0 & 0 & 0 \end{pmatrix} \quad (4.3)$$

and is visualised by the interconnection graph $\mathcal{G}_\mathrm{K}$ shown in Fig. 4.4a. A more detailed description of the plant is outlined in Section 4.4.3.

The processing of Algorithm 4.1 is visualised in Figs. 4.4b to 4.4l. The subfigures emphasise three aspects. First, design agent D_i on which the function `getModel` is performed at iteration κ is highlighted by the bold framed vertex. Second, the local interconnection graph $\mathcal{G}_{\mathrm{K}i}$ is shown, which can be constructed with the model set $\mathcal{M}_i$ known by design agent D_i (bold framed vertex) at the iteration κ. Third, the grey shadowed vertices indicate the corresponding subsystems that are categorised as strongly connected to subsystem S_1 (i.e., $stat_i = 1$ in D_i).

In particular, the local search is performed as follows: Initially, at $\kappa = 0$ Algorithm 4.1 is started at design agent D_1 by running `getModel(1,1)`. S_1 is initially categorised as strongly connected to S_1 (shadowed vertex in Fig. 4.4b). In the first step D_1 requests the uninvestigated design agent with the lowest index within the set $\mathcal{P}_1 = \{3, 4\}$, i.e., D_3 $(\kappa = 1)$. Accordingly, the function `getModel` is now executed on D_3 (bold vertex in Fig. 4.4c) and D_1 waits for a response. From the perspective of design agent D_3, it only knows the interaction relations shown by the local interconnection graph $\mathcal{G}_{\mathrm{K}3}(1)$ (Fig. 4.4c). As $\mathcal{P}_3 = \{2\}$, D_3 requests design agent D_2 at $\kappa = 2$ which starts with a request to D_1 $(\kappa = 3)$. As S_1 is categorised as strongly connected to S_1, subsystem S_2 is also categorised as strongly connected to S_1 (node 2 is shadowed in Fig. 4.4e). Thereafter, D_2 requests design agent D_5 at iteration $\kappa = 4$. The local interconnection graph $\mathcal{G}_{\mathrm{K}5}$ of D_5 is shown Fig. 4.4f. Since, there are no predecessors of subsystem S_5 (i.e., $\mathcal{P}_5 = \emptyset$), it is backtracked to design agent D_2 at $\kappa = 5$. As S_5 is not classified as strongly connected to S_1, the node 1, D_5 only transmits this information to design agent D_2, but no model information. A further backtracking from D_2 to D_3 is performed at $\kappa = 6$, since all predecessors of design agent D_2 are investigated $(\mathcal{B}_2 = \mathcal{P}_2)$. As subsystem S_2 is categorised as strongly connected to S_1, the model set $\mathcal{M}_2 = \{F_2, K_2\}$ is sent to D_3. With the additional information at hand, D_3 can determine the local interconnection graph $\mathcal{G}_{\mathrm{K}3}$ shown in Fig. 4.4h.

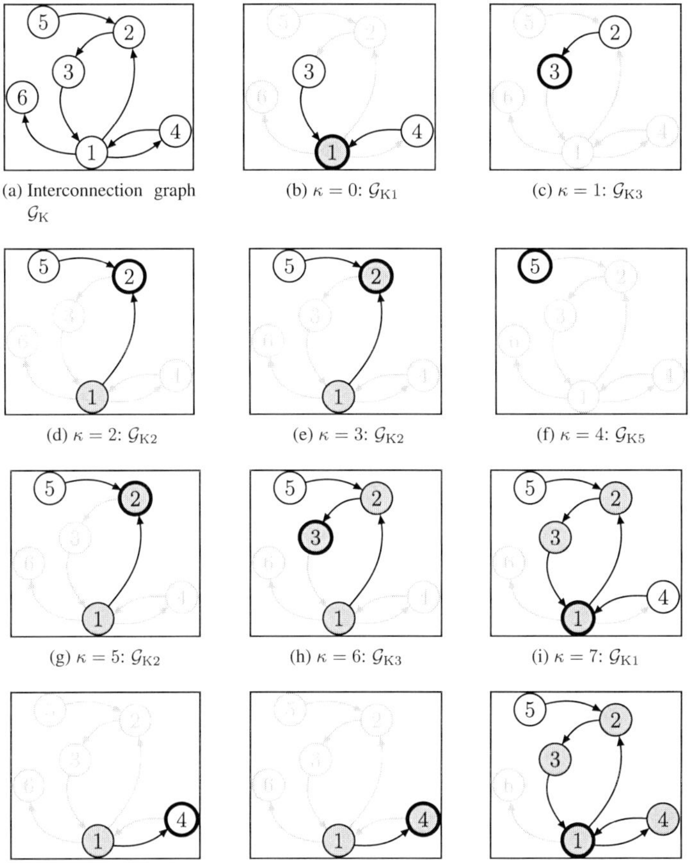

(a) Interconnection graph $\mathcal{G}_{\mathrm{K}}$ (b) $\kappa = 0$: $\mathcal{G}_{\mathrm{K1}}$ (c) $\kappa = 1$: $\mathcal{G}_{\mathrm{K3}}$ (d) $\kappa = 2$: $\mathcal{G}_{\mathrm{K2}}$ (e) $\kappa = 3$: $\mathcal{G}_{\mathrm{K2}}$ (f) $\kappa = 4$: $\mathcal{G}_{\mathrm{K5}}$ (g) $\kappa = 5$: $\mathcal{G}_{\mathrm{K2}}$ (h) $\kappa = 6$: $\mathcal{G}_{\mathrm{K3}}$ (i) $\kappa = 7$: $\mathcal{G}_{\mathrm{K1}}$ (j) $\kappa = 8$: $\mathcal{G}_{\mathrm{K4}}$ (k) $\kappa = 9$: $\mathcal{G}_{\mathrm{K4}}$ (l) $\kappa = 10$: $\mathcal{G}_{\mathrm{K1}}$

Figure 4.4: Processing of Alg. 4.1: Evaluation of the local interconnection graphs $\mathcal{G}_{\mathrm{K}i}$. The bold vertex shows the design agent on which the algorithm currently performed and the grey vertices mark subsystems strongly connected to S_1.

Moreover, subsystem S_3 is also categorised as strongly connected to S_1 (shadowed vertex in Fig. 4.4h). At iteration $\kappa = 7$, it is furthermore backtracked from D_3 to D_1 with the transmission of the model sets $\mathcal{M}_2$ and $\mathcal{M}_3$ such that D_1 can now construct the graph $\mathcal{G}_{\mathrm{K1}}$ shown in Fig. 4.4i. Afterwards, design agent D_4 is requested by D_1 at $\kappa = 8$. S_4 is then categorised as strongly connected to S_1 at $\kappa = 9$ since D_4 requests D_1. Finally, D_4 transmits its local model set $\mathcal{M}_4 = \{F_4, K_4\}$ to D_1 at the last iteration $\kappa = 10$.

The final local interconnection graph $\mathcal{G}_{K1}$ available to D_1 after ten iterations is illustrated in Fig. 4.4l. The shadowed vertices represent the set $\mathcal{N}_{S1} = \{1, 2, 3, 4\}$. Eventually, design agent D_1 stores the models $\mathcal{M}_1 = \{S_1, K_1\} \cup \{F_i, K_i, i = 1, ..., 4\}$. It has to be emphasised that design agent D_6 is not requested during the execution of Algorithm 4.1, since subsystem S_6 is no predecessor of any other subsystem. That is, there exists no path from S_6 to S_1 in the interconnection graph $\mathcal{G}_K$ (Fig. 4.4a).

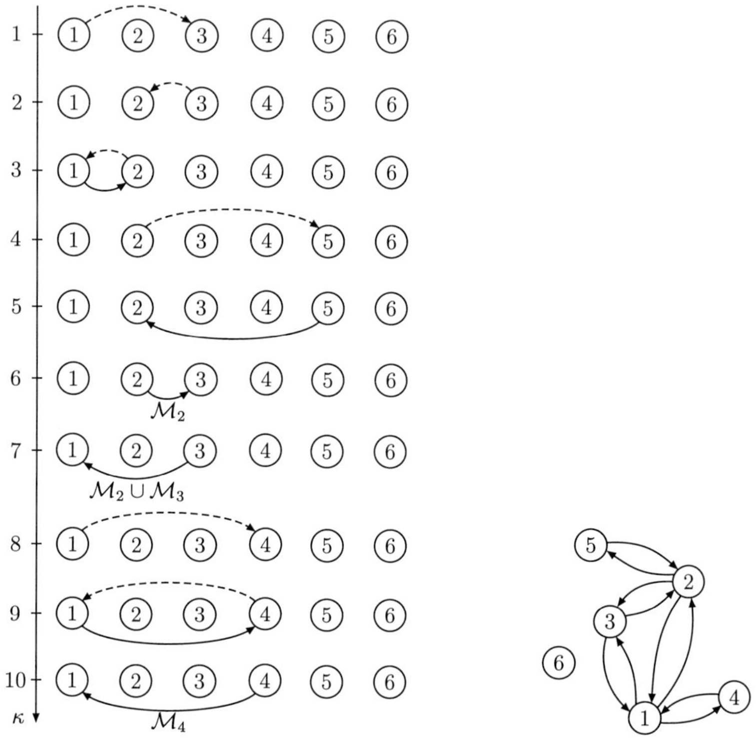

(a) Communication graph $\mathcal{G}_D(\kappa)$: Dashed edges denote requests and solid edges denote responses. If a request ist denied, the response is received at the same iteration

(b) Cumulated communication graph $\mathcal{G}_D$

Figure 4.5: Execution of Alg. 4.1: Communication graph $\mathcal{G}_D(\kappa)$ and cumulated communication graph $\mathcal{G}_D$.

During the procurement phase, at each iteration κ only two design agents are in contact (due to the serial processing of the algorithm) which is emphasised by the communication graph shown in Fig. 4.5a. At iterations $\kappa = 3$ and $\kappa = 9$ design agents D_2 and D_4, respectively, request model information from design agent D_1. Since D_1 has already been marked as requested, it denies the request immediately and transmits no models (no weight on the solid edges in Fig. 4.5a). Model information is transmitted at iterations $\kappa = 6$, $\kappa = 7$ and $k = 10$ as illustrated by the weighted edges in Fig. 4.5a. The cumulated communication graph is shown in Fig. 4.5b. Since the requested design agent D_i is chosen from the set $\mathcal{P}_i$ that collects the indices of subsystems S_j which influence S_i directly, the established communication links are inherent to the physical interconnection structure. As a consequence, the cumulated communication graph yields a subgraph of the bidirectional representation of the interconnection graph (compare $\mathcal{G}_{\rm K}$ in Fig. 4.4a and $\mathcal{G}_{\rm D}$ in Fig. 4.5b).

Although the function `getModel` is implemented on all design agents, it is only processed by those design agents that receive a request message. This fact is visualised in Fig. 4.4, where the bold vertex indicates which design agent currently runs the function `getModel`. In this example, design agent D_6 does not execute the function at all (cf. Fig. 4.5b, where no edge to node 6 exists in the cumulated communication graph).

The proof of Algorithm 4.1 (see Appendix B.1) reveals that the required number of iterations scales linearly with the number of vertices and edges of the reachability graph $\mathcal{G}_{\rm R1}(\mathcal{G}_{\rm K}^\top)$. That is, the time complexity is $O(|\mathcal{E}_{\rm R1}| + |\mathcal{V}_{\rm R1}|)$, where $\mathcal{V}_{\rm R1}$ collects the nodes $i \in \mathcal{V}_{\rm K}$ reachable from vertex 1 and $\mathcal{E}_{\rm R1} = \bigcup_{i\in\mathcal{V}_{\rm R1}} \bigcup_{j\in\mathcal{P}_i} (i,j)$ (see Section 2.1.6). In particular, the number of iterations yields

$$\kappa = |\mathcal{E}_{\rm R1}| + |\mathcal{V}_{\rm R1}| - 1.$$

Referring to the example, the reachability graph $\mathcal{G}_{\rm R1}(\mathcal{G}_{\rm K}^\top)$ is depicted in Fig. 4.6. Accordingly, $\kappa = 6 + 5 - 1 = 10$ iterations are required.

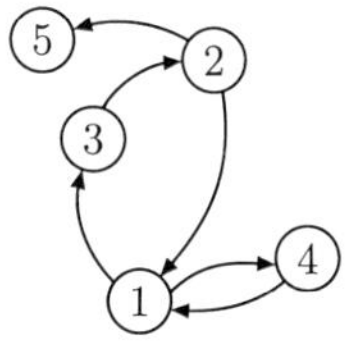

Figure 4.6: Processing of Alg. 4.1: Reachability graph $\mathcal{G}_{\rm R1}(\mathcal{G}_{\rm K}^\top)$.

Moreover, the number of request messages and response messages exchanged among the design agents results to $2 \cdot |\mathcal{E}_{\rm R1}|$. Thus, the complexity of Algorithm 4.1 in accordance to the transmitted messages scales linearly with the number of edges, i.e., $O(|\mathcal{E}_{\rm R1}|)$. Regarding Fig. 4.5a, twelve massages are communicated among the design agents.

4.4 Application scenario: Fault-tolerant control

The situation is considered in which an actuator in subsystem S_1 fails, and control station C_1 shall be reconfigured, although the fault has effect on all other controlled subsystems through the physical couplings. The reconfiguration must guarantee the retrieval of overall closed-loop asymptotic stability by means of a virtual actuator. To focus on the reconfiguration process, the following two assumptions are made:

A 4.1 *A diagnosis system detects the fault and uniquely identifies the model S_{1f} of the faulty subsystem. As a result, the model S_{1f} is known by design agent D_1. The diagnostic process does not consume any time.*

A 4.2 *There exist N control stations C_i, $(\forall i \in \mathcal{N})$ such that the overall closed-loop system is asymptotically stable.*

Concerning assumption A 4.1, a local diagnosis system is proposed in Chapter 8.

A virtual actuator is proposed that uses the models $S_1^{\mathcal{N}_{S1}}$ and $S_{1f}^{\mathcal{N}_{S1}}$ of the healthy and faulty subsystem together with its strongly connected controlled subsystems, respectively. An LMI-based design method is presented in Corollary 4.1. The whole reconfiguration procedure that is managed by design agent D_1 is summarised in Algorithm 4.2.

4.4.1 Local reconfiguration with a virtual actuator using the model of the physical interaction

This section, first, presents the virtual actuator that is used to recover asymptotic stability of the overall closed-loop system. It is shown that only the model $S_1^{\mathcal{N}_{S1}}$ is required for the design of the virtual actuator as well as during the execution of the virtual actuator. Finally, an H_∞-design of the virtual actuator is proposed.

At time instant $t_\mathrm{f} \geq 0$ an actuator in subsystem S_1 fails. The model S_{1f} of the faulty subsystem is, thus, represented by (2.26), where the corresponding column in $\boldsymbol{B}_1$ that represents the faulty actuator is set to zero. Accordingly, the model $S_{1f}^{\mathcal{N}_{S1}}$ that shall be set up yields

$$S_{1f}^{\mathcal{N}_{S1}} = \mathrm{con}\left(S_{1f} \cup \left\{ F_i|_{\boldsymbol{w}_i=\boldsymbol{0}}, i \in \mathcal{N}_{S1} \setminus \{1\}\right\} \cup \{K_i, i \in \mathcal{N}_{S1}\}, \ \boldsymbol{u}_{1f}, \ \boldsymbol{y}_{1f}\right) \qquad (4.4)$$

and is represented in state space by

$$S_{1f}^{\mathcal{N}_{S1}}: \begin{cases} \dot{\boldsymbol{x}}_{1f}^{\mathcal{N}_{S1}}(t) = \boldsymbol{A}_1^{\mathcal{N}_{S1}} \boldsymbol{x}_{1f}^{\mathcal{N}_{S1}}(t) + \boldsymbol{B}_{1f}^{\mathcal{N}_{S1}} \boldsymbol{u}_{1f}(t), \quad \boldsymbol{x}_{1f}^{\mathcal{N}_{S1}}(t_\mathrm{f}) = \boldsymbol{0} \\ \boldsymbol{y}_{1f}(t) = \boldsymbol{C}_1^{\mathcal{N}_{S1}} \boldsymbol{x}_{1f}^{\mathcal{N}_{S1}}(t), \end{cases}$$

where $\boldsymbol{x}_{1f}^{\mathcal{N}_{S1}}(t)$ accumulates the state $\boldsymbol{x}_{1f}(t)$ of the faulty subsystem and the states $\boldsymbol{x}_{Fi}(t)$ of the controlled subsystems F_i, $(\forall i \in \mathcal{N}_{S1} \setminus \{1\})$ that are strongly connected to S_1. The structure of the model $S_{1f}^{\mathcal{N}_{S1}}$ is highlighted yellow in Fig. 4.7.

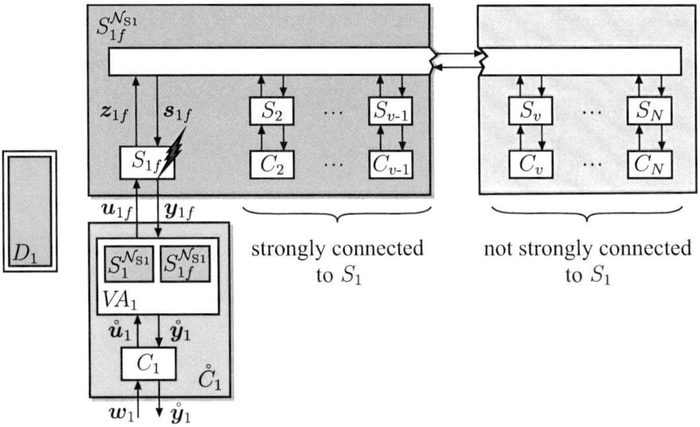

Figure 4.7: Structure of the reconfigured closed-loop system: Local reconfiguration of control station C_1 using a virtual actuator VA_1. The result is the reconfigured control station $\mathring{C}_1$. For the design and the execution of the virtual actuator the models $S_1^{\mathcal{N}_{S1}}$ and $S_{1f}^{\mathcal{N}_{S1}}$ (yellow) are required which are set up using Alg. 4.1.

The virtual actuator uses the models $S_1^{\mathcal{N}_{S1}}$ and $S_{1f}^{\mathcal{N}_{S1}}$ of the fault-free and faulty subsystem under consideration of the physical interaction in order to hide the effect of the fault from the nominal control station C_1 (cf. Section 2.4.2). In particular, the virtual actuator is represented by the state-space model

$$VA_1 : \begin{cases} \dot{\boldsymbol{x}}_{\delta 1}(t) &= \boldsymbol{A}_{\delta 1}\boldsymbol{x}_{\delta 1}(t) + \boldsymbol{B}_1^{\mathcal{N}_{S1}}\mathring{\boldsymbol{u}}_1(t), \quad \boldsymbol{x}_{\delta 1}(t_f) = \boldsymbol{0} \\ \boldsymbol{u}_{1f}(t) &= \boldsymbol{M}_1\boldsymbol{x}_{\delta 1}(t) \\ \mathring{\boldsymbol{y}}_1(t) &= \boldsymbol{y}_{1f}(t) + \boldsymbol{C}_1^{\mathcal{N}_{S1}}\boldsymbol{x}_{\delta 1}(t), \end{cases} \tag{4.5}$$

where $\boldsymbol{x}_{\delta 1}(t) = \boldsymbol{x}_{1f}^{\mathcal{N}_{S1}}(t) - \boldsymbol{x}_1^{\mathcal{N}_{S1}}(t)$ is the difference state and $\boldsymbol{A}_{\delta 1} = \boldsymbol{A}_1^{\mathcal{N}_{S1}} - \boldsymbol{B}_{1f}^{\mathcal{N}_{S1}}\boldsymbol{M}_1$. The feedback gain $\boldsymbol{M}_1$ has to be designed so that the virtual actuator (4.5) stabilises the faulty overall closed-loop system

$$F_f = \mathrm{con}\left(\{S_{1f}, C_1, K_1\} \cup \{F_i, K_i,\ i \in \mathcal{N} \setminus \{1\}\},\ \boldsymbol{w},\ \boldsymbol{y}_f\right), \tag{4.6}$$

where $\boldsymbol{y}_f = \begin{pmatrix} \boldsymbol{y}_{1f}^\top & \boldsymbol{y}_2^\top & \cdots & \boldsymbol{y}_N^\top \end{pmatrix}^\top$. An LMI-based design is presented as follows.

Corollary 4.1 (Stabilising reconfiguration after an actuator failure in S_1 using a virtual actuator) *The virtual actuator* (4.5) *recovers asymptotic stability of the faulty overall closed-loop system F_f if and only if there exit feasible solutions $\boldsymbol{X} = \boldsymbol{X}^\top \prec 0$ and $\boldsymbol{Y}$ to the LMI*

$$\boldsymbol{A}_1^{\mathcal{N}_{S1}} \boldsymbol{X} + \boldsymbol{X} \boldsymbol{A}_1^{\mathcal{N}_{S1}\top} + \boldsymbol{B}_{1f}^{\mathcal{N}_{S1}} \boldsymbol{Y} + \boldsymbol{Y}^\top \boldsymbol{B}_{1f}^{\mathcal{N}_{S1}\top} \prec 0. \tag{4.7}$$

The feedback gain of the virtual actuator results to $\boldsymbol{M}_1 = -\boldsymbol{Y}\boldsymbol{X}^{-1}$. Feasible solutions to the LMI exist if and only if the pair $\left(\boldsymbol{A}_1^{\mathcal{N}_{S1}},\ \boldsymbol{B}_{1f}^{\mathcal{N}_{S1}}\right)$ is stabilisable.

Proof. See Appendix B.2. □

Corollary 4.1 states an asymptotic stability of reconfigured closed-loop system

$$\mathring{F}_1^{\mathcal{N}_{S1}} = \operatorname{con}\left(\left\{S_{1f}^{\mathcal{N}_{S1}}, VA_1, C_1, K_1\right\} \cup \left\{F_i|_{\boldsymbol{w}_i=\boldsymbol{0}}, K_i,\ i \in \mathcal{N}_{S1} \setminus \{1\}\right\},\ \boldsymbol{w}_1,\ \mathring{\boldsymbol{y}}_1\right)$$

is necessary and sufficient to recover asymptotic stability of faulty overall closed-loop system F_f according to assumption A 4.2. Hence, with the model $S_{1f}^{\mathcal{N}_{S1}}$ available to D_1, the virtual actuator (4.5) can be designed with respect to a necessary and sufficient condition. Note that C_1 can be designed independently from the other control stations. This is due to the fact that the faulty interacting subsystem $S_{1f}^{\mathcal{N}_{S1}}$ has to be stabilised through the I/O pair ($\boldsymbol{u}_{1f}$,$\boldsymbol{y}_{1f}$) as claimed by the stabilisability of the pair $\left(\boldsymbol{A}_1^{\mathcal{N}_{S1}},\ \boldsymbol{B}_{1f}^{\mathcal{N}_{S1}}\right)$. The virtual actuator VA_1 together with the nominal control station C_1 represents the reconfigured control station $\mathring{C}_1$ shown in Fig. 4.7.

In contrast to the classical virtual actuator (2.27) presented in Section 2.4.2, the virtual actuator (4.5) has the main advantages that, both, during the design phase and the execution phase only the model $S_{1f}^{\mathcal{N}_{S1}}$ is required and not the interacting subsystem model $S_{1f}^{\mathcal{N}}$. Due to the naturally sparse interconnection between the subsystems usually the relation $\mathcal{N}_{S1} \subset \mathcal{N}$ holds. Hence, the computational workload is reduced. The virtual actuator represents a decentralised realisation of the global virtual actuator and can, therefore, be implemented on control station C_1. Whenever the pair $\left(\boldsymbol{A}_1^{\mathcal{N}_{S1}},\ \boldsymbol{B}_{1f}^{\mathcal{N}_{S1}}\right)$ is not stabilisable, distributed solutions as proposed in [32, 12] have to be preferred.

4.4.2 Plug-and-play reconfiguration algorithm

This section shows how plug-and-play control can be applied to the reconfiguration of control station C_1 by design agent D_1. Therefore, the results from Section 4.2 and Section 4.4.1 are combined to Algorithm 4.2.

If an actuator in subsystem S_1 fails, asymptotic stability of the overall closed-loop system is jeopardised due to the physical couplings to other subsystem. The whole reconfiguration process is organised by the corresponding design agent D_1. Accordingly, design agent D_1 procures relevant model information through the communication network from other design agents D_i (i.e., solve Problem 1.2) to accomplish the reconfiguration according to Corollary 4.1 (i.e., solve Problem 1.1). This procedure is summarised in the Algorithm 4.2.

Algorithm 4.2: Plug-and-play reconfiguration of control station C_1

Given: Control stations C_i, $(\forall i \in \mathcal{N})$ exist such that F is asymptotically stable,
model set $\mathcal{M}_1 = \{S_1, S_{f1}, C_1, K_1\}$ available to D_1 and
model set $\mathcal{M}_i = \{F_i, K_i\}$ available to D_i, $(\forall i \in \mathcal{N} \setminus \{1\})$

Processing on D_1:

1. Run Algorithm 4.1 to gather the model set
$\mathcal{M}_1 = \{S_1, S_{f1}, C_1, K_1\} \cup \{F_i, K_i,\ i \in \mathcal{N}_{\mathrm{S1}} \setminus \{1\}\}$ in order to set up the models $S_1^{\mathcal{N}_{\mathrm{S1}}}$, $S_{1f}^{\mathcal{N}_{\mathrm{S1}}}$ and $F_{1f}^{\mathcal{N}_{\mathrm{S1}}}$
2. **if** $F_{1f}^{\mathcal{N}_{\mathrm{S1}}}$ is unstable **then** design the virtual actuator (4.5) by means of Corollary 4.1, **else** STOP (no reconfiguration necessary)
3. Implement the virtual actuator VA_1 into the control equipment

Result: Asymptotically stable reconfigured closed-loop system

---

At the end of Step 1, design agent D_1 has available to set up the model $S_1^{\mathcal{N}_{\mathrm{S1}}}$ of the fault-free interacting subsystem (cf. model (4.1)), the model $S_{1f}^{\mathcal{N}_{\mathrm{S1}}}$ of the faulty interacting subsystem (cf. model (4.4)), and the model $F_{1f}^{\mathcal{N}_{\mathrm{S1}}} = \mathrm{con}\left(\left\{S_{1f}^{\mathcal{N}_{\mathrm{S1}}},\ C_1\right\},\ \boldsymbol{w}_1,\ \boldsymbol{y}_{1f}\right)$ of the faulty controlled interacting subsystem. The latter model is used to verify whether or not a reconfiguration is necessary in Step 2. If $F_{1f}^{\mathcal{N}_{\mathrm{S1}}}$ is unstable, the virtual actuator is designed (Step 2), implemented and started up (Step 3).

Note that alternatively the proposed LMI-based design, the reconfiguration can be performed by arbitrary automated procedures from literature [119, 120]. From assumption A 3.4 it follows that the reconfiguration is processed instantaneously.

4.4.3 Example: Interconnected tank system

The plug-and-play reconfiguration algorithm is applied to the interconnected tank system depicted in Fig. 4.3, which has been introduced in Section 4.3. The simulation highlights that asymptotic stability is preserved by the design of a virtual actuator which only uses the models of the controlled subsystems that are strongly connected to the faulty tank.

Process description. As presented in Fig. 4.3, the interconnected tank system consists of six tanks T_i, $(i = 1, ..., 6)$ and one auxiliary tank T_A physically coupled through pipes. The tank T_1 and the auxiliary tank T_A comprise subsystem S_1 with the state

$$\boldsymbol{x}_1(t) = \begin{pmatrix} l_\mathrm{A}(t) \\ l_1(t) \end{pmatrix}$$

representing the level $l_\mathrm{A}(t)$ of the auxiliary tank and the level $l_1(t)$ of the tank T_1. The subsystem is described by the state-space model (2.6) defined by

$$\boldsymbol{A}_1 = \begin{pmatrix} -0.65 & 0 \\ 0.81 & -1 \end{pmatrix}, \quad \boldsymbol{B}_1 = \begin{pmatrix} 0 & 1.2 \\ 2 & 0 \end{pmatrix}, \quad \boldsymbol{E}_1 = \begin{pmatrix} 0 \\ 0.33 \end{pmatrix},$$
$$\boldsymbol{C}_1 = \begin{pmatrix} 0 & 1 \end{pmatrix}, \qquad \boldsymbol{C}_{\mathrm{z}1} = \begin{pmatrix} 0 & 3.14 \end{pmatrix}.$$

Subsystem S_1 is controlled by a PI-controller using both valves V_A and V_1, represented by the state-space model (2.11) with

$$\boldsymbol{A}_{\mathrm{C}1} = \mathbf{O}, \qquad \boldsymbol{B}_{\mathrm{C}1} = \begin{pmatrix} 1 \\ 1 \end{pmatrix}, \qquad \boldsymbol{C}_{\mathrm{C}1} = \begin{pmatrix} 1 \\ 0 \end{pmatrix}, \qquad \boldsymbol{D}_{\mathrm{C}1} = \begin{pmatrix} 0.55 \\ 0.55 \end{pmatrix}.$$

Accordingly, the level dynamics of the other tanks T_i, $(i = 2, ..., 6)$ are represented by the respective subsystems model (2.6), with

$$\begin{aligned}
&\boldsymbol{A}_2 = -23.7, && \boldsymbol{B}_2 = 0.2, && \boldsymbol{E}_2 = 1, && \boldsymbol{C}_2 = 1, && \boldsymbol{C}_{\mathrm{z}2} = 23.7,\\
&\boldsymbol{A}_3 = -11, && \boldsymbol{B}_3 = 0.5, && \boldsymbol{E}_3 = 0.5, && \boldsymbol{C}_3 = 1, && \boldsymbol{C}_{\mathrm{z}3} = 22,\\
&\boldsymbol{A}_4 = -1.3, && \boldsymbol{B}_4 = 0.6, && \boldsymbol{E}_4 = 0.2, && \boldsymbol{C}_4 = 1, && \boldsymbol{C}_{\mathrm{z}4} = 6.5,\\
&\boldsymbol{A}_5 = -0.2, && \boldsymbol{B}_5 = 0.2, && \boldsymbol{E}_5 = 0.2, && \boldsymbol{C}_5 = 1, && \boldsymbol{C}_{\mathrm{z}5} = 1.4,\\
&\boldsymbol{A}_6 = -0.4, && \boldsymbol{B}_6 = 0.2, && \boldsymbol{E}_6 = 0.2, && \boldsymbol{C}_6 = 1, && \boldsymbol{C}_{\mathrm{z}6} = 2.1
\end{aligned}$$

and are interconnected according to the interconnection matrix (4.3), where the local interconnection model K_i can be derived. The tanks are controlled by PI-controllers C_i, $(i = 2, ..., 6)$, defined by

$$\boldsymbol{A}_{\mathrm{C}i} = 0, \qquad \boldsymbol{B}_{\mathrm{C}i} = 1, \qquad \boldsymbol{C}_{\mathrm{C}i} = 1, \qquad \boldsymbol{D}_{\mathrm{C}i} = 0.55$$

manipulating the angle of the respective inflow valve V_i. The control stations are designed to result in an asymptotically stable closed-loop system. The step response of the nominal controlled system

$$F_1^{\mathcal{N}_{S1}} = \mathrm{con}\left(\{C_1,\, S_1^{\mathcal{N}_{S1}}\},\, \boldsymbol{w}_1,\, \boldsymbol{y}_1\right)$$

is shown in Fig. 4.8.

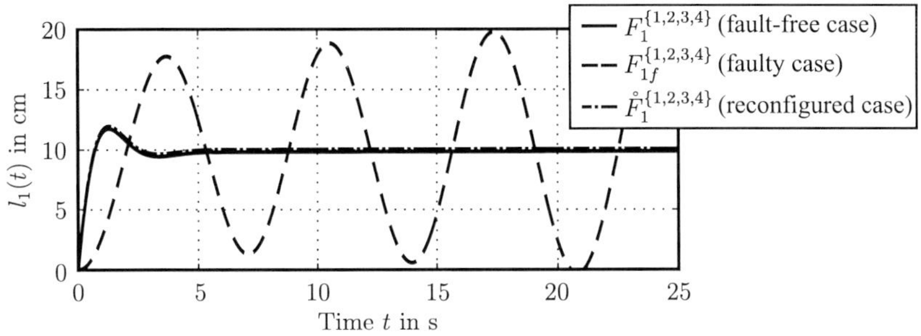

Figure 4.8: Application: Response of the level $l_1(t)$ to the reference signal $w_1(t) = \sigma(t) \cdot 10\,\mathrm{cm}$ for the nominal, faulty and reconfigured case.

Fault occurrence in subsystem S_1. At $t = t_\mathrm{f} = 0\,\mathrm{s}$ the valve V_1 stuck in its operating point. Hence, the model S_{1f} of the faulty subsystem with

$$\boldsymbol{B}_{\mathrm{f}1} = \begin{pmatrix} 0 & 1.2 \\ 0 & 0 \end{pmatrix}$$

is immediately identified by the diagnosis system and is available to design agent D_1 (cf. assumption A 4.1).

Algorithm 4.2 is performed by D_1 to reconfigure control station C_1 in order that asymptotic stability of the overall closed-loop system is recovered by using the remaining valve V_A.

Step 1: The processing of Algorithm 4.1 is analysed in Section 4.3. Finally, design agent D_1 has the models $\mathcal{M}_1 = \{S_1, S_{1f}\} \cup \{F_i, K_i,\, i = 2,3,4\}$ available. The connection of the models results to $S_1^{\{1,2,3,4\}}$ and $S_{1f}^{\{1,2,3,4\}}$, where $\boldsymbol{x}_1^{\{1,2,3,4\}}, \boldsymbol{x}_{1f}^{\{1,2,3,4\}} \in \mathbb{R}^8$. Moreover, the faulty controlled interacting subsystem $F_{1f}^{\{1,2,3,4\}} = \mathrm{con}\left(\left\{S_{1f}^{\{1,2,3,4\}}\right\},\, C_1,\, \boldsymbol{w}_1,\, \boldsymbol{y}_{1f}\right)$ is set up.

Step 2: As it can be seen by the step response of $\boldsymbol{y}_{1f}(t)$ in Fig. 4.8, the faulty controlled interacting subsystem $F_{1f}^{\{1,2,3,4\}}$ is unstable. Accordingly, design agent D_1 designs the virtual actuator by solving the LMI (4.7). The feedback gain yields

$$\boldsymbol{M}_1 = \begin{pmatrix} 0 & 0 & 0 & 0 & 0 & 0 & 0 & 0 \\ 0.05 & 0.35 & 0.73 & 0.04 & 0.45 & 0.02 & 0.25 & 0.18 \end{pmatrix}.$$

The first row of $\boldsymbol{M}_1$ is set to zero since it is related to the opening angle of the failed valve V_1.
Step 3: The virtual actuator is implemented into the control hardware and is subsequently started up. Figure 4.8 visualises that stability of the reconfigured controlled interacting subsystem $\mathring{F}_1^{\{1,2,3,4\}}$ is recovered.

Figure 4.8 shows that the reconfigured system $\mathring{F}_1^{\{1,2,3,4\}}$ has the same behaviour than the nominal system $F_1^{\{1,2,3,4\}}$. That is, the virtual actuator hides the effect of the fault from the nominal control station. Since the fault affects the system in its operating point and the reconfiguration is assumed to be immediate, the fault has no effect on the overall closed-loop system.

It has to be emphasised that $\boldsymbol{x}_{\delta 1} \in \mathbb{R}^8$. That is, the number of online computed state $\boldsymbol{x}_{\delta 1}$ by the virtual actuator is reduced by four (correspond to $33\,\%$) since the controlled subsystems F_5 ($\boldsymbol{x}_{\mathrm{F}5} \in \mathbb{R}^2$) and F_6 ($\boldsymbol{x}_{\mathrm{F}6} \in \mathbb{R}^2$) are not taken into account. This reduces the computational workload of the virtual actuator compared to the classical virtual actuator from Section 2.4.2.

5 Plug-and-play control using the local model based on restricted operating sets

This chapter is addressed to the design of control station C_1 by design agent D_1 using only the initially known models S_1 and K_1 of the subsystem and the local couplings, respectively. Nevertheless, the design has to guarantee overall closed-loop objectives despite the unknown dynamics of the physical interaction. As a main result of this chapter, local design conditions are presented that guarantee boundedness of the overall closed-loop system and a certain overall closed-loop performance. These conditions are derived by restricting the state space of each controlled subsystem. Furthermore, two design methods are proposed that rely on control in invariant sets. First, an LMI-based design of a static state-feedback and, second, the design of a tube-based model predictive controller (MPC) from [127] is presented. The last part of this chapter applies the proposed design method on the thermofluid process to guarantee fault-tolerance. An experiment shows the successful recovery of the overall close-loop performance by reconfiguring C_1 using a tube-based MPC.

5.1 Problem formulation

This chapter focusses on the design of control station C_1 using only the subsystem model S_1 and the model K_1 of the physical couplings that are initially available to D_1. This design must guarantee the adherence of a global control aim, even though a model which represents the physical influence of all other controlled subsystems on subsystem S_1 is unknown by D_1. The aim of this chapter is to derive locally verifiable design conditions to enable D_1 to perform the synthesis of C_1. These design conditions need to be conservative to compensate the lack of knowledge about the physical interaction $P_1^{\mathcal{N}}$. Nevertheless, the design with local models is preferable for the following reasons:

- Satisfaction of privacy constraints, i.e., considering the issue that design agents are not willing to share their model information (cf. motivation example in Section 1.2.1).
- Reduced complexity of the design due to a low dimensional model so that the design can be embedded in hardware with low computational power.
- Scalability of the design in the sense that the control stations can be designed independently from each other.

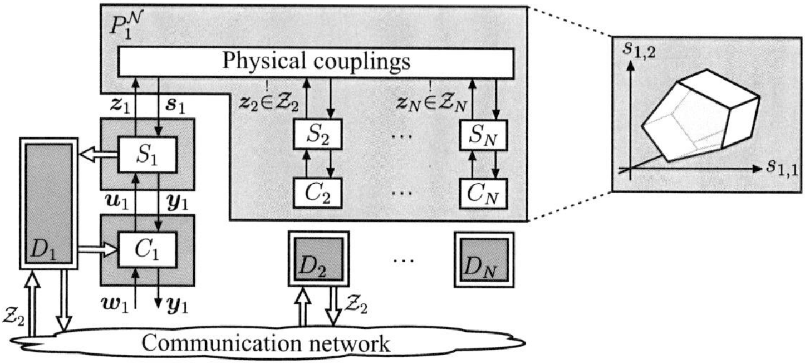

Figure 5.1: Structure and idea of plug-and-play control using the local model based on restricted operating sets. If each coupling output z_i remains in their desired operating set $\mathcal{Z}_i$, the coupling input s_1 becomes bounded. This bound is used for the design of C_1.

The model that represents the physical interaction with S_1, that is considered to be unknown for the design of C_1 consists, of all controlled subsystems F_i, $(i = 2, ..., N)$ as highlighted by the grey box in Fig. 5.1. The method presented in this chapter aims at restricting the influence of the physical interaction $P_1^{\mathcal{N}_{S1}}$ on subsystem S_1 by a suitable design of the control stations C_i, $(i = 2, ..., N)$. The main idea is to force the coupling output z_i of each controlled subsystem F_i to remain within a desired operating set $\mathcal{Z}_i$ in spite of the influence through the physical couplings (highlighted by the claim $z_i \overset{!}{\in} \mathcal{Z}_i$ in Fig. 5.1). As a result, the dynamics of the physical interaction $P_1^{\mathcal{N}}$ will also operate within a constrained set (right-hand side in Fig. 5.1). This set is composed of the targeted operating sets $\mathcal{Z}_i$ of the direct neighbours so that D_1 has to procure only these sets (for example $\mathcal{Z}_2$ as shown in Fig. 5.1).

These considerations are formalised in the following problem:

Problem 5.1 (Plug-and-play control using local model information)

Problem: *Although the dynamics of subsystem S_1 depends upon the physical interactions with all other controlled subsystems F_i, $(i = 2, ..., N)$, D_1 has to perform the design of C_1 (i.e., solve Problem 1.1) using its initially available model information only. This design must guarantee the satisfaction of overall closed-loop specifications.*

Given:
- *D_i, $(\forall i \in \mathcal{N})$ store the corresponding model set $\mathcal{M}_i = \{S_i, K_i\}$*
- *Control objectives of the overall closed-loop system*

Find: *Local design conditions which guarantee the adherence of the overall closed-loop specifications.*

Approach: *Restrict the state space of the controlled subsystems in spite of the physical influence.*

A solution to Problem 5.1 yields a solution to Problem 1.2. Moreover, the issue how an appropriate control method can look like and how it can be designed is deepened in this chapter. In particular, if a controller can be found that makes the operating region *robustly positively invariant* for the controlled subsystem, then its state will never leave this region. Methods are proposed to design a static state-feedback and a tube-based MPC. These methods are solutions to Problem 1.1.

Section 5.2 formulates the main idea of this chapter in order to enable the controller design without a model of the physical interaction. It will be shown that only the desired operating regions of neighbouring design agents are required. Based on this idea, Section 5.3 proposed local design conditions that guarantee boundedness of the overall closed-loop system state and a certain overall closed-loop performance. The following Section 5.4 is devoted to the particular synthesis of the control station in order to satisfy the local objectives. Plug-and-play control is finally applied to the thermofluid process to evaluate the proposed methods by means of simulations and experiments for the fault-tolerant control scenario in Section 5.5.

5.2 Conditions to enable the design with the local model

This section outlines the effect of restricting the operating region of each controlled subsystem and states the required amount of information.

Consider the subsystems S_i, ($\forall i \in \mathcal{N}$) that are modelled in state space by (2.6), where the state $\boldsymbol{x}_i(t)$ is assumed to be measurable. Each subsystem is controlled by a static state-feedback represented by the model

$$C_i: \ \boldsymbol{u}_i(t) = \boldsymbol{K}_i \boldsymbol{x}_i(t). \tag{5.1}$$

with the aim to hold the subsystem state in its origin (i.e., $\boldsymbol{w}_i(t) = \boldsymbol{0}$). For the moment, assume that linear feedback gains $\boldsymbol{K}_i$, ($\forall i \in \mathcal{N}$) exists such that the solutions $\boldsymbol{z}_i(t)$, ($\forall i \in \mathcal{N}$) of the controlled subsystems

$$F_i: \begin{cases} \dot{\boldsymbol{x}}_i(t) &= (\boldsymbol{A}_i + \boldsymbol{B}_i \boldsymbol{K}_i)\, \boldsymbol{x}_i(t) + \boldsymbol{E}_i \boldsymbol{s}_i(t), \quad \boldsymbol{x}_i(0) = \boldsymbol{x}_{i,0} \\ \boldsymbol{z}_i(t) &= \boldsymbol{C}_{zi} \boldsymbol{x}_i(t), \end{cases} \tag{5.2}$$

are bounded within the corresponding given operating set $\mathcal{Z}_i$ according to

$$\boldsymbol{z}_i(t) \in \mathcal{Z}_i, \quad \forall t \geq 0, \quad \forall i \in \mathcal{N}$$

in spite of the influence through the coupling input $\boldsymbol{s}_i(t)$. Then, based on the equality (2.7)

$$\boldsymbol{s}_i(t) = \sum_{j \in \mathcal{P}_i} \boldsymbol{L}_{ij} \boldsymbol{C}_{zj} \boldsymbol{x}_j(t),$$

the coupling input $\boldsymbol{s}_i(t)$ becomes bounded according to

$$\boldsymbol{s}_i(t) \in \mathcal{S}_i = \bigoplus_{j \in \mathcal{P}_i} \boldsymbol{L}_{ij} \mathcal{Z}_j, \quad \forall t \geq 0. \tag{5.3}$$

With the focus on the controlled subsystem F_1, it can be concluded that the physical interaction

$$P_1^{\mathcal{N}}: \quad \boldsymbol{s}_1(t) = \boldsymbol{P}_1^{\mathcal{N}}(t) * \boldsymbol{z}_1(t) \tag{5.4}$$

which comprises all controlled subsystems F_i, ($i = 2, ..., N$) becomes an *unknown-but-bounded input signal* represented by

$$\bar{P}_1^{\mathcal{P}_1 \cup \{1\}}: \quad \boldsymbol{s}_1(t) \in \mathcal{S}_1 \tag{5.5}$$

as highlighted in Fig. 5.2[1]. The superscribe of the model $\bar{P}_1^{\mathcal{P}_1 \cup \{1\}}$ indicates that only information from D_1 and D_i, ($\forall i \in \mathcal{P}_1$) is required to construct this system.

The crucial point of these considerations is that the coupling outputs $\boldsymbol{z}_i(t)$, ($\forall i \in \mathcal{N}$) have to remain in the corresponding set $\mathcal{Z}_i$ despite the influence of the physical interaction. As a consequence, design agent D_i has to procure only the sets $\mathcal{Z}_i$ of its direct neighbours D_j, ($\forall j \in \mathcal{P}_j$)

[1] Although the bound represented by (5.5) is not a comparison system (cf. Definitions 2.13 and 2.14), the notation of a comparison system is used to indicate that the dynamics of the physical interaction are bounded

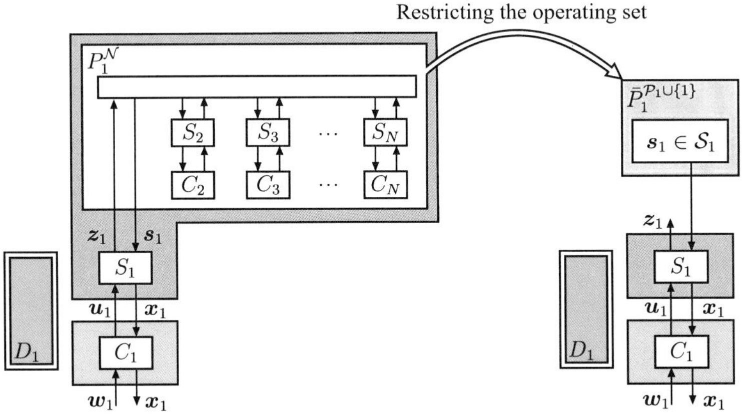

Figure 5.2: Effect of restricting the operating sets: The physical interaction $P_1^{\mathcal{N}}$ can be considered as unknown-but-bounded input signal represented by $\bar{P}_1^{\mathcal{P}_1\cup\{1\}}$.

and no information from other design agents. For instance, consider a locally interconnected system, where S_i has direct influence on S_{i+1} and on S_{i-1} and vice versa as depicted by the interconnection graph in Fig. 3.7a on page 60. From the local view of subsystem S_1, the influence of all other controlled subsystems F_i, $(i = 2, ..., N)$ on S_1 is represented by the operating set $\mathcal{Z}_2$. According to eqn (5.3), to calculate the constrained set $\mathcal{S}_1$, design agent D_1 has to procure only the set $\mathcal{Z}_2$ from its direct neighbour D_2 and no information from other design agents.

However, the actual dynamics of the physical interaction are only roughly represented by the model (5.5) for the following reasons. First, this model is composed of desired sets $\mathcal{Z}_i$ and not of the actual operating set (cf. eqn (5.3)) and, second, the system (5.5) is static in the sense that the bound $\mathcal{S}_i$ represents the maximal value the signal $s_i(t)$ is allowed to have for all times. Indeed, in dependence on the actual behaviour of the physical interaction $P_i^{\mathcal{N}}$, the bound $\mathcal{S}_i$ is rough or precise as illustrated in Fig. 5.3. A rough bound results if the desired sets $\mathcal{Z}_i$ roughly bound the actual dynamics as illustrated in Fig. 5.3a. Conversely, a tight bound results if all desired sets $\mathcal{Z}_i$ are tight bounds of the coupling output signals, highlighted in Fig. 5.3b.

Due to the fact that control station C_1 is designed using only the operating set $\mathcal{S}_1$ instead of the exact model $P_1^{\mathcal{N}}$, conservative design conditions are necessary (see Section 5.3). The above mentioned facts reflect the reason for this conservatism.

Remark 5.1 *Whenever constraints on the states of all subsystems* S_i, $(\forall i \in \mathcal{N})$ *are considered that are satisfied by an appropriate controller, the coupling input signal* s_1 *becomes bounded. According to this situation a model predictive control solution to satisfy these state constraints*

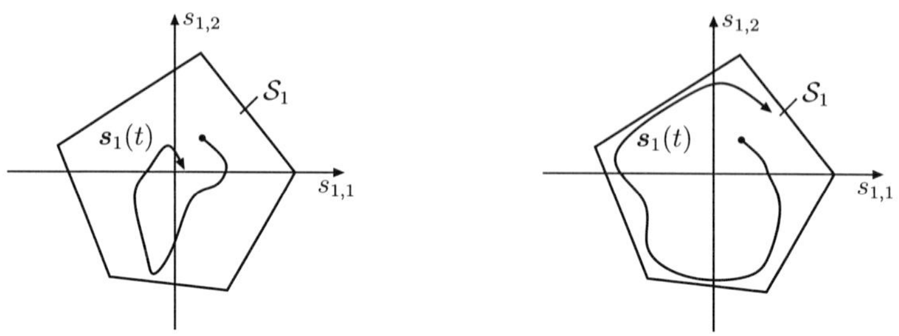

(a) Bound $\mathcal{S}_1$ is rough since the distance between the signal $s_1(t)$ and the boundary of $\mathcal{S}_1$ is large

(b) Bound $\mathcal{S}_1$ is precise since the distance between the signal $s_1(t)$ and the boundary of $\mathcal{S}_1$ is short

Figure 5.3: Example of a rough and a precise bound $\mathcal{S}_1$: The precision of the bound depends on how precise the desired sets $\mathcal{Z}_i$ reflect the actual bound on the coupling output signals $z_i(t)$.

has been proposed by [145]. The particular model predictive control algorithm can, thus, be designed using the local subsystem model S_1 and the bound $\mathcal{S}_1$ of the coupling input signal.

Similar to this approach, the main idea of this chapter is to limit the signal space of all coupling output signal z_i, $(\forall i \in \mathcal{N})$ by an appropriate controller design. Local design conditions are proposed that are not related to a particular control method (i.e., solution to Problem 1.2). Based on this design conditions controller design methods are proposed to obtain a static state-feedback (Section 5.4.1) and a model predictive controller (Section 5.4.2) (i.e., solution to Problem 1.1).

Communication among the design agents. The calculation of the set $\mathcal{S}_1$ by design agent D_1 according to (5.3) requires the available model set $\mathcal{M}_1 = \{S_1, K_1\}$ as well as the operating sets $\mathcal{Z}_i$ from all direct neighbours D_i, $(i \in \mathcal{P}_1)$. This information is requested by D_1 in parallel as highlighted by the communication graph in Fig. 5.4 for $\mathcal{P}_1 = \{3, 4\}$.

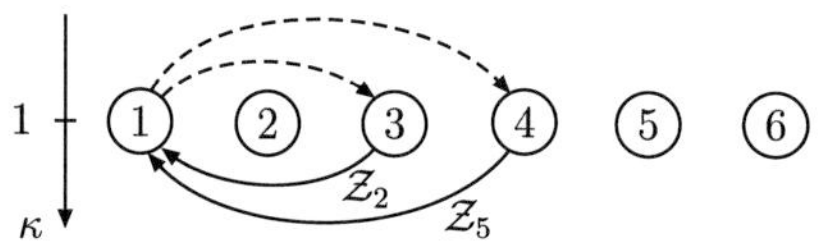

Figure 5.4: Communication graph for the request of D_1 with $\mathcal{P}_1 = \{3, 4\}$.

After the communication phase, the model set

$$\mathcal{M}_1 = (S_1, K_1, \{\mathcal{Z}_i, i \in \mathcal{P}_1\}).$$

is available to D_1.

5.3 Local design conditions

This section, first, presents local design conditions to guarantee boundedness of the overall closed-loop system state. Thereafter, local design conditions are proposed to additionally adhere actuator limitations.

5.3.1 Boundedness conditions

This section states local design requirements to guarantee the boundedness of the overall closed-loop system state.

Consider the local local stability condition $\mathcal{A}_i$ that reads as follows:

$$\begin{aligned} \mathcal{A}_i : \quad & \text{the state of the controlled subsystem } F_i \text{ is bounded by the given set} \\ & \mathcal{X}_i \subset \mathbb{R}^{n_i} \text{ according to} \\ & \qquad \boldsymbol{x}_i(t) \in \mathcal{X}_i, \quad \forall t \geq 0, \quad \forall \boldsymbol{s}_i \in \mathcal{S}_i. \end{aligned} \tag{5.6}$$

This local design condition requires not only the restriction of the coupling output signal space (required for the boundedness of the coupling input signal) but, more restrictive, the constraint of the state space. Accordingly, the satisfaction of the claim (5.6) implies the inclusion of the coupling output signal $\boldsymbol{z}_i(t)$ within the set

$$\mathcal{Z}_i = \boldsymbol{C}_{\mathrm{z}i}\mathcal{X}_i.$$

Note that for an observable pair $(\boldsymbol{A}_i, \boldsymbol{C}_{\mathrm{z}i})$ the boundedness of the subsystem state is equivalent to the boundedness of the coupling output.

As deduced in Section 5.2, the dynamics of all controlled subsystems F_i, $(\forall i \in \mathcal{N})$ can be analysed independently from each other if all controlled subsystems satisfy the respective local claim (5.6). Hence, the analysis of the overall closed-loop behaviour directly follows from the analysis of each controlled subsystem's motion. Thus, the overall closed-loop state is bounded according to

$$\boldsymbol{x}(t) \in \mathcal{X}, \quad \forall t \geq 0$$

with $\mathcal{X} = \mathcal{X}_1 \times \mathcal{X}_2 \times \cdots \times \mathcal{X}_N$.

The following proposition summarises the relation between the local design condition $\mathcal{A}_i$ and the boundedness of the overall closed-loop state $\boldsymbol{x}(t)$.

Proposition 5.1 (Local condition for boundedness of the overall closed-loop system) *If all controlled subsystems F_i, $(\forall i \in \mathcal{N})$ satisfy the corresponding local design condition $\mathcal{A}_i$ defined in (5.6), then the solution $\boldsymbol{x}(t)$ of the overall closed-loop system is bounded.*

5.3.2 Performance conditions

This section proposes local performance conditions in view of the experiments in Sec. 5.5.

The control of essentially all technical processes has to regard actuator limitations. For example, limitations on the available heating power of the heating rods or the admissible opening angle of the valves have to be taken into account in the control of the thermofluid process considered in this thesis. Hence, the global performance condition $\mathcal{A}_\mathrm{D}$ reads as follows:

$$\mathcal{A}_\mathrm{D}: \begin{cases} 1.\ \text{the state of the overall closed-loop system } F \text{ is bounded} & (5.7\text{a}) \\ 2.\ \text{the control signal of all controlled subsystems } F_i \text{ satisfy given} & \\ \quad \text{actuator limitation } \mathcal{U}_i \subset \mathbb{R}^{m_i} \text{ according to} & (5.7\text{b}) \\ \qquad \boldsymbol{u}_i(t) \in \mathcal{U}_i, \quad \forall t \geq 0, \quad \forall i \in \mathcal{N}. & \end{cases}$$

Condition (5.7a) holds if the local condition $\mathcal{A}_i$ is satisfied. Then, the state of the controlled subsystem F_i operates within the set $\mathcal{X}_i$ so that the control action $\boldsymbol{u}_i(t)$ is constrained by

$$\boldsymbol{u}_i(t) \in \boldsymbol{K}_i \mathcal{X}_i.$$

Obviously, to guarantee that the actuators operate within their allowed range $\mathcal{U}_i$ according to (5.7b), the feedback gains $\boldsymbol{K}_i$, $(\forall i \in \mathcal{N})$ has to be chosen such that

$$\boldsymbol{K}_i \mathcal{X}_i \subseteq \mathcal{U}_i, \quad \forall i \in \mathcal{N}.$$

In conclusion, the local performance condition $\mathcal{A}_{\mathrm{D}i}$ encompasses the following two claims:

$$\mathcal{A}_{\mathrm{D}i}: \begin{cases} 1.\ \text{the state of the controlled subsystem } F_i \text{ is bounded by the given set} & \\ \quad \mathcal{X}_i \subset \mathbb{R}^{n_i} \text{ according to} & (5.8\text{a}) \\ \qquad \boldsymbol{x}_i(t) \in \mathcal{X}_i, \quad \forall t \geq 0, \quad \forall \boldsymbol{s}_i \in \mathcal{S}_i & \\ 2.\ \text{the control signal of the controlled subsystem } F_i \text{ satisfies the given} & \\ \quad \text{actuautor limitation } \mathcal{U}_i \subset \mathbb{R}^{m_i} \text{ according to} & (5.8\text{b}) \\ \qquad \boldsymbol{K}_i \mathcal{X}_i \subseteq \mathcal{U}_i. & \end{cases}$$

The following proposition summarises the relation between the local performance condition $\mathcal{A}_{\mathrm{D}i}$ and the desired performance of the overall closed-loop system.

Proposition 5.2 (Local condition for the adherence of an overall closed-loop performance) *If all controlled subsystems F_i, ($\forall i \in \mathcal{N}$) satisfy the corresponding local design condition $\mathcal{A}_{\mathrm{D}i}$ defined in* (5.8), *then the overall closed-loop system has a performance described by $\mathcal{A}_{\mathrm{D}}$ defined in* (5.7).

Performance conditions on the free motion. This paragraph briefly discusses additional performance claims on the free motion of the controlled subsystem F_i.

As mentioned in Section 5.2, the coupling input signal becomes an unknown-but-bounded input signal for the controlled subsystem. Hence, additional claims can be stated individually on the "disturbed" controlled subsystem. In detail, the free motion of the controlled subsystem (5.2) under consideration of the physical interaction (5.4) is described by

$$\boldsymbol{y}_i(t) = \boldsymbol{F}_{\mathrm{yx0}i}(t) * \boldsymbol{x}_{i,0} + \boldsymbol{F}_{\mathrm{ys}i}(t) * \boldsymbol{\Psi}_i(t) * \boldsymbol{F}_{\mathrm{zx0}i}(t) * \boldsymbol{x}_{i,0} \tag{5.9}$$

with

$$\boldsymbol{\Psi}_i(t) = \boldsymbol{P}_i^{\mathcal{N}}(t) + \boldsymbol{P}_i^{\mathcal{N}}(t) * \boldsymbol{F}_{\mathrm{zs}i}(t) * \boldsymbol{\Psi}_i(t)$$

and

$$\begin{aligned} \boldsymbol{F}_{\mathrm{yx0}i}(t) &= \boldsymbol{C}_i \mathrm{e}^{\boldsymbol{A}_i + \boldsymbol{B}_i \boldsymbol{K}_i}, & \boldsymbol{F}_{\mathrm{zx0}i}(t) &= \boldsymbol{C}_{\mathrm{z}i} \mathrm{e}^{\boldsymbol{A}_i + \boldsymbol{B}_i \boldsymbol{K}_i}, \\ \boldsymbol{F}_{\mathrm{ys}i}(t) &= \boldsymbol{C}_i \mathrm{e}^{\boldsymbol{A}_i + \boldsymbol{B}_i \boldsymbol{K}_i} \boldsymbol{E}_i, & \boldsymbol{F}_{\mathrm{zs}i}(t) &= \boldsymbol{C}_{\mathrm{z}i} \mathrm{e}^{\boldsymbol{A}_i + \boldsymbol{B}_i \boldsymbol{K}_i} \boldsymbol{E}_i. \end{aligned}$$

The second addend of eqn (5.9) represents the effect of the other controlled subsystems through the physical couplings. As the dynamics $\boldsymbol{P}_i^{\mathcal{N}}(t)$ are unknown, the solution $\boldsymbol{y}_i(t)$ is unknown too. However, due to the satisfaction of $\mathcal{A}_i$, ($\forall i \in \mathcal{N}$), it is known that

$$\boldsymbol{s}_i(t) = \boldsymbol{\Psi}_i(t) * \boldsymbol{F}_{\mathrm{zx0}i}(t) * \boldsymbol{x}_{i,0} \in \mathcal{S}_i$$

whenever $\boldsymbol{x}_{i,0} \in \mathcal{X}_i$. Accordingly, eqn (5.9) can be rewritten into

$$\boldsymbol{y}_i(t) = \boldsymbol{F}_{\mathrm{yx0}i}(t) * \boldsymbol{x}_{i,0} + \boldsymbol{F}_{\mathrm{ys}i}(t) * \boldsymbol{s}_i(t), \quad \forall \boldsymbol{x}_{i,0} \in \mathcal{X}_i,$$

where $\boldsymbol{s}_i(t)$ represents an unknown input but is known to be bounded in $\mathcal{S}_i$ (see Fig. 5.2). As a consequence, performance conditions on the free motion (5.9) can be formalised in terms of disturbance attenuation [127, 140].

5.4 Controller design based on robustly positively invariant operating sets

This section proposes two different control methods to satisfy the local design conditions $\mathcal{A}_i$ and $\mathcal{A}_{\mathrm{D}i}$. The first method is based on Lyapunov's stability criterion, where the solution is a static state-feedback (5.1). The second method constitutes the well known tube-based model predictive controller proposed in [127]. Both techniques follow the idea to transform the operating set $\mathcal{X}_i$ into a robustly positively invariant (RPI) set for the controlled subsystem.

5.4.1 LMI-based design of a static state-feedback

This section proposes two LMI-based method to design the static state-feedback gain of the control station (5.1) with respect to the local design conditions $\mathcal{A}_i$ and $\mathcal{A}_{\mathrm{D}i}$. The first design method aims at making a given ellipsoidal operating set RPI for the controlled subsystem (5.2), whereas the second design method focusses on polytopic operating sets.

Ellipsoids are very popular as candidates for positive invariant sets since the symmetric positive-definite matrix $\boldsymbol{P}$, which is a solution of the well known Lyapunov inequality

$$\boldsymbol{x}^\top(\boldsymbol{A}^\top\boldsymbol{P}+\boldsymbol{P}\boldsymbol{A})\boldsymbol{x}\leq 0,$$

defines an ellipsoid $\mathcal{I}=\left\{\boldsymbol{x}\in\mathbb{R}^n\colon \boldsymbol{x}^\top\boldsymbol{P}\boldsymbol{x}\leq 1\right\}$ that is positively invariant for the system

$$\dot{\boldsymbol{x}}(t)=\boldsymbol{A}\boldsymbol{x}(t),\ \ \boldsymbol{x}(0)=\boldsymbol{x}_0.$$

See [50] for more details. A condition to design a static state-feedback gain has been derived in [182] with the aim to make a given ellipsoidal set RPI for the controlled subsystem (5.2).

Lemma 5.1 (LMI condition to make an ellipsoid RPI [182]) *Let the constraint set $\mathcal{S}_i$ and the ellipsoidal operating set $\mathcal{X}_i=\left\{\boldsymbol{x}_i\in\mathbb{R}^{n_i}\colon \boldsymbol{x}_i^\top\boldsymbol{X}^{-1}\boldsymbol{x}_i\leq 1\right\}$ with $\boldsymbol{X}=\boldsymbol{X}^\top\succ 0$ be given. If there exists a feasible solution $\boldsymbol{Y}$ to the LMI*

$$\boldsymbol{X}\boldsymbol{A}_i^\top+\boldsymbol{Y}^\top\boldsymbol{B}_i^\top+\boldsymbol{A}_i\boldsymbol{X}+\boldsymbol{B}_i\boldsymbol{Y}+\alpha\boldsymbol{X}+\frac{\beta}{\alpha}\boldsymbol{E}_i\boldsymbol{E}_i^\top\preceq 0$$

for an arbitrary $\alpha>0$, where $\beta=\max_{\boldsymbol{s}_i\in\mathcal{S}_i}\ \boldsymbol{s}_i^\top\boldsymbol{s}_i$, then

- *the matrix $\boldsymbol{A}_i+\boldsymbol{B}_i\boldsymbol{K}_i$ is Hurwitz and*
- *the set $\mathcal{X}_i$ is RPI for the controlled subsystem* (5.2)*,*

where $\boldsymbol{K}_i=\boldsymbol{Y}\boldsymbol{X}^{-1}$.

Accordingly, if the static state-feedback gain $\boldsymbol{K}_i$ is designed in accordance with Lemma 5.1, the controlled subsystem (5.2) satisfies the local condition $\mathcal{A}_i$. Based on this result, LMI-based conditions are formulated to satisfy the local performance condition $\mathcal{A}_{\mathrm{D}i}$.

Theorem 5.1 (LMI condition to satisfy the local performance condition $\mathcal{A}_{\mathrm{D}i}$) *Let the constraint set $\mathcal{S}_i$, the ellipsoidal operating set $\mathcal{X}_i = \left\{\boldsymbol{x}_i \in \mathbb{R}^{n_i} \colon \boldsymbol{x}_i^\top \boldsymbol{X}^{-1}\boldsymbol{x}_i \leq 1\right\}$ with $\boldsymbol{X} = \boldsymbol{X}^\top \succ 0$ and the admissible control action range $\mathcal{U}_i = \{\boldsymbol{u}_i \in \mathbb{R}^{m_i} \colon |\boldsymbol{u}_i| \leq \boldsymbol{\varrho}\}$ be given. If there exists a feasible solution $\boldsymbol{Y}$ to the LMIs*

$$\begin{cases} \boldsymbol{X}\boldsymbol{A}_i^\top + \boldsymbol{A}_i\boldsymbol{Y} + \boldsymbol{Y}^\top\boldsymbol{B}_i^\top + \boldsymbol{B}_i\boldsymbol{Y} + \alpha\boldsymbol{X} + \dfrac{\beta}{\alpha}\boldsymbol{E}_i\boldsymbol{E}_i^\top \preceq 0 & (5.10a) \\ \begin{pmatrix} \varrho_j^2 & \boldsymbol{y}_j^\top \\ \star & \boldsymbol{X} \end{pmatrix} \succeq 0, \quad j = 1, ..., m_1 & (5.10b) \end{cases}$$

for an arbitrary $\alpha > 0$, where $\beta = \max_{\boldsymbol{s}_i \in \mathcal{S}_i} \boldsymbol{s}_i^\top \boldsymbol{s}_i$ and $\boldsymbol{y}_j^\top$ is the j-th row of $\boldsymbol{Y}$ and ϱ_j is the j-th element of $\boldsymbol{\varrho}$, then

- *the matrix $\boldsymbol{A}_i + \boldsymbol{B}_i\boldsymbol{K}_i$ is Hurwitz,*
- *the set $\mathcal{X}_i$ is RPI for the controlled subsystem (5.2) and*
- *the control action is limited according to $\boldsymbol{u}_i \in \mathcal{U}_i$.*

The feedback gain results to $\boldsymbol{K}_i = \boldsymbol{Y}\boldsymbol{X}^{-1}$.

Proof. See Appendix B.3. □

Note that additional performance conditions which can be expressed as LMIs can simply be added to the LMIs (5.10). See for example [58, 62, 154] for the formulation of H_∞-optimal requirements, optimal output energy conditions or pole placement methods.

Extensions to polytopic operating sets. From a technical viewpoint, the operating sets are often represented by linear combinations of the subsystem state, e.g., admissible temperature range of a work piece or a certain ratio between two volume flows. These linear combinations result in polytopic operating sets. This paragraph extends the above presented result towards the consideration of polytopic operation sets $\mathcal{X}_i$. The main idea is to find the biggest ellipsoidal RPI set $\mathcal{I}_i$ that is contained in the polytope $\mathcal{X}_i$.

The following result is a combination of Lemma 5.1 and results from [58]:

Corollary 5.1 (LMI conditions for constructing an ellipsoidal RPI set that is contained in a polytope) *Let the constraint set $\mathcal{S}_i$, the polytopic operating set $\mathcal{X}_i = \{\boldsymbol{x}_i \in \mathbb{R}^{n_i} : |\boldsymbol{p}_j \boldsymbol{x}_i| \leq b_j,\ j = 1, ..., g\}$ and the initial state $\boldsymbol{x}_{i,0}^\top$ of the controlled subsystem (5.2) be given. If there exist solutions $\boldsymbol{X} = \boldsymbol{X}^\top \succ 0$ and $\boldsymbol{Y}$ to the optimisation problem*

$$\begin{cases} \min\limits_{\boldsymbol{X}} \ -\log\big(\det(\boldsymbol{X})\big) & (5.11a) \\ s.t.\ \ \boldsymbol{X}\boldsymbol{A}_i^\top + \boldsymbol{A}_i\boldsymbol{X} + \boldsymbol{Y}^\top\boldsymbol{B}_i^\top + \boldsymbol{B}_i\boldsymbol{Y} + \alpha\boldsymbol{X} + \frac{\beta}{\alpha}\boldsymbol{E}_i\boldsymbol{E}_i^\top \preceq 0 & (5.11b) \\ \qquad \boldsymbol{p}_j^\top \boldsymbol{X} \boldsymbol{p}_j - b_j^2 \leq 0,\ j = 1, ..., g & (5.11c) \\ \qquad \begin{pmatrix} 1 & \boldsymbol{x}_{i,0}^\top \\ \star & \boldsymbol{X} \end{pmatrix} \succeq 0 & (5.11d) \end{cases}$$

for an arbitrary $\alpha > 0$, where $\beta = \max_{\boldsymbol{s}_i \in \mathcal{S}_i} \boldsymbol{s}_i^\top \boldsymbol{s}_i$, then

- *the matrix $\boldsymbol{A}_i + \boldsymbol{B}_i\boldsymbol{K}_i$ is Hurwitz,*
- *the set $\mathcal{I}_i = \{\boldsymbol{x}_i \in \mathbb{R}^{n_i} : \boldsymbol{x}_i^\top \boldsymbol{X}^{-1} \boldsymbol{x}_i \leq 1\}$ is the biggest ellipsoidal RPI set for the controlled subsystem (5.2) that is contained in $\mathcal{X}_i$ and*
- *the initial state $\boldsymbol{x}_{i,0}$ is contained in $\mathcal{I}_i$.*

The feedback gain results to $\boldsymbol{K}_i = \boldsymbol{Y}\boldsymbol{X}^{-1}$.

Proof. See Appendix B.4. □

The optimisation problem (5.11) expands the volume of the ellipsoid (cost function (5.11a)) while the ellipsoid shall be RPI (LMI (5.11b)), shall be contained in $\mathcal{X}_i$ (LMIs (5.11c)) and shall contain the initial state $\boldsymbol{x}_{i,0}$ (LMI (5.11d)). In addition, to take actuator limitations into account (i.e., condition (5.8b)) the LMIs (5.10b) from Theorem 5.1 have to be added as additional constraints to the optimisation problem (5.11).

Nevertheless, the proposed design in Corollary 5.1 does not guarantee the satisfaction of the local design condition $\mathcal{A}_i$ or $\mathcal{A}_{\mathrm{D}i}$, respectively. This is due to the fact that the resulting ellipsoidal RPI set $\mathcal{I}_i$ yields a proper subset of the desired polytopic operating set $\mathcal{X}_i$, i.e., $\mathcal{I}_i \subset \mathcal{X}_i$. As a consequence, it cannot be guaranteed for the initial states that are contained in $\mathcal{X}_i$ but not in $\mathcal{I}_i$ the solution $\boldsymbol{x}_i(t)$ will remain in $\mathcal{X}_i$. Hence, the initial state is explicitly taken into account within the optimisation problem (5.11) (this becomes relevant for the controller reconfiguration scenario in Section 5.5.1). However, if the initial state is located at a vertex of $\mathcal{X}_i$, the optimisation problem will never have a solution. The following Section 5.4.2 presents an MPC-based control scheme that enables the transformation of a polytopic operating set $\mathcal{X}_i$ into an RPI set.

Remark 5.2 *Results, similar to Corollary 5.1 have been proposed in [106, 114] for discrete time systems. The publication [106] has stated LMI conditions to determine the biggest RPI ellipsoid that is contained in a given polytope. In [114] time-varying systems with noisy measurements has been studied.*

Remark 5.3 *To find a control law for a linear system that makes the whole polytop (robustly) positively invariant for F_i is a well discussed issue in literature (see [50, 51] for surveys). Basically, for linear systems it is sufficient to associate a control law to each vertex of a given polytope $\mathcal{X}_i$ that pushes the state inside $\mathcal{X}_i$. Due to the linearity of the system and the convexity of the polytope, it is guaranteed that each point on the boundary of $\mathcal{X}_i$ is pushed inside. Solutions to the so called linear constrained regulation problem in the presence of additive disturbances (DLCRP) have been presented in [49, 132, 171] for instance, where the control law is considered to be a linear static state-feedback. However, the DLCRP often does not provide a solution so that non-linear control strategies are exploited like piecewise linear controllers [87, 141] or model predictive control schemes [43, 127].*

5.4.2 Design of a tube-based model predictive controller

This section applies the tube-based model predictive controller (tube-based MPC) from [127] to transform a given polytopic operating set $\mathcal{X}_1$ into an RPI set for the controlled subsystem. First, the tube-based MPC from [127] is briefly recapitulated. Thereafter, the issue how the tube-based MPC has to be parametrised is studied and a design algorithm (Algorithm 5.2) is proposed.

Consider the subsystems S_i represented by the linear discrete time model

$$S_i : \begin{cases} \boldsymbol{x}_i(k+1) &= \boldsymbol{A}_i\boldsymbol{x}_i(k) + \boldsymbol{B}_i\boldsymbol{u}_i(k) + \boldsymbol{E}_i\boldsymbol{s}_i(k), \quad \boldsymbol{x}_i(0) = \boldsymbol{x}_{i,0} \\ \boldsymbol{z}_i(k) &= \boldsymbol{C}_{zi}\boldsymbol{x}_i(k), \end{cases} \tag{5.12}$$

where $\boldsymbol{x}_i(k)$ is assumed to be measurable. The tube-based MPC scheme can be interpreted as a two degree of freedom controller as is illustrated in Fig. 5.5 for the control of subsystem S_1. The first degree of freedom yields a model predictive controller $\mathfrak{P}_1(\boldsymbol{x}_1(k), S_1|_{\boldsymbol{s}_1=\boldsymbol{0}})$ that is used to track the state of the isolated subsystem

$$S_1|_{\boldsymbol{s}_1=\boldsymbol{0}} : \quad \boldsymbol{x}_1(k+1) = \boldsymbol{A}_1\boldsymbol{x}_1(k) + \boldsymbol{B}_1\boldsymbol{u}_1(k), \;\; \boldsymbol{x}_1(0) = \boldsymbol{x}_{1,0}$$

back to the origin, whereas the second degree of freedom, a static state-feedback $\boldsymbol{K}_1$, is used to compensate the influence of the physical interaction.

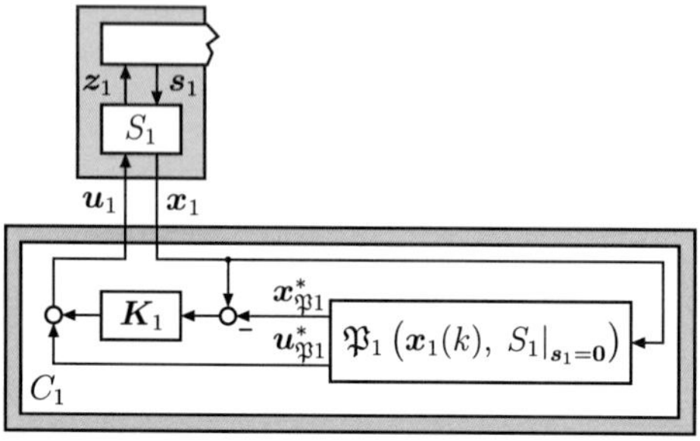

Figure 5.5: Structure of the tube-based MPC applied to subsystem S_1.

The tube-based model predictive control station C_1 is described by the model

$$C_1 : \begin{cases} \text{solve} \quad \mathfrak{P}_1(\boldsymbol{x}_1(k),\, S_1|_{\boldsymbol{s}_1=\boldsymbol{0}}) \\ \boldsymbol{u}_1(k) \;= \boldsymbol{u}^*_{\mathfrak{P}1}(k) + \boldsymbol{K}_1\big(\boldsymbol{x}_1(k) - \boldsymbol{x}^*_{\mathfrak{P}1}(k)\big), \end{cases} \tag{5.13}$$

where $\boldsymbol{u}^*_{\mathfrak{P}1}(k)$ and $\boldsymbol{x}^*_{\mathfrak{P}1}(k)$ are the optimal solution of the model predictive controller

$$\mathfrak{P}_1(\boldsymbol{x}_1(k),\, S_1|_{\boldsymbol{s}_1=\boldsymbol{0}}) :$$

$$\begin{cases} \min\limits_{\mathscr{X}_{\mathfrak{P}1},\mathscr{U}_{\mathfrak{P}1}} \; V_1(\mathscr{X}_{\mathfrak{P}1},\mathscr{U}_{\mathfrak{P}1}) = \sum\limits_{j=k}^{k+h-1} \boldsymbol{x}_{\mathfrak{P}1}(j)^\top \boldsymbol{Q}_1 \boldsymbol{x}_{\mathfrak{P}1}(j) + \boldsymbol{u}_{\mathfrak{P}1}(j)^\top \boldsymbol{R}_1 \boldsymbol{u}_{\mathfrak{P}1}(j) & (5.14\text{a}) \\ s.t. \;\; \boldsymbol{x}_{\mathfrak{P}1}(j{+}1) = \boldsymbol{A}_1 \boldsymbol{x}_{\mathfrak{P}1}(j) + \boldsymbol{B}_1 \boldsymbol{u}_{\mathfrak{P}1}(j), \quad j = k,..,k{+}h{-}1 & (5.14\text{b}) \\ \quad \boldsymbol{x}_{\mathfrak{P}1}(j) \in \mathcal{X}_{\mathfrak{P}1}, \quad j = k,..,k{+}h{-}1 & (5.14\text{c}) \\ \quad \boldsymbol{u}_{\mathfrak{P}1}(j) \in \mathcal{U}_{\mathfrak{P}1}, \quad j = k,..,k{+}h{-}1 & (5.14\text{d}) \\ \quad \boldsymbol{x}_{\mathfrak{P}1}(k+h) = \boldsymbol{0} & (5.14\text{e}) \\ \quad \boldsymbol{x}_1(k) \in \boldsymbol{x}_{\mathfrak{P}1}(k) \oplus \mathcal{I}_1 & (5.14\text{f}) \end{cases}$$

The model predictive control scheme intends to solve the optimisation problem (5.14) for each time instant k over the prediction horizon h. The aim is to predict the optimal behaviour of the isolated subsystem (5.14b) while minimising a linear quadratic cost function (5.14a) with the positive definite weight matrices $\boldsymbol{Q}_1$ and $\boldsymbol{R}_1$. The optimisation is subject to additional constraints (5.14b)–(5.14f) which are explained in the following. The result of optimisation is, on the one hand, the optimal state sequence

$$\mathscr{X}^*_{\mathfrak{P}1} := \{\boldsymbol{x}^*_{\mathfrak{P}1}(k),\, \boldsymbol{x}^*_{\mathfrak{P}1}(k+1) \ldots,\, \boldsymbol{x}^*_{\mathfrak{P}1}(k+h-1)\},$$

and, on the other hand, the optimal control sequence

$$\mathscr{U}^*_{\mathfrak{P}1} := \{\boldsymbol{u}^*_{\mathfrak{P}1}(k),\, \boldsymbol{u}^*_{\mathfrak{P}1}(k+1),\, \ldots,\, \boldsymbol{u}^*_{\mathfrak{P}1}(k+h-1)\}.$$

The first elements $\boldsymbol{x}^*_{\mathfrak{P}1}(k)$ and $\boldsymbol{u}^*_{\mathfrak{P}1}(k)$ of the respective optimal sequences are applied to subsystem S_1 according to (5.13).

All admissible control sequences for the initial state $\boldsymbol{x}_{\mathfrak{P}1}(k)$ of the optimisation problem are collected in the set

$$\mathscr{U}_1(\boldsymbol{x}_{\mathfrak{P}1}(k)) = \{\mathscr{U}_{\mathfrak{P}1} \colon \boldsymbol{u}_{\mathfrak{P}1}(i) \in \mathscr{U}_{\mathfrak{P}1},\, \boldsymbol{x}_{\mathfrak{P}\mathscr{U}}(i; \boldsymbol{x}_{\mathfrak{P}1}(k)) \in \mathcal{X}_1 \text{ for } i = k, .., k+h\},$$

where $\boldsymbol{x}_{\mathfrak{P}\mathscr{U}}(i; \boldsymbol{x}_{\mathfrak{P}1}(k))$ represents the predicted state $\boldsymbol{x}_{\mathfrak{P}1}(i)$, $(i = k, ..., k+h)$ which corresponds to the control sequence $\mathscr{U}_{\mathfrak{P}1}$ and the initial state $\boldsymbol{x}_{\mathfrak{P}1}(k)$. All initial states $\boldsymbol{x}_{\mathfrak{P}1}(k)$ of the optimisation problem for which an admissible control sequence $\mathscr{U}_1(\boldsymbol{x}_{\mathfrak{P}1}(k))$ exist are collected in the feasible set

$$\mathcal{F}_{\mathfrak{P}1} = \{\boldsymbol{x}_{\mathfrak{P}1}(k) \colon \exists \boldsymbol{x}_1(k) \in \boldsymbol{x}_{\mathfrak{P}1}(k) \oplus \mathcal{I}_1 \text{ such that } \mathscr{U}_1(\boldsymbol{x}_{\mathfrak{P}1}(k)) \neq \emptyset\}.$$

That means that all predicted states $\boldsymbol{x}_{\mathfrak{P}1}(k) \in \mathcal{F}_{\mathfrak{P}1}$ are transferred to the origin in accordance with constraint (5.14e).

Since the optimisation is subject to the isolated subsystem (cf. constraint (5.14b)), the constraint (5.14f) assures that the actual subsystem state $\boldsymbol{x}_1(k)$ is located in the set $\mathcal{I}_1$ centered around the optimal predicted state $\boldsymbol{x}^*_{\mathfrak{P}1}(k)$, i.e., $\boldsymbol{x}_1(k) \in \boldsymbol{x}_{\mathfrak{P}1}(k) \oplus \mathcal{I}_1$. The set $\mathcal{I}_1$ represents the RPI set for an controlled subsystem

$$F_1 : \begin{cases} \boldsymbol{x}_1(k+1) &= (\boldsymbol{A}_1 + \boldsymbol{B}_1\boldsymbol{K}_1)\,\boldsymbol{x}_1(k) + \boldsymbol{E}_1\boldsymbol{s}_1(k), \quad \boldsymbol{x}_1(0) = \boldsymbol{x}_{1,0} \\ \boldsymbol{z}_1(k) &= \boldsymbol{C}_{\mathrm{z}1}\boldsymbol{x}_1(k) \end{cases} \tag{5.15}$$

with the static state-feedback gain $\boldsymbol{K}_1$ from (5.13) and the bounded coupling input $\boldsymbol{s}_1(k) \in \mathcal{S}_1$. Hence, due to the repeatedly solving of the optimisation problem (5.14), it is guaranteed that during the transfer from $\boldsymbol{x}_{1,0}$ into the set $\mathcal{I}_1$ the current state $\boldsymbol{x}_1(k)$ is located in $\boldsymbol{x}^*_{\mathfrak{P}1}(k) \oplus \mathcal{I}_1$ for all times $k \geq 0$ (cf. Proposition 1 of [126]). From this fact, a tube around the optimal predicted state trajectory results which encloses the actual state trajectory of the controlled subsystem (5.12) and (5.13) as visualised in Fig. 5.6. The constraint (5.14e) guarantees that all states

$\boldsymbol{x}_1(k) \in \mathcal{F}_{\mathfrak{P}1} \oplus \mathcal{I}_1$ are be transferred into $\mathcal{I}_1$ by the tube-based MPC. That is, the set

$$\mathcal{I}_{\mathfrak{P}1} := \mathcal{F}_{\mathfrak{P}1} \oplus \mathcal{I}_1$$

represents the RPI set of the subsystem (5.12) controlled by the tube-based MPC (5.13).

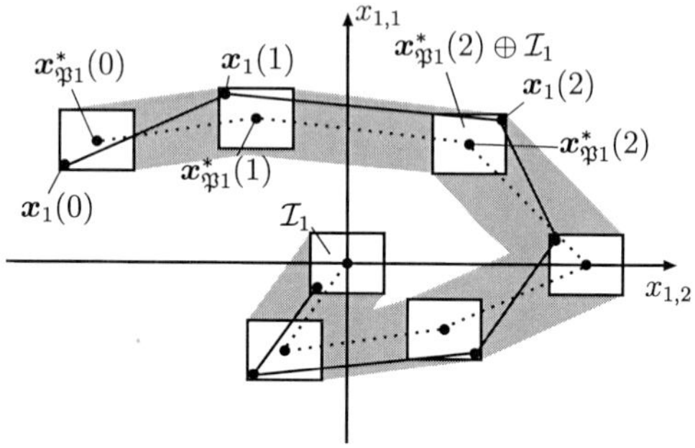

Figure 5.6: Example of a resulting tube for a subsystem (5.12) with $\boldsymbol{x}_1 = (x_{1,1}\ x_{1,2})^\top$ controlled by a tube-based MPC (5.13). The dotted line denotes $\boldsymbol{x}^*_{\mathfrak{P}1}(k)$ and the solid line represents $\boldsymbol{x}_1(k)$.

To guarantee that the controlled subsystem state is located within the operating set $\mathcal{X}_1$ (i.e., $\boldsymbol{x}_1(k) \in \mathcal{X}_1$) and the applied control sequence does adhere the actuator constraints $\mathcal{U}_1$ (i.e., $\boldsymbol{u}_1 \in \mathcal{U}_1$), the predicted state $\boldsymbol{x}_{\mathfrak{P}1}(k)$ and predicted control signal $\boldsymbol{u}_{\mathfrak{P}1}(k)$ has to respect the tightened constraints $\mathcal{X}_{\mathfrak{P}1}$ (cf. constraint (5.14c)) and $\mathcal{U}_{\mathfrak{P}1}$ (cf. constraint (5.14d)) respectively. These tightened constraints are calculated by

$$\mathcal{X}_{\mathfrak{P}1} = \mathcal{X}_1 \ominus \mathcal{I}_1, \qquad \mathcal{U}_{\mathfrak{P}1} = \mathcal{U}_1 \ominus \boldsymbol{K}_1\mathcal{I}_1. \tag{5.16}$$

Note that due to the convex cost function, the linear model and the polytopic sets $\mathcal{I}_1, \mathcal{U}_1$ and $\mathcal{X}_1$, the whole optimisation problem (5.14) is convex and, hence, can be efficiently implemented as quadratic program [85]. It is worth noting that also ellipsoidal sets can be used, especially if real-time issues need to be considered [190]. In [127] it has been proved (by Theorem 1) that the tube-based MPC drives the state in the set $\mathcal{I}_1$ for all $\boldsymbol{x}_1(k) \in \mathcal{I}_{\mathfrak{P}1}$ and for all $\boldsymbol{s}_1(k) \in \mathcal{S}_1$.

Remark 5.4 *Without loss of generality, the proposed tube-based controller aims at driving the system state into the origin (i.e., $\boldsymbol{w}_1 = \boldsymbol{0}$ and $\boldsymbol{u}_1 = \boldsymbol{0}$). Arbitrary desired fixed operating point ($\boldsymbol{w}_1 = \bar{\boldsymbol{w}}_1$) can be reached by easily shifting the origin of the state space into the desired operating point and process the optimisation with the shifted state $\tilde{x}_{\mathfrak{P}1} = \boldsymbol{x}_{\mathfrak{P}1} - \bar{\boldsymbol{w}}_1$ and the corresponding shifted control signal $\tilde{\boldsymbol{u}}_{\mathfrak{P}1} = \boldsymbol{u}_{\mathfrak{P}1} - \bar{\boldsymbol{u}}_1$. Furthermore, the tracking of trajectories has been studied by [108].*

Parametrisation of the tube-based MPC. This paragraph considers the appropriate parametrisation of the tube-based MPC so that the operating set $\mathcal{X}_1$ becomes RPI for the controlled subsystem (5.12), (5.13).

As mentioned before, the resulting RPI set of the subsystem (5.12) that is controlled by the tube-based MPC (5.13) yields

$$\mathcal{I}_{\mathfrak{P}1} = \mathcal{F}_{\mathfrak{P}1} \oplus \mathcal{I}_1.$$

Accordingly, the aim is to parametrise the tube-based MPC so that the equality

$$\mathcal{I}_{\mathfrak{P}1} = \mathcal{X}_1 \tag{5.17}$$

is fulfilled. Indeed, the satisfaction of (5.17) depends on the shape of the feasible set $\mathcal{F}_{\mathfrak{P}1}$ and the RPI set $\mathcal{I}_1$ of the controlled subsystem (5.15) as outlined in the following:

1. The feasible set $\mathcal{F}_{\mathfrak{P}1}$: The size of the feasible set depends upon the weight matrices $\boldsymbol{Q}_1$ and $\boldsymbol{R}_1$ as well as the prediction horizon h. For instance, a larger horizon leads to a the expansion of the feasible set. The biggest feasible set is represented by the tightened constraint itself, i.e.,

$$\mathcal{F}_{\mathfrak{P}1} = \mathcal{X}_{\mathfrak{P}1}, \tag{5.18}$$

   which is necessary to guarantee (5.17). Note that if no actuator limitations are considered (as in the local design condition $\mathcal{A}_i$), the equality (5.18) always holds.

2. The RPI set $\mathcal{I}_1$: As the Minkowski sum is not the complement of the Pontryagin difference (see [155]), only the relation

$$\underbrace{(\mathcal{X}_1 \ominus \mathcal{I}_1)}_{\mathcal{X}_{\mathfrak{P}1}} \oplus \mathcal{I}_1 \subseteq \mathcal{X}_1 \tag{5.19}$$

   holds. The equality of (5.19) holds if the set $\mathcal{I}_1$ has a specific shape. Since the RPI set $\mathcal{I}_1$ is not unique, the appropriate choice of this set is subject of discussion in the following.

The RPI set of a controlled subsystem is not unique (cf. condition (2.5)). In the following, an algorithm is proposed to find an RPI set $\mathcal{I}_1$ for the controlled subsystem (5.15) such that the relation (5.19) holds with the equality sign, i.e.,

$$(\mathcal{X}_1 \ominus \mathcal{I}_1) \oplus \mathcal{I}_1 = \mathcal{X}_1. \tag{5.20}$$

As a consequence, the operating set $\mathcal{X}_1$ becomes RPI for the subsystem (5.12) controlled by the tube-based MPC (5.13), providing that eqn (5.18) holds. First, a condition is proposed which guarantees the satisfaction of the equality (5.20) for the case that $\mathcal{I}_1$ is not RPI for F_1. Thereafter, additional claims are stated which consider $\mathcal{I}_1$ to be RPI.

A necessary and sufficient condition for the satisfaction of the equality (5.20) for arbitrary polytopes $\mathcal{X}$ and $\mathcal{Y}$ is given in [155] by Lemma 3.1.11 and Theorem 3.2.11. Both results are summarised in the following theorem:

Theorem 5.2 (Complement of Minkowski sum and Pontryagin difference [155]) *Let $\mathcal{X}, \mathcal{Y} \subset \mathbb{R}^n$ be polytopes. The relation*

$$(\mathcal{X} \ominus \mathcal{Y}) \oplus \mathcal{Y} = \mathcal{X}$$

holds if and only if the support set $\mathcal{F}_{\mathcal{X}}(\boldsymbol{v})$ contains a translate of the support set $\mathcal{F}_{\mathcal{Y}}(\boldsymbol{a})$, whenever $\mathcal{F}_{\mathcal{Y}}(\boldsymbol{a})$ is an edge of $\mathcal{Y}$ and $\boldsymbol{a} \in \mathbb{R}^n$, $|\boldsymbol{a}| = 1$. Then $\mathcal{Y}$ is called a summand *of $\mathcal{X}$.*

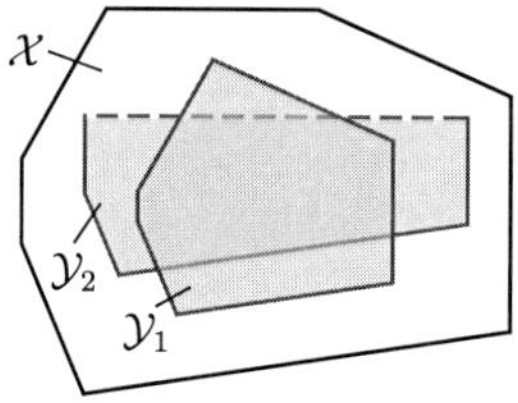

Figure 5.7: Examples of summands: $\mathcal{Y}_1$ is a summand of $\mathcal{X}$ since each edge of $\mathcal{Y}$ is a transient of an edge of $\mathcal{X}$. $\mathcal{Y}_2$ is not a summand since the dashed edge is longer than the respective edge of $\mathcal{X}$.

Roughly speaking, if each edge in $\mathcal{Y}$ is a transient of an edge of $\mathcal{X}$ and, additionally, is not longer than the edge of $\mathcal{X}$, then the equation (5.20) is satisfied. An example is shown in Fig. 5.7, where the set $\mathcal{Y}_1$ represents a summand of $\mathcal{X}$, whereas $\mathcal{Y}_2$ is not a summand. Nevertheless, the summand $\mathcal{Y}_1$ is not necessarily RPI for the controlled subsystem F_1. In the following an algorithm is deduced that can be used to find a set that is a summand of $\mathcal{X}$ and, additionally, RPI. Therefore, the notions of a minimal RPI set and of a minimal summand are introduced. According to [140] the *minimal RPI set* is defined as follows:

Definition 5.1 (Minimal robust positive invariant set [140]) *The minimal RPI (mRPI) set $\mathcal{I}_{\min,1}$ of the controlled subsystem* (5.15) *is the RPI set in $\mathbb{R}^{n_1}$ that is contained in every RPI set of the controlled subsystem* (5.15).

From Definition 5.1, it can be concluded that $\mathcal{I}_1$ is RPI for F_1 only if it contains the mRPI set

$\mathcal{I}_{\min,1}$, i.e., $\mathcal{I}_1 \supseteq \mathcal{I}_{\min,1}$. As presented in [103], the mRPI set $\mathcal{I}_{\min,1}$ of F_1 is calculated by

$$\mathcal{I}_{\min,1} = \bigoplus_{i=0}^{\infty} (\boldsymbol{A}_1 + \boldsymbol{B}_1 \boldsymbol{K}_1)^i \cdot \boldsymbol{E}_1 \mathcal{S}_1. \tag{5.21}$$

Since it is generally difficult to exactly determine the result of the infinite Minkowski sum of sets, the set $\mathcal{I}_{\min,1}$ can be approximated for example by a polytopic set [139, 140] or by an ellipsoidal set [58, 190]. Referring to Definition 5.1, the *minimal summand* is introduced.

Definition 5.2 (Minimal summand containing a convex set) *Let $\mathcal{X}, \mathcal{Z} \subset \mathbb{R}^n$ be convex sets with $\mathcal{Z} \subseteq \mathcal{X}$. The summand $\mathcal{Y}_{\min}$ of $\mathcal{X}$ that contains $\mathcal{Z}$ (i.e., $\mathcal{Y}_{\min} \supseteq \mathcal{Z}$) is called* minimal summand *if it is contained in all other summands of $\mathcal{X}$ that contain $\mathcal{Z}$.*

A corollary of Theorem 5.2 presents a necessary and sufficient condition for a set to be a minimal summand and to contain a polytope.

Corollary 5.2 (Minimal summand containing a convex set) *Let $\mathcal{X}, \mathcal{Z} \subset \mathbb{R}^n$ be polytopes with $\mathcal{Z} \subseteq \mathcal{X}$. The polytope $\mathcal{Y}_{\min} \subset \mathbb{R}^n$ is the* minimal summand *of $\mathcal{X}$ that contains $\mathcal{Z}$ in accordance with Definition 5.2 if and only if*

1. *the support set $\mathcal{F}_{\mathcal{X}}(\boldsymbol{a})$ contains a translate of the support set $\mathcal{F}_{\mathcal{Y}_{\min}}(\boldsymbol{a})$ and*

2. *the distance $h_{\mathcal{Y}_{\min}}(\boldsymbol{a}) - h_{\mathcal{Z}}(\boldsymbol{a})$ is minimal and non negative*

whenever $\mathcal{F}_{\mathcal{Y}_{\min}}(\boldsymbol{a})$ is an edge of $\mathcal{Y}_{\min}$ and $\boldsymbol{a} \in \mathbb{R}^n$, $|\boldsymbol{a}| = 1$.

Proof. See Appendix B.5 □

Let the set to be contained in the minimal summand be the minimal RPI set, then the following necessary condition is concluded: Only if a subset of $\mathcal{X}$ contains the minimal summand, it can be RPI and a summand of $\mathcal{X}$. This relation states a starting point for the construction of an RPI set $\mathcal{I}_1$ for which the equality (5.20) holds. These construction steps are formalised in Algorithm 5.1.

The first part (Lines 1–7) of the algorithm presents the construction of the minimal summand that contains the minimal RPI set $\mathcal{I}_{\min,1}$ according to Corollary 5.2. Based on this minimal summand, the second part (Lines 8–13) of the algorithm describes a way to find an RPI set $\mathcal{I}_1$ while preserving this set to be a summand of $\mathcal{X}_1$.

The construction steps of the minimal summand are explained by means of the example shown in Fig. 5.8a. Recapitulate from Section 2.1.3 that a polytope can be represented as the

Algorithm 5.1: Determination of an RPI summand of $\mathcal{X}_1$

Given: Polytopic operating set $\mathcal{X}_1 \subset \mathbb{R}^{n_1}$,
mRPI set $\mathcal{I}_{\mathrm{min},1}$ of the controlled subsystem (5.15) with $\mathcal{I}_{\mathrm{min},1} \subseteq \mathcal{X}_1$
step width δ

Initialise: $\mathcal{V}_{\mathrm{cor}} = \emptyset$, $\mathcal{V}_{\mathrm{end}} = \emptyset$

// Construct the minimal summand of $\mathcal{X}_1$ which contains $\mathcal{I}_{\mathrm{min},1}$

1. Set $\mathcal{V} = \{\boldsymbol{a} \in \mathbb{R}^{n_1} : \mathcal{F}_{\mathcal{X}_1}(\boldsymbol{a}) \text{ is an edge of } \mathcal{X}_1,\ |\boldsymbol{a}| = 1\}$
2. Set $\mathcal{I}_1 = \left\{\boldsymbol{x}_1 \in \mathbb{R}^{n_1} : \boldsymbol{x}_1 \in \bigcap_{\boldsymbol{a} \in \mathcal{V}} \mathcal{H}_{\mathcal{I}_{\mathrm{min},1}}(\boldsymbol{a})\right\}$
3. Set $\mathcal{V}_{\mathrm{cor}} = \{\boldsymbol{a} \in \mathbb{R}^{n_1} : \mathcal{F}_{\mathcal{I}_1}(\boldsymbol{a}) \text{ is longer than } \mathcal{F}_{\mathcal{X}_1}(\boldsymbol{a}),\ \boldsymbol{a} \in \mathcal{V}\}$
4. **do**
5. Translate the halfspaces $\mathcal{H}_{\mathcal{I}_1}(\boldsymbol{a})$, $(\boldsymbol{a} \in \mathcal{V}_{\mathrm{cor}})$ back along $\boldsymbol{a}$ until the length of $\mathcal{F}_{\mathcal{I}_1}(\boldsymbol{a})$ is equal to the length of $\mathcal{F}_{\mathcal{X}_1}(\boldsymbol{a})$
6. Set $\mathcal{V}_{\mathrm{cor}} = \{\boldsymbol{a} \in \mathbb{R}^{n_1} : \mathcal{F}_{\mathcal{I}_1}(\boldsymbol{a}) \text{ is longer than } \mathcal{F}_{\mathcal{X}_1}(\boldsymbol{a}),\ \boldsymbol{a} \in \mathcal{V}\}$
7. **while** $\mathcal{V}_{\mathrm{cor}} \neq \emptyset$

// Make the summand RPI

8. **while** $\mathcal{I}_1$ is not RPI **do**
9. **if** $\mathcal{V}_{\mathrm{end}} = \mathcal{V}$ **then** STOP (no RPI set exists)
10. $\boldsymbol{a} \leftarrow \mathcal{V} \backslash \mathcal{V}_{\mathrm{end}}$
11. Translate the halfspaces $\mathcal{H}_{\mathcal{I}_1}(\boldsymbol{a})$ back along $\boldsymbol{a}$ by the step width δ
12. **if** $\mathcal{H}_{\mathcal{I}_1}(\boldsymbol{a}) = \mathcal{H}_{\mathcal{X}_1}(\boldsymbol{a})$ **then** $\mathcal{V}_{\mathrm{end}} \leftarrow \boldsymbol{a}$
13. **end**

Result: RPI set $\mathcal{I}_1$ for the controlled subsystem (5.15) that is a summand of $\mathcal{X}_1$

intersection of a finite number of closed halfspaces. A halfspace of $\mathcal{X}_1$ is denoted by $\mathcal{H}_{\mathcal{X}_1}(\boldsymbol{a})$, where $\boldsymbol{a}$ is the orthogonal vector.

In Line 1 all vectors $\boldsymbol{a}$ are collected that correspond to a halfspace of $\mathcal{X}_1$ (see Section 2.1.3). Then, the initial set is constructed in Line 2. Therefore, all halfspaces are translated in direction to the origin, until they are transient to $\mathcal{I}_{\mathrm{min},1}$ (i.e., $h_{\mathcal{I}_1}(\boldsymbol{a}) - h_{\mathcal{I}_{\mathrm{min},1}}(\boldsymbol{a}) = 0$). The resulting set is highlighted in Fig. 5.8b by the grey polytope. This set is not a summand, since the dashed edge is longer than the corresponding edge of $\mathcal{X}_1$. The vector $\boldsymbol{a}$ that corresponds to this edge is stored in the set $\mathcal{V}_{\mathrm{corr}}$ (Line 3). Then the halfspace $\mathcal{H}_{\mathcal{I}_1}(\boldsymbol{a})$ is translated back until the resulting edge has the same length than the edge of $\mathcal{X}_1$. The resulting set is highlighted as grey polytope in Fig. 5.8c. This set is the minimal summand of the operating set $\mathcal{X}_1$ that contains the minimal RPI set $\mathcal{I}_{\mathrm{min},1}$, i.e., $\mathcal{I}_1 = \mathcal{Y}_{\mathrm{min},1}$. Note that if an edge of $\mathcal{I}_1$ is longer than the corresponding edge of $\mathcal{X}_1$ as a result from the translation in direction to the origin, the reverse effect occurs if this halfspace $\mathcal{H}_{\mathcal{X}_1}(\boldsymbol{a})$ is translated back to its original position. The shortest distance $h_{\mathcal{I}_1}(\boldsymbol{a}) - h_{\mathcal{I}_{\min,1}}(\boldsymbol{a})$ results if the halfspace is translated back until the edge $\mathcal{F}_{\mathcal{I}_1}(\boldsymbol{a})$ has the same length than the edge $\mathcal{F}_{\mathcal{X}_1}(\boldsymbol{a})$ (Line 5). In the worst case, the loop (Lines 4–7) will end when $\mathcal{H}_{\mathcal{I}_1}(\boldsymbol{a}) = \mathcal{H}_{\mathcal{X}_1}(\boldsymbol{a})$ for all $\boldsymbol{a} \in \mathcal{V}_{\text{cor}}$. That is the minimal summand is $\mathcal{X}_1$ itself.

Based on this minimal summand, it is checked in Line 8 whether or not $\mathcal{I}_1$ is RPI for F_1 according to (2.5). If this is not the case, the halfspaces $\mathcal{H}_{\mathcal{I}_1}(\boldsymbol{a})$, $(\forall \boldsymbol{a} \in \mathcal{V})$ are translated back step by step to their original positions. If $\mathcal{I}_1 = \mathcal{X}_1$ is not RPI, then the algorithm aborts (Line 9). That particular choice of $\boldsymbol{a}$ (Line 10) is not specified. Possible strategies are, on the one hand, the translation of a particular halfspace $\mathcal{H}_{\mathcal{I}_1}(\boldsymbol{a})$ until $\mathcal{H}_{\mathcal{I}_1}(\boldsymbol{a}) = \mathcal{H}_{\mathcal{X}_1}(\boldsymbol{a})$ holds before another halfspace is chosen or, on the other hand, the change of the halfspace by each iteration of the loop (Lines 8–13).

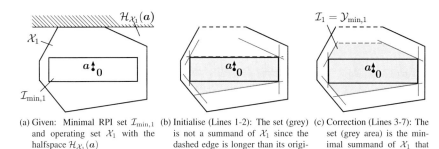

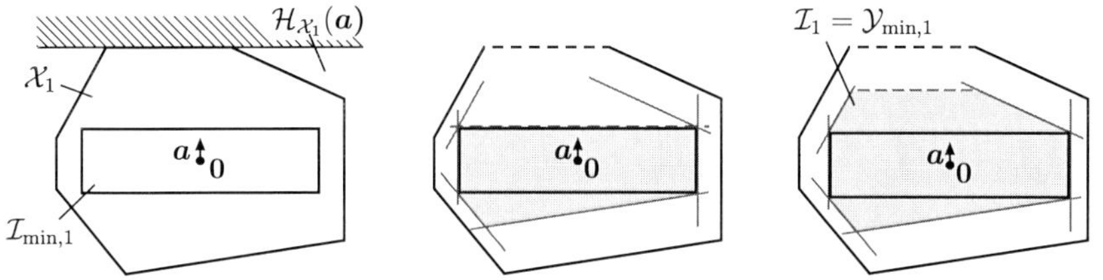

(a) Given: Minimal RPI set $\mathcal{I}_{\min,1}$ and operating set $\mathcal{X}_1$ with the halfspace $\mathcal{H}_{\mathcal{X}_1}(\boldsymbol{a})$

(b) Initialise (Lines 1-2): The set (grey) is not a summand of $\mathcal{X}_1$ since the dashed edge is longer than its original.

(c) Correction (Lines 3-7): The set (grey area) is the minimal summand of $\mathcal{X}_1$ that contains $\mathcal{I}_{\min,1}$

Figure 5.8: Processing of Line 1–7 of Alg. 5.1: Example to construct a minimal summand $\mathcal{Y}_{\min,1}$ of $\mathcal{X}_1$ that contains the minimal RPI set $\mathcal{I}_{\min,1}$.

Design algorithm of the tube-based MPC. In the following the design steps of the tube-based model predictive control station (5.13) regarding the local performance conditions $\mathcal{A}_{\mathrm{D}i}$ are summarised in Algorithm 5.2.

To find a feedback gain $\boldsymbol{K}_1$ (Step 1), methods from literature like optimal controller design or pole placement techniques can be applied. Then, the mRPI set is determined in Step 2. As mentioned in the previous paragraph, if the mRPI set is not a subset of the considered operating set, then there exists no RPI set that is contained in $\mathcal{X}_1$. Hence, the design is aborted (Step 3). The steps 4 and 5 concern the aim to make the whole operating set $\mathcal{X}_1$ RPI for the controlled subsystem (5.12), (5.13). In Step 4 the RPI summand $\mathcal{I}_1$ of $\mathcal{X}_1$ is determined by means of Algorithm 5.1. If Algorithm 5.1 fails, then the mRPI set $\mathcal{I}_{\min,1}$ is chosen (i.e., $\mathcal{I}_1 = \mathcal{I}_{\min,1}$ in Step 5). The design algorithm ends with the calculation of the tightened constraints (Step 6)

Algorithm 5.2: Tube-based MPC design for subsystem S_1

Given: Model set $\mathcal{M}_1 = \{S_1, K_1\}$ and set $\mathcal{S}_1$ available to D_1,
local condition $\mathcal{A}_{D1}$ with polytopic operating set $\mathcal{X}_1$ and actuator constraints $\mathcal{U}_1$ of polytopic shape available to D_1

Processing on D_1:

1. Design the static state-feedback gain $\boldsymbol{K}_1$ such that F_1 from (5.15) is stable
2. Calculate the mRPI set $\mathcal{I}_{\text{min},1}$ in accordance with (5.21)
3. **if** $\mathcal{I}_{\text{min},1} \not\subseteq \mathcal{X}_1$ **then** STOP (design failed)
4. Determine an RPI set $\mathcal{I}_1$ that is a summand of $\mathcal{X}_1$ by running Algorithm 5.1
5. **if** Algorithm 5.1 fails **then** set $\mathcal{I}_1 = \mathcal{I}_{\text{min},1}$
6. Compute the tightened constraints $\mathcal{X}_{\mathfrak{P}1}$ and $\mathcal{U}_{\mathfrak{P}1}$ according to (5.16)
7. Choose the prediction horizon h and the symmetric positive matrices $\boldsymbol{R}_1$ and $\boldsymbol{Q}_1$

Result: Control station (5.13) in order that the controlled subsystem (5.12), (5.13) satisfies the local performance condition $\mathcal{A}_{D1}$

and the choice of the weight matrices and the prediction horizon (Step 7).

In Step 5, any other RPI set that is contained in $\mathcal{X}_1$ can be chosen. If the mRPI set $\mathcal{I}_{\text{min},1}$ is selected, the tube becomes small as well as the final divergence to the origin. Note that, due to the consideration of actuator constraints, the prediction horizon h has to be chosen large enough so that the equality (5.18) holds. Then the operating set $\mathcal{X}_1$ is RPI for the controlled subsystem (5.12), (5.13), providing that $\mathcal{I}_1$ is a summand of $\mathcal{X}_1$. Nevertheless, the larger the prediction horizon is chosen, the higher becomes the computational workload to solve the optimisation problem.

5.5 Application scenario: Fault-tolerant control

The presented plug-and-play control concept is applied to the thermofluid process (Section 2.5.1) to achieve fault tolerance. The situation is considered in which a fault in reactor TB has occurred and only the corresponding control station shall be reconfigured, although the fault has effect on all other tanks of the process through the physical interaction.

For the following investigation, the existence of a nominal controller is assumed:

A 5.1 *There exist N control stations C_i, ($\forall i \in \mathcal{N}$) such that each controlled subsystem F_i satisfies the corresponding local performance condition $\mathcal{A}_{Di}$ defined in* (5.8).

Algorithm 5.3 summarises the processing of design agent D_1 to reconfigure control station C_1 in a plug-and-play control manner. An experiment highlights that the desired overall closed-loop performance (in accordance to $\mathcal{A}_{\mathrm{D}}$) is preserved in spite of a fault. Therefore, a tube-based MPC is applied.

5.5.1 Plug-and-play reconfiguration algorithm

At time instant $k_{\mathrm{f}} \geq 0$, subsystem S_1 is affected by a process fault or an actuator fault. Thus, the faulty subsystem is modelled by

$$S_{1f}: \begin{cases} \boldsymbol{x}_{1f}(k+1) &= \boldsymbol{A}_{1f}\boldsymbol{x}_{1f}(k) + \boldsymbol{B}_{1f}\boldsymbol{u}_{1f}(k) + \boldsymbol{E}_1\boldsymbol{s}_{1f}(k), \quad \boldsymbol{x}_{1f}(k_{\mathrm{f}}) = \boldsymbol{x}_1(k_{\mathrm{f}}) \\ \boldsymbol{z}_{1f}(k) &= \boldsymbol{C}_{\mathrm{z}1}\boldsymbol{x}_{1f}(k). \end{cases} \tag{5.22}$$

As a consequence of the fault, the RPI set $\mathcal{I}_1 \subseteq \mathcal{X}_1$ of the nominal controlled subsystem F_1 is probably no longer RPI for the faulty subsystem S_{1f} controlled by the nominal control station C_1, i.e.,

$$(\boldsymbol{A}_{1f} + \boldsymbol{B}_{1f}\boldsymbol{K}_1)\mathcal{I}_1 \oplus \boldsymbol{E}_1\mathcal{S}_1 \nsubseteq \mathcal{I}_1.$$

Accordingly, the faulty state $\boldsymbol{x}_{1f}(k)$ will leave the desired operating set $\mathcal{X}_1$ and, thus, jeopardises the boundedness of the overall closed-loop system state and violates the actuator constraints. Hence, the existing control station C_1 must be reconfigured. To perform the reconfiguration on time, the following assumption is made.

A 5.2 *A diagnosis system detects the fault and uniquely identifies the model S_{1f} of the faulty subsystem at time instant $k_{\mathrm{I}} \geq k_{\mathrm{f}}$, where $\boldsymbol{x}_{1f}(k_{\mathrm{I}}) \in \mathcal{X}_1$. As a result, the model S_{1f} is known by design agent D_1 at time instant $k_{\mathrm{I}} \geq k_{\mathrm{f}}$.*

A local diagnosis system is presented in Chapter 8.

Initially each design agent D_1 knows the models $\mathcal{M}_1 = (S_1, S_{1f}, C_1, K_1)$ that consists of the model of the subsystem, the model of the control station and the local couplings (cf. assumption A 3.2). If a fault has occurred in subsystem S_1, the design agent, first, gathers the operating sets $\mathcal{Z}_i$, $(\forall i \in \mathcal{P}_1)$ from the neighbouring design agents through the network, second, redesigns the control station and, third, implements it into the control hardware. These reconfiguration steps performed by D_1 are summarised in Algorithm 5.3.

In Step 1, design agent D_1 requests the operating sets $\mathcal{Z}_i$ from of its direct neighbours D_i, $(\forall i \in \mathcal{P}_1)$. After the interaction of the design agents according to the communication graph

Algorithm 5.3: Plug-and-play reconfiguration of control station C_1

Given: Control stations C_i, $(i \in \mathcal{N})$ exist such that F_i satisfy $\mathcal{A}_{\mathrm{D}i}$ for all $i \in \mathcal{N}$,
model set $\mathcal{M}_1 = \{S_1, S_{1f}, C_1, K_1\}$ available to D_1,
model set $\mathcal{M}_i = \{F_i, K_i\}$ available to D_i, $(\forall i \in \mathcal{N} \setminus \{1\})$ and
local design condition $\mathcal{A}_{\mathrm{D}1}$ with operating set $\mathcal{X}_1$, actuator limitation $\mathcal{U}_1$ and design parameter ε_1 available to D_1

Processing on D_1**:**

1. Gather the operating set $\mathcal{Z}_i$ from D_i, $(\forall i \in \mathcal{P}_1)$ and calculate $\mathcal{S}_1$ according to (5.3)
2. **if** $\mathcal{A}_{\mathrm{D}1}$ is violated **then** design
   a) the static state-feedback gain $\boldsymbol{K}_1$ by means of solving the LMIs (5.10) or the LMIs (5.11) with $\boldsymbol{x}_{1,0} = \boldsymbol{x}_{1f}(k_{\mathrm{I}})$ or
   b) the tube-based MPC (5.13) by means of Algorithm 5.2
   else STOP (no reconfiguration necessary)
3. Implement the reconfigured control station into the control equipment

Result: The reconfigured closed-loop system satisfies the global performance condition $\mathcal{A}_{\mathrm{D}}$

illustrated in Fig. 5.4 the information

$$\mathcal{M}_1 = (S_1, S_{1f}, C_1, K_1, \{\mathcal{Z}_i, \, i \in \mathcal{P}_1\})$$

are at hand of D_1. Then, D_1 is able to calculate the constraint set $\mathcal{S}_1$ of the coupling input signal and to check whether a reconfiguration is necessary in Step 2. If a reconfiguration is required, either a static state-feedback (5.1) can be designed (Step 2a) or a tube-based MPC (5.13) can be parametrised (Step 2b), which is finally implemented into the control hardware (Step 3).

It is worth noting that the proposed reconfiguration algorithm can be used in case of multiple fault occurrence in several subsystems. That property is attributable to the fact that the reconfiguration of control stations can be designed independently from each other. Moreover, note that due to assumption A 3.4 the processing of Algorithm 5.3 does not consume time.

5.5.2 Experiment: Interconnected thermofluid process

The plug-and-play reconfiguration algorithm (Algorithm 5.3) is applied to the thermofluid process introduced in Section 2.5.1 after a fault in reactor TB has occurred. Tube-based MPC is the solution for the reconfiguration. By means of simulations and experiments, the behaviour of the fault-tolerant overall system is analysed. The experimental results highlight, on the one hand, the conservatism of the local design condition and, on the other hand, the applicability of the conservative design to a real world process.

The subsystems are modelled by (5.12) with the parameters from (A.7a)–(A.7c). The physical coupling among the tanks is modelled by (2.7) with the coupling gains from (A.6a)–(A.6c). The sets of predecessors yield

$$\mathcal{P}_1 = \{2, 3\}, \qquad \mathcal{P}_2 = \emptyset, \qquad \mathcal{P}_3 = \{1, 3\}.$$

In the operating point, the control action is physically limited by

$$\begin{aligned}
\boldsymbol{u}_1(t) = \begin{pmatrix} u_{\mathrm{TB}}(t) \\ u_{\mathrm{HB}}(t) \end{pmatrix} \in \mathcal{U}_1 = \left\{ \boldsymbol{u}_1 \in \mathbb{R}^2 \colon \begin{pmatrix} -0.35 \\ -0.24 \end{pmatrix} \leq \boldsymbol{u}_1 \leq \begin{pmatrix} 0.65 \\ 0.76 \end{pmatrix} \right\} \\
u_2(t) = u_{\mathrm{T1}}(t) \in \mathcal{U}_2 = \{ u_2 \in \mathbb{R} \colon -0.23 \leq u_2 \leq 0.77 \} \\
\boldsymbol{u}_3(t) = \begin{pmatrix} u_{\mathrm{TS}}(t) \\ u_{\mathrm{HS}}(t) \end{pmatrix} \in \mathcal{U}_3 = \left\{ \boldsymbol{u}_3 \in \mathbb{R}^2 \colon \begin{pmatrix} -0.18 \\ -0.1 \end{pmatrix} \leq \boldsymbol{u}_3 \leq \begin{pmatrix} 0.82 \\ 0.9 \end{pmatrix} \right\}
\end{aligned}$$

and the level as well as the temperature in the tanks has to be within the polytopic operating sets

$$\begin{aligned}
\boldsymbol{x}_1(t) = \begin{pmatrix} l_{\mathrm{TB}}(t) \\ \vartheta_{\mathrm{TB}}(t) \end{pmatrix} \in \mathcal{X}_1 = \left\{ \boldsymbol{x}_1 \in \mathbb{R}^2 \colon |\boldsymbol{x}_1| \leq \begin{pmatrix} 0.02\,\mathrm{m} \\ 1.5\,\mathrm{K} \end{pmatrix} \right\} \\
x_2(t) = l_{\mathrm{T1}}(t) \in \mathcal{X}_2 = \{ x_2 \in \mathbb{R} \colon |x_2| \leq 0.05\,\mathrm{m} \} \\
\boldsymbol{x}_3(t) = \begin{pmatrix} l_{\mathrm{TS}}(t) \\ \vartheta_{\mathrm{TS}}(t) \end{pmatrix} \in \mathcal{X}_3 = \left\{ \boldsymbol{x}_3 \in \mathbb{R}^2 \colon |\boldsymbol{x}_3| \leq \begin{pmatrix} 0.05\,\mathrm{m} \\ 1.5\,\mathrm{K} \end{pmatrix} \right\}.
\end{aligned}$$

The existing nominal control stations are static state-feedbacks, modelled by

$$C_i : \ \boldsymbol{u}_i(k) = \boldsymbol{K}_i \boldsymbol{x}_i(k) \tag{5.23}$$

with

$$\boldsymbol{K}_1 = 10^{-1} \begin{pmatrix} 21.91 & -0.03 \\ 0.06 & -1.17 \end{pmatrix}, \qquad \boldsymbol{K}_2 = -6, \qquad \boldsymbol{K}_3 = 10^{-1} \begin{pmatrix} 22.11 & -0.03 \\ 0.07 & -0.5 \end{pmatrix}$$

in order that the controlled subsystems (5.12), (5.23) are asymptotically stable and adhere the state and actuator constraints. The control stations are designed by solving a standard LQ optimisation problem with

$$\boldsymbol{Q}_1 = \boldsymbol{Q}_3 = \begin{pmatrix} 5 & 0 \\ 0 & 0.5 \end{pmatrix}, \quad \boldsymbol{Q}_2 = 0.5, \quad \boldsymbol{R}_1 = \boldsymbol{R}_3 = \begin{pmatrix} 1 & 0 \\ 0 & 20 \end{pmatrix}, \quad \boldsymbol{R}_2 = 20. \tag{5.24}$$

Fault occurrence in subsystem S_1. At $t_{\mathrm{f}} = 0\,\mathrm{s}$ the valve from tank T2 to reactor TB gets stuck in an open position modelled by a greater valve angle

$$u_{2\mathrm{B}f} = 0.7.$$

This fault represents a process fault. The corresponding model

$$S_{1f}: \begin{cases} \boldsymbol{x}_{1f}(k+1) = \begin{pmatrix} 0.9999 & 0 \\ -0.0012 & 0.9964 \end{pmatrix} \boldsymbol{x}_{1f}(k) + \boldsymbol{B}_1 \boldsymbol{u}_{1f}(k) + \boldsymbol{E}_1 \boldsymbol{s}_{1f}(k), \\ \boldsymbol{x}_{1f}(k_{\mathrm{f}}) = \boldsymbol{x}_1(k_{\mathrm{f}}) \\ \boldsymbol{y}_{1f}(k) = \boldsymbol{x}_{1f}(k) \\ \boldsymbol{z}_{1f}(k) = \boldsymbol{x}_{1f}(k), \end{cases}$$

of the faulty subsystem is identified at $t_{\mathrm{I}} = 32.3\,\mathrm{s}$ (i.e., $k_{\mathrm{I}} = 161$), where

$$\boldsymbol{x}_1(k_{\mathrm{I}}) = \begin{pmatrix} 1.5\,\mathrm{cm} \\ 0.22\,\mathrm{K} \end{pmatrix} \in \mathcal{X}_1$$

as shown in Fig. 5.11a. The local diagnosis system forwards the model S_{1f} to design agent D_1, which subsequently starts the reconfiguration of control station C_1 by running Algorithm 5.3:
Step 1: Design agent D_1 requests the model information $\mathcal{Z}_i$ from its neighbouring design agents D_i, $\mathcal{P}_1 = \{2, 3\}$ and immediately receives their information. After the communication, the locally available model information yields

$$\mathcal{M}_1 = \{S_1,\, S_{1f},\, C_1,\, K_1,\, \mathcal{Z}_2,\, \mathcal{Z}_3\}\,.$$

The constraint signal space of the coupling input signal $\boldsymbol{s}_1$ results to $\mathcal{S}_1 = \boldsymbol{L}_{12}\mathcal{Z}_2 \oplus \boldsymbol{L}_{13}\mathcal{Z}_3$, shown in Fig. 5.9a.
Step 2: The local performance condition is violated since the controlled subsystem F_1 is unstable. That is, both eigenvalues of the system matrix $\boldsymbol{A}_{1f} + \boldsymbol{B}_1\boldsymbol{K}_1$ are outside the unit circle. Accordingly, design agent D_1 performs Algorithm 5.2 to design a tube-based MPC.

Step 2.1: The redesign is accomplished by solving an LQ optimisation problem with the

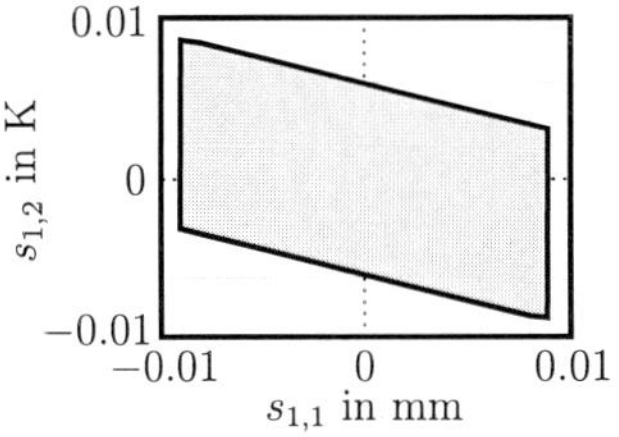

(a) Constraint signals space $\mathcal{S}_1$

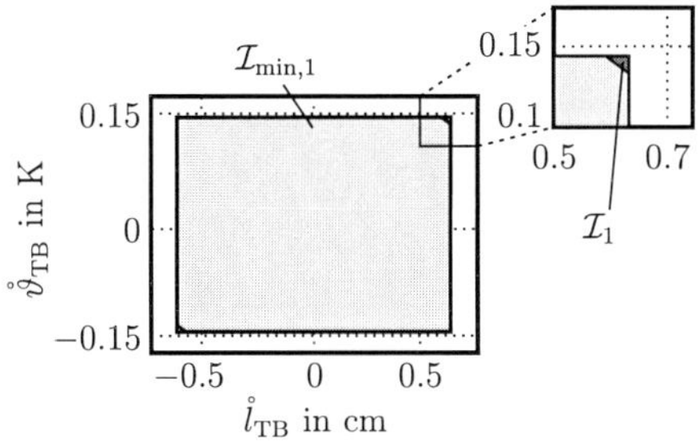

(b) mRPI set $\mathcal{I}_{\min,1}$ and RPI summand $\mathcal{I}_1$

Figure 5.9: Application: Constraint signals space $\mathcal{S}_1$ and calculated RPI set $\mathcal{I}_1$ of the reconfigured controlled subsystem $\mathring{F}_1$.

weight matrices from (5.24). The result is the static state-feedback gain

$$\mathring{\boldsymbol{K}}_1 = 10^{-1}\begin{pmatrix} 21.85 & -0.04 \\ 0.1 & -1.11 \end{pmatrix}$$

such that all eigenvalues of $\boldsymbol{A}_{1f} + \boldsymbol{B}_1\mathring{\boldsymbol{K}}_1$ are inside the unit circle. The superscript $\mathring{(\cdot)}$ denotes the reconfigured case.

Step 2.2: The mRPI set $\mathcal{I}_{\min,1}$ for the reconfigured controlled subsystem

$$\mathring{F}_1 : \begin{cases} \mathring{\boldsymbol{x}}_1(k+1) = \left(\boldsymbol{A}_{1f} + \boldsymbol{B}_1\mathring{\boldsymbol{K}}_1\right)\mathring{\boldsymbol{x}}_1(k) + \boldsymbol{E}_1\boldsymbol{s}_1(k), \quad \mathring{\boldsymbol{x}}_1(k_{\mathrm{I}}) = \boldsymbol{x}_{1f}(k_{\mathrm{I}}) \\ \mathring{\boldsymbol{z}}_1(k) \quad = \mathring{\boldsymbol{x}}_1(k), \end{cases}$$

is determined using Algorithm 1 from [140]. The resulting mRPI set $\mathcal{I}_{\min,1}$ is depicted in Fig. 5.9b.

Step 2.3: As $\mathcal{I}_{\min,1} \subset \mathcal{X}_1$ holds, a tube-based MPC exists that satisfies the local performance condition $\mathcal{A}_{\mathrm{D}1}$.

Steps 2.4 and 2.5: Design agent D_1 performs Algorithm 5.1 which terminates successfully. The The resulting RPI summand yields the rectangle

$$\mathcal{I}_1 = \left\{ \boldsymbol{x}_1 \in \mathbb{R}^2 \colon |\boldsymbol{x}_1| \leq \begin{pmatrix} 6.08\,\mathrm{mm} \\ 0.15\,\mathrm{K} \end{pmatrix} \right\},$$

where $\mathcal{I}_1 \supset \mathcal{I}_{\min,1}$ as highlighted in Fig. 5.9b.

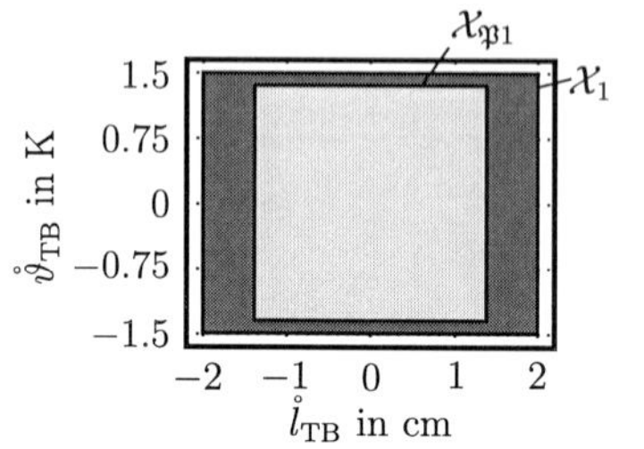

(a) Operating set $\mathcal{X}_1$ and tightened constraints $\mathcal{X}_{\mathfrak{P}1}$

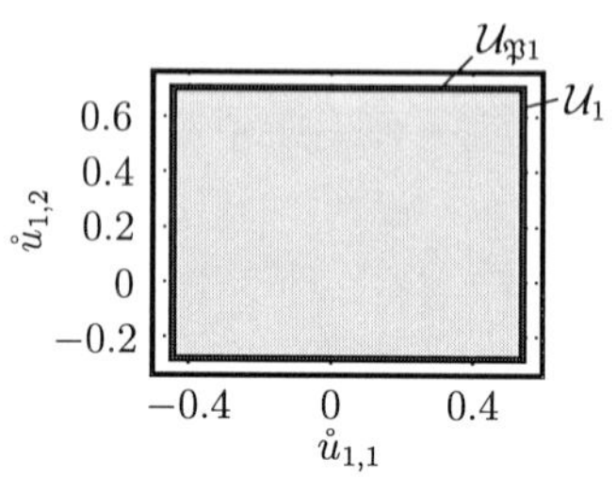

(b) Admissible actuator action $\mathcal{U}_1$ and tightened constraints $\mathcal{U}_{\mathfrak{P}1}$

Figure 5.10: Application: Tightened constraints $\mathcal{X}_{\mathfrak{P}1}$ and $\mathcal{U}_{\mathfrak{P}1}$ in comparison with the operating set $\mathcal{X}_1$ and the actuator constraints $\mathcal{U}_1$.

Step 2.6: The tightened constraints result to

$$\mathcal{X}_{\mathfrak{P}1} = \left\{ \boldsymbol{x}_1 \in \mathbb{R}^2 \colon |\boldsymbol{x}_1| \leq \begin{pmatrix} 0.013\,\mathrm{m} \\ 1.35\,\mathrm{K} \end{pmatrix} \right\},$$

$$\mathcal{U}_{\mathfrak{P}1} = \left\{ \boldsymbol{u}_1 \in \mathbb{R}^2 \colon \begin{pmatrix} -0.44 \\ -0.27 \end{pmatrix} \leq \boldsymbol{u}_1 \leq \begin{pmatrix} 0.54 \\ 0.69 \end{pmatrix} \right\}$$

as shown in Fig. 5.10.

Step 2.7: The prediction horizon is chosen to $h = 32$ and the weight matrices read

$$\boldsymbol{Q}_1 = \begin{pmatrix} 5 & 0 \\ 0 & 0.5 \end{pmatrix}, \qquad \boldsymbol{R}_1 = \begin{pmatrix} 1 & 0 \\ 0 & 20 \end{pmatrix},$$

adapted from (5.24).

Step 3: The reconfigured control station

$$\mathring{C}_1 : \begin{cases} \text{solve} \quad \mathfrak{P}_1(\mathring{\boldsymbol{x}}_1(k), S_{1f}|_{\boldsymbol{s}_{1f}=\boldsymbol{0}}) \\ \mathring{\boldsymbol{u}}_1(k) = \boldsymbol{u}_{\mathfrak{P}1}(k) + \mathring{\boldsymbol{K}}_1(\mathring{\boldsymbol{x}}_1(k) - \boldsymbol{x}_{\mathfrak{P}1}(k))) \end{cases} \tag{5.25}$$

is subsequently implemented into the control hardware and is switched on. The reconfiguration is finished.

Figure 5.11a shows the level $l_{\mathrm{TB}}(k)$ and the temperature $\vartheta_{\mathrm{TB}}(k)$ in reactor TB in simulation (dashed lines) and experiment (solid lines). The corresponding actuator control signals are shown in Fig. 5.11b. The dots mark the time of fault occurrence and the time of fault identifica-

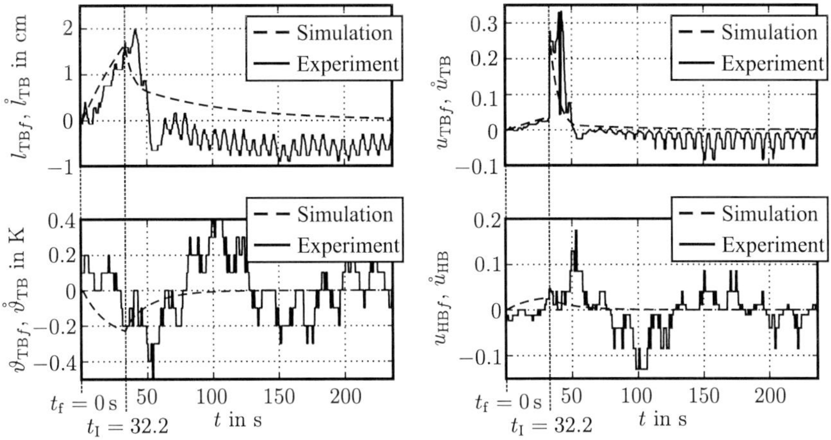

(a) Level l_{TB} and temperature ϑ_{TB}, where $|l_{TB}| \leq 2\,\mathrm{cm}$ and $|\vartheta_{TB}| \leq 1.5\,\mathrm{K}$

(b) Valve opening u_{TB} and heat power u_{HB}, where $-0.35 \leq u_{TB} \leq 0.65$ and $-0.24 \leq u_{HB} \leq 0.76$

Figure 5.11: Application: Behaviour of level l_{TB} and temperature ϑ_{TB} as well as the control signals u_{TB} and u_{HB} in reactor TB.

tion, respectively. Instantaneously, the reconfigured control station $\mathring{C}_1$ is started up and drives the reconfigured state back to the origin while the actuator limitations are satisfied.

Figure 5.12 represents the motion of the simulated state $\boldsymbol{x}_1(k)$ in its state space in the time interval between fault occurrence ($t_f = 0\,\mathrm{s}$) and fault isolation ($t_I = 32.2\,\mathrm{s}$) as dashed lines as well as after the reconfiguration ($t > t_I$) by solid lines. Moreover, the predicted state $\boldsymbol{x}_{\mathfrak{p}1}(k)$ of the tube-based MPC is shown together with the RPI set $\mathcal{I}_1$. It can be seen that after $k = 223$ ($t = 44.6\,\mathrm{s}$) the predicted state reached the origin ($\boldsymbol{x}_{\mathfrak{p}1}(k) = \mathbf{0}$) such that only the state-feedback remains active to keep the state within its RPI set $\mathcal{I}_1$. At that moment, the repeated solving of the optimisation problem stops.

Figure 5.13 shows the corresponding result for the experiment. For the sake of visualisation, the measured level $l_{TB}(k)$ is plotted against the motion of the simulated temperature $\vartheta_{TB}(k)$. After the reconfiguration, the tube-based MPC drives the state successfully back into the RPI set $\mathcal{I}_1$. The translation takes much longer ($200.2\,\mathrm{s}$ in comparison to $44.6\,\mathrm{s}$ in the simulation) due to the measurement uncertainties. Nevertheless, the experiment highlights that despite noisy measurements and the uncertainty of the model in terms of disregarded nonlinear effects such as the actuation of the valves by pulse-width modulation, the tube based MPC guarantees that the state remains in its operating set $\mathcal{X}_1$, while regarding the actuator limitation $\mathcal{U}_1$.

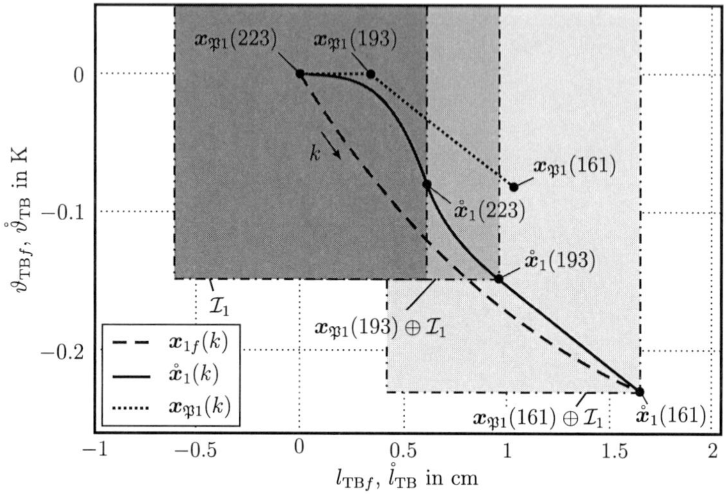

Figure 5.12: Application: Trajectory of the simulated state $x_1(k)$ of reactor TB in state space.

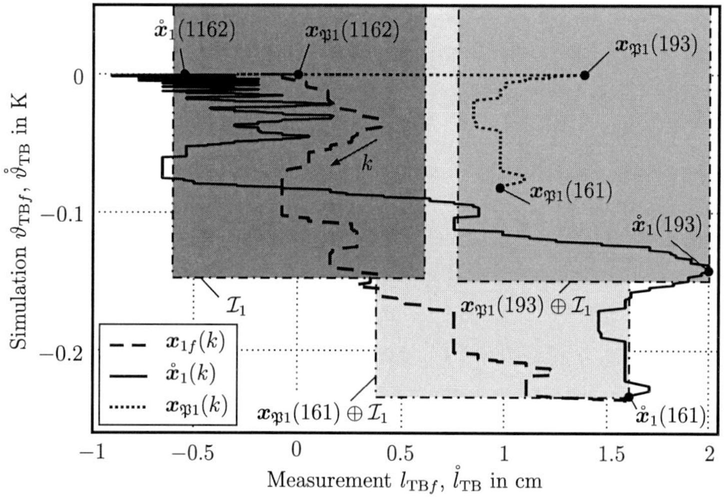

Figure 5.13: Application: Trajectory of the measured level $l_{\mathrm{TB}}(k)$ and simulated temperature $\vartheta_{\mathrm{TB}}(k)$ of reactor TB in state space.

This experiment emphasises that even though the local design conditions are conservative (cf. Section 5.2), the reconfiguration was successful. This success is particularly owed to the small couplings the theromfluid process provides. That is, the constraint set $\mathcal{S}_1$ (cf. Fig. 5.9a) is tiny in comparison to the operating set $\mathcal{X}_1$ (cf. Fig. 5.10a). In can be concluded that the challenge of the proposed tube-based MPC is rather to be robust against the uncertainty of the used subsystem model than to compensate the ignored dynamics of the physical interaction.

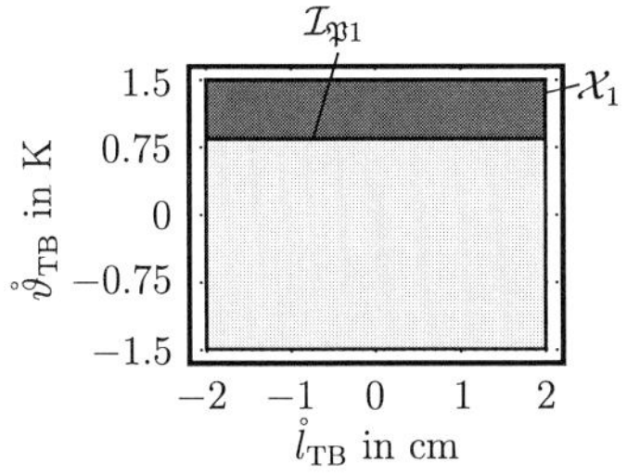

(a) RPI set $\mathcal{I}_{\mathfrak{P}1}$ of the reconfigured controlled reactor TB in comparison with the operating reagion $\mathcal{X}_1$

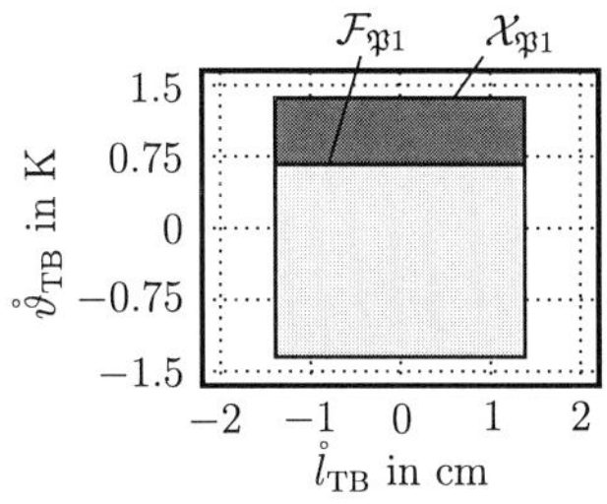

(b) Feasible set $\mathcal{F}_{\mathfrak{P}1}$ together with the operating reagion $\mathcal{X}_1$

Figure 5.14: Application: RPI set $\mathcal{I}_{\mathfrak{P}1}$ and feasible set $\mathcal{F}_{\mathfrak{P}1}$ of the reactor TB controlled by the tube-based MPC.

Size of the RPI set. For the reconfigured controlled subsystem (5.22), (5.25), the whole operating set $\mathcal{X}_1$ is not RPI as highlighted in Fig. 5.14a. As shown in Fig. 5.14b, the feasible set is smaller than the tightened constraints, i.e., $\mathcal{F}_{\mathfrak{P}1} \subset \mathcal{X}_{\mathfrak{P}1}$. This is due to the fact that the thermofluid process does not provide a separate cooling unit to cool down the fluid. It is cooled only indirectly by the cooler inflows from the other tanks which takes much longer compared to active cooling. Hence, the tube-based MPC cannot steer the subsystem into the origin in $h = 32$ steps (i.e. $6.4\,\mathrm{s}$). Hence, the prediction horizon $h = 32$ (i.e. $6.4\,\mathrm{s}$) has to be widened in order to take the low dynamic into account. However, this entails a higher computational workload. With the used equipment, it can no longer be guaranteed that the optimisation problem (5.14) can be solved between two samples (i.e., $200\,\mathrm{ms}$).

5.5.3 Further application scenarios

The presented controller design can be used without changes to process an ordinary design of a single control station. Since only the local models and the desired sets from the neighbouring design agents are necessary, the design can be accomplished independently from other control stations.

Moreover, Algorithm 6.1 presented in Section 6.5.1 can be applied with slight changes to incorporate new subsystems to the interconnected systems. The particular changes encompass the additional procurement of the sets $\mathcal{Z}_i$, the calculation of $\mathcal{S}_i$ and the usage of a design method proposed in this chapter. That is, the Steps 1 and 2 as well as the Steps 4 and 5 in Algorithm 6.1 have to be replaced by the Steps 1 and 2 of Algorithm 5.3, respectively.

6 Plug-and-play control using the local model based on limited amplifications

This chapter considers the design of control station C_1 based on the subsystem model S_1 and local coupling model K_1 which are initially available to D_1. This design must ensure the adherence of a global control aim, even though no information about the physical influence form other subsystems to subsystem S_1 is known by D_1. The main result of this chapter are local design conditions that guarantee overall closed-loop I/O stability and a certain overall closed-loop performance. The basis of these conditions is to attenuate the amplification from the coupling input to the coupling output of each controlled subsystem by design. Moreover, an H_∞-design is proposed. This chapter closes with the demonstration of the proposed design method on the thermofluid process. An experiment is shown in which a new reactor is integrated to the process at runtime.

6.1 Problem formulation

The situation is considered in which control station C_1 shall be designed using only the local models S_1 and K_1 of the subsystem and the local couplings, respectively, that are initially available to design agent D_1. Although the physical interaction of subsystem S_1 with all other subsystems remain unknown, the design must guarantee I/O stability and a certain performance of the overall closed-loop system. The focus of this chapter is to derive locally checkable design conditions so as to enable design agent D_1 to accomplish the synthesis of C_1. As the physical interaction $P_1^{\mathcal{N}}$ is unknown, restrictive design conditions are needed which, however, will pay-off whenever privacy constraints need to be considered, only low computational power is available or the scalability of the design is desired (see Section 5.1).

The unknown model of the physical interaction $P_1^{\mathcal{N}}$ comprises all other controlled subsystems F_i, $(i = 2, ..., N)$ as shown by the grey box in Fig. 6.1. The main idea of this chapter is to bound the effect of the physical interaction on subsystem S_1 by an appropriate design of the control

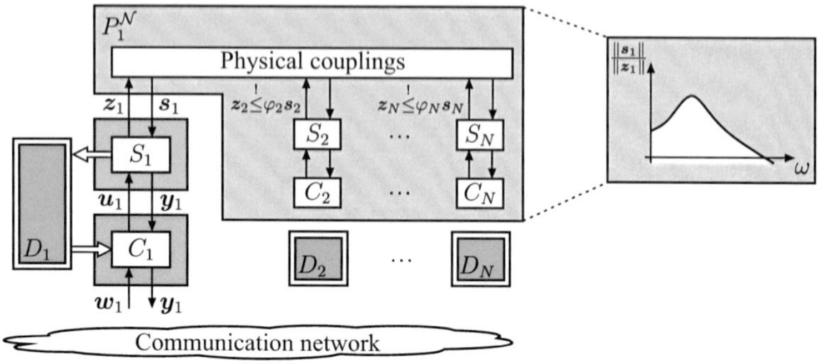

Figure 6.1: Structure and idea of plug-and-play control using local model information based on limited amplifications. Whenever all controlled subsystems limit their amplification of the coupling signals ($z_i \leq \varphi_i s_i$), the physical interaction with S_1 become bounded. This bound can be determined locally by D_1 and is used for the design of C_1.

stations C_i, $(i = 2, ..., N)$. The aim of this design is the attenuation of the coupling input s_i on the coupling output z_i of each controlled subsystem F_i, $(i = 2, ..., N)$ illustrated by the claim $z_i \overset{!}{\leq} \varphi_i s_i$ in Fig. 6.1. The suitable choice of the local bounds φ_i, $(i = 2, ..., N)$, enables design agent D_1 to determine the upper bound of the physical interaction (shown on the right-hand side of Fig. 6.1) with the initially available model information. That is, no additional models have to be requested from other design agents. Based on these considerations local design requirements are derived that regard I/O stability and a certain I/O performance of the overall closed-loop system. This problem to be solved in this chapter is summarised as follows:

Problem 6.1 (Plug-and-play control using local model information)

Problem: *Although the dynamics of subsystem S_i depend upon the physical interaction, design agent D_i has to perform the design of C_i with local model information only (i.e., solve Problem 1.1). This local design has to guarantee I/O stability and a certain I/O performance of the overall closed-loop system.*

Given:
- *D_i, $(\forall i \in \mathcal{N})$ store the corresponding model set $\mathcal{M}_i = \{S_i, K_i\}$*
- *Global design conditions for overall closed-loop I/O stability and an overall closed-loop performance*

Find: *Local design conditions which guarantee the adherence of the global conditions.*

Approach: *Limiting the amplification from the coupling input signal $\boldsymbol{s}_i$ to the coupling output signal $\boldsymbol{z}_i$ of each controlled subsystem F_i.*

A solution to Problem 6.1 states a solution to Problem 1.2 and, hence, enables D_1 to accomplish the controller design, i.e, solve the design problem Problem 1.1. Therefore, this chapter presents an H_∞-design of control station C_i regarding the local condition.

Section 6.2 presents conditions that enable D_1 to design C_1 without using models from other design agents. Based on these conditions, Section 6.3 decomposes a global I/O stability condition and a global I/O performance condition into locally verifiable conditions. The H_∞-design method is proposed in Section 6.3. Finally, the proposed method for plug-and-play control is applied to the scenario that a new subsystem has to be integrated at runtime. By means of simulations and experiments on the thermofluid process the method is evaluated in Section 6.5.

6.2 Conditions to enable the design with the local model

This section proposes conditions on the controlled subsystems so that an upper bound of the physical interaction can be determined with the locally available model information only.

Consider the controlled subsystems F_i, $(\forall i \in \mathcal{N})$ that is represented in frequency domain by

$$F_i: \begin{cases} \boldsymbol{y}_i(s) = \boldsymbol{F}_{\mathrm{yw}i}(s)\,\boldsymbol{w}_i(s) + \boldsymbol{F}_{\mathrm{ys}i}(s)\,\boldsymbol{s}_i(s) \\ \boldsymbol{z}_i(s) = \boldsymbol{F}_{\mathrm{zw}i}(s)\,\boldsymbol{w}_i(s) + \boldsymbol{F}_{\mathrm{zs}i}(s)\,\boldsymbol{s}_i(s). \end{cases}$$

Accordingly, the model $P_i^{\mathcal{N}}$ of physical interaction that results from the connection (3.9) yields

$$P_i^{\mathcal{N}}: \quad \boldsymbol{s}_i(s) = \boldsymbol{P}_i^{\mathcal{N}}(s)\,\boldsymbol{z}_i(s) \tag{6.1}$$

with the dynamics

$$\begin{aligned} \boldsymbol{P}_i^{\mathcal{N}}(s) = \boldsymbol{L}_{\{i\},\mathcal{N}\setminus\{i\}} &\Big(\mathbf{I} - \mathrm{diag}(\boldsymbol{F}_{\mathrm{zs}j}(s))_{j\in\mathcal{N}\setminus\{i\}}\,\boldsymbol{L}_{\mathcal{N}\setminus\{i\}}\Big)^{-1} \\ &\mathrm{diag}(\boldsymbol{F}_{\mathrm{zs}j}(s))_{j\in\mathcal{N}\setminus\{i\}}\,\boldsymbol{L}_{\mathcal{N}\setminus\{i\},\{i\}}. \end{aligned} \tag{6.2}$$

This transfer function matrix only depends upon the coupling gains $\boldsymbol{L}_{ij}$ as well as the transfer functions $\boldsymbol{F}_{\mathrm{zs}j}(s)$ from the controlled subsystems F_j, $(\forall j \in \mathcal{N}\setminus\{i\})$. Intuitively, the limitation of the magnitudes of these transfer functions $\boldsymbol{F}_{\mathrm{zs}j}(s)$ for all $j \in \mathcal{N}\setminus\{i\}$ directly leads to the limitation of the effect of the physical interaction on subsystem S_i. Based on this idea, the following lemma is formalised.

Lemma 6.1 (Local conditions to constrain the effect of the physical interaction) *If the relations*

$$\|\boldsymbol{F}_{\mathrm{zs}i}(\mathrm{j}\omega)\| \sum_{j\in\mathcal{P}_i} \|\boldsymbol{L}_{ij}\| < \varphi(\omega), \quad \forall\omega\in\mathbb{R}, \quad \forall i\in\mathcal{N} \tag{6.3}$$

are satisfied for a given function $\varphi(\omega)\colon \mathbb{R}\to\mathbb{R}_+$ *with* $\varphi(\omega)\leq 1$, *then the system*

$$\bar{P}_i^{\{i\}}: \quad \bar{s}_i(\omega) = \underbrace{\textstyle\sum_{j\in\mathcal{P}_i} \|\boldsymbol{L}_{ij}\|\varphi(\omega)}_{\bar{P}_i^{\{i\}}(\omega)} \|\boldsymbol{z}_i(\mathrm{j}\omega)\| \tag{6.4}$$

is a comparison system of the physical interaction $P_i^{\mathcal{N}}$ *represented by* (6.1) *for all* $i\in\mathcal{N}$.

Proof. See Appendix B.6. □

The superscribe of the comparison system $\bar{P}_i^{\{i\}}$ indicates that only the model information from D_i is required to determine this system. As a consequence of Lemma 6.1, the physical interaction (6.1) that is composed of all controlled subsystems can be handled as *unknown-but-bounded model uncertainty* represented by the comparison system (6.4) as illustrated in Fig. 6.2 from the local view of subsystem S_1.

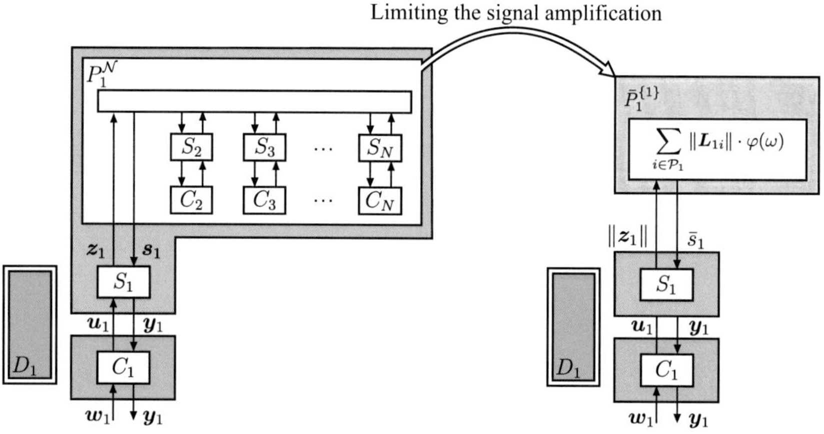

Figure 6.2: Effect of limiting the signal amplification: The physical interaction $P_1^{\mathcal{N}}$ is considered as unknown-but-bounded model uncertainty. A comparison system $\bar{P}_1^{\{1\}}$ of the unknown physical interaction $P_1^{\mathcal{N}}$ can be determined by D_1.

The crucial point of this approach is that the amplification of each transfer function $\boldsymbol{F}_{\mathrm{zs}i}(s)$, $(i \in \mathcal{N})$ has to be bounded by the same desired function $\varphi(\omega)$. As a result, an upper bound of the dynamics (6.2) becomes representable by the local coupling gains $\boldsymbol{L}_{ij}$ and this boundary $\varphi(\omega)$ only (cf. (6.4)). That is, the exact dynamics $\boldsymbol{F}_{\mathrm{zs}i}(s)$, $(\forall i \in \mathcal{N})$ become irrelevant to set up a comparison system. Under the consideration that the function $\varphi(\omega)$ is known by D_i, the comparison system $\bar{P}_i^{\{i\}}$ can be constructed with the a priory known model set $\mathcal{M}_i = \{S_i, K_i\}$.

However, the accuracy by which $\varphi(\omega)$ represents each dynamic $\|\boldsymbol{F}_{\mathrm{zs}i}(\mathrm{j}\omega)\|$ depends upon the individual dynamics. An example of $\varphi(\omega)$ for three controlled subsystems is highlighted in Fig. 6.3. Accordingly, the function $\varphi(\omega)$ is a poor description of $\|\boldsymbol{F}_{\mathrm{zs}1}(\mathrm{j}\omega)\|$ and $\|\boldsymbol{F}_{\mathrm{zs}3}(\mathrm{j}\omega)\|$, but a good representation of $\|\boldsymbol{F}_{\mathrm{zs}2}(\mathrm{j}\omega)\|$. As a consequence, the more the individual dynamics $\|\boldsymbol{F}_{\mathrm{zs}i}(\mathrm{j}\omega)\| \sum_{j\in\mathcal{P}_i} \|\boldsymbol{L}_{ij}\|$ differ from each other, the more imprecise becomes the upper bound $P_i^{\{i\}}(\omega)$ representing the physical interaction $\boldsymbol{P}_1^{\mathcal{N}}(s)$.

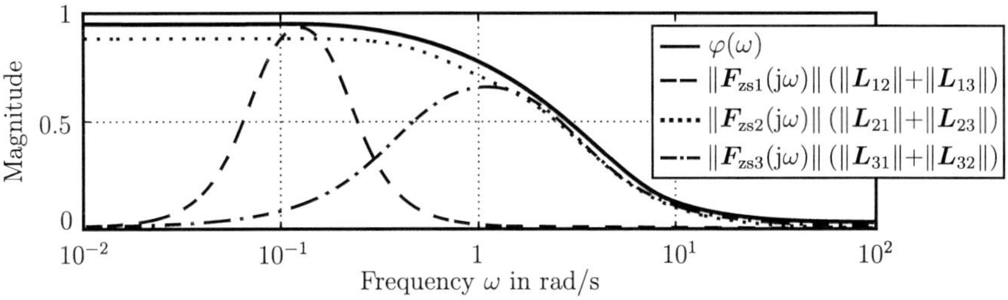

Figure 6.3: Example of $\varphi(\omega)$: Common upper bound of all controlled subsystem dynamics. The plots $\|\boldsymbol{F}_{\mathrm{zs}1}(\mathrm{j}\omega)\|$ and $\|\boldsymbol{F}_{\mathrm{zs}3}(\mathrm{j}\omega)\|$ are roughly described by $\varphi(\omega)$, whereas $\varphi(\omega)$ yields a precise representation of the plot $\|\boldsymbol{F}_{\mathrm{zs}2}(\mathrm{j}\omega)\|$.

The following example emphasises this issue.

Example 6.1 *Accuracy of the physical interaction model*

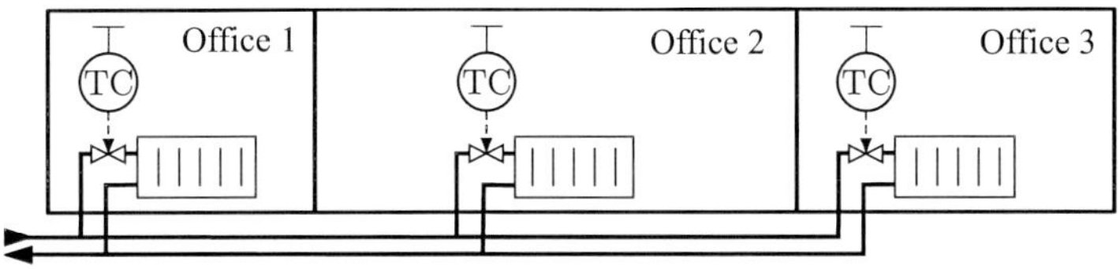

Figure 6.4: Example of a temperature control in three rooms.

Consider the temperature control of three adjacent offices as shown in Fig. 6.4. The offices are of different size and are equipped with different heaters. Each room represents a subsystem with the state

$$x_1(t) = \vartheta_1(t), \quad x_2(t) = \vartheta_2(t), \quad x_3(t) = \vartheta_3(t).$$

The control input $u_i(t)$ is the angle of the heating valve. The temperature dynamics are modelled by (2.6), where

$$\begin{aligned} A_1 &= -2, & B_1 &= E_1 = 1.4, & C_1 &= C_{z1} = 1 \\ A_2 &= -0.05, & B_2 &= E_2 = 0.02, & C_2 &= C_{z2} = 1 \\ A_3 &= -0.33, & B_3 &= E_3 = 0.18, & C_3 &= C_{z3} = 1. \end{aligned}$$

The heat transfer through the thin walls is reflected by (2.8) with

$$L_{12} = L_{21} = L_{23} = L_{32} = -3.$$

Each office is equipped with a thermostat valve (TC in Fig. 6.4) which have proportional characteristics and, hence, are modelled by (2.11) with

$$D_{C1} = 3, \qquad D_{C2} = 6 \qquad D_{C3} = 6$$

in order that each controlled subsystem is I/O stable and the local claims (6.3) are satisfied with $\varphi(\omega)=1$. According to Lemma 6.1, the comparison system

$$\bar{P}_1^{\{1\}}: \quad \bar{s}_1(\mathrm{j}\omega) = 3 \cdot |z_1(\omega)| \tag{6.5}$$

results, where $\bar{P}_1^{\{1\}}(\omega) = \|L_{12}\| = 3$ is determined with the locally known models by D_1.

Figure 6.5 highlights the upper bound $\bar{P}_1^{\{1\}}(\omega)$ in comparison with the actual dynamics $\|P_1^{\{1,2,3\}}(\mathrm{j}\omega)\|$ of the physical interaction. The upper bound only represents a very rough description of the actual dynamics. □

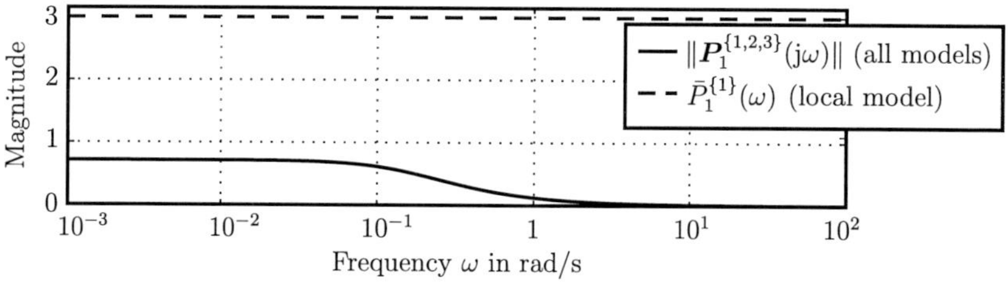

Figure 6.5: Room temperature control: Locally determined upper bound.

Determination of the design function $\varphi(\omega)$. This paragraph states some remarks on the determination of the function $\varphi(\omega)$ if it is not initially known to D_1.

First, focus on the situation in which a nominal controller exists as considered in the application scenarios proposed in Section 1.2.2. Moreover, let the nominal controller satisfy the relation $\|F_{zsi}(\mathrm{j}\omega)\| \sum_{j\in\mathcal{P}_i} \|L_{ij}\| < 1$ for all $i \in \mathcal{N}$. For this situation, the function $\varphi(\omega)$ can be determined online by the particular design agent D_1 using the local search algorithm Algorithm 4.1. Only the functions $\|\boldsymbol{F}_{\mathrm{zs}i}(\mathrm{j}\omega)\| \sum_{j\in\mathcal{P}_i} \|\boldsymbol{L}_{ij}\|$ need to be exchanged so that the function $\varphi(\omega)$ can be determined to

$$1 \geq \varphi(\omega) \geq \sup_{\omega}\Big(\|\boldsymbol{F}_{\mathrm{zs}i}(\mathrm{j}\omega)\| \sum_{j\in\mathcal{P}_i} \|\boldsymbol{L}_{ij}\|,\ \forall i \in \mathcal{N}_{\mathrm{S1}}\Big),$$

where $\mathcal{N}_{\mathrm{S1}}$ collects all indices of the controlled subsystems F_i that are strongly connected to subsystem S_1 (cf. eqn (4.2)). If no initial controller exists, a possible approach is that each design agent individually fixes a local function $\varphi_i(\omega) \leq 1$, $(i \in \mathcal{N})$ on the basis of the subsystem dynamics $\|\boldsymbol{S}_{\mathrm{zs}i}(\mathrm{j}\omega)\| \sum_{j\in\mathcal{P}_i} \|\boldsymbol{L}_{ij}\|$. Hence, to determine $\varphi(\omega)$ for the design of the particular control station C_1, design agent D_1 runs Algorithm 4.1 to procure the local bounds $\varphi_i(\omega)$ from all subsystems that are strongly connected to subsystem S_1 and calculates $\varphi(\omega)$ according to

$$1 \geq \varphi(\omega) \geq \sup_{\omega}\Big(\varphi_i(\omega),\ \forall i \in \mathcal{N}_{\mathrm{S1}}\Big).$$

This approach is reasonable from the technical point of view. Due to the consideration of technical limitations, the dynamics of a controlled process are usually close to the dynamics of the process itself.

Throughout this thesis less attention is paid on the determination of $\varphi(\omega)$, which leads to the following assumption.

A 6.1 *The function $\varphi(\omega)$ is known by all design agents D_i, $(\forall i \in \mathcal{N})$.*

6.3 Local design conditions

This section, first, proposed local design conditions to guarantee I/O stability of the overall closed-loop system. Thereafter, local performance conditions are derived to guarantee a certain performance of the overall closed-loop system.

6.3.1 I/O stability conditions

This section presents local design conditions to guarantee I/O stability of the overall closed-loop system.

Consider the overall closed-loop system

$$F = \mathrm{con}\left(\{F_i,\, i \in \mathcal{N}\} \cup \{K_i,\, i \in \mathcal{N}\},\ \boldsymbol{w},\ \boldsymbol{y}\right)$$

that is represented by the frequency domain model

$$F: \begin{array}{l} \boldsymbol{y}(s) = \operatorname{diag}\left(\boldsymbol{F}_{\mathrm{yw}i}(s)\right)_{i\in\mathcal{N}} \boldsymbol{w}(s) + \\ \quad \operatorname{diag}\left(\boldsymbol{F}_{\mathrm{ys}i}(s)\right)_{i\in\mathcal{N}} \boldsymbol{L}\left(\mathbf{I} - \operatorname{diag}\left(\boldsymbol{F}_{\mathrm{zs}i}(s)\right)_{i\in\mathcal{N}} \boldsymbol{L}\right)^{-1} \operatorname{diag}\left(\boldsymbol{F}_{\mathrm{zw}i}(s)\right)_{i\in\mathcal{N}} \boldsymbol{w}(s). \end{array} \tag{6.6}$$

The first addend of (6.6) represents the behaviour of the isolated overall closed-loop system, whereas the second addend reflects the effect of the physical interaction among the controlled subsystems. Essentially, the return difference matrix $\mathbf{I} - \operatorname{diag}\left(\boldsymbol{F}_{\mathrm{zs}i}(s)\right)_{i\in\mathcal{N}} \boldsymbol{L}$ indicates existing feedback structures among the controlled subsystems through the physical couplings. Due to this feedback structure I/O stability of the overall closed loop can be jeopardised although the individual controlled subsystems F_i are I/O stable. According to [116], a necessary and sufficient condition $\mathcal{A}$ for I/O stability of the overall closed-loop system comprises the following two requirements:

$$\mathcal{A}: \begin{cases} 1.\ \text{I/O stability of the controlled subsystems } F_i,\ \forall i \in \mathcal{N} & (6.7a) \\ 2.\ \text{the Nyquist plot} \\ \qquad \det\left(\mathbf{I} - \operatorname{diag}\left(\boldsymbol{F}_{\mathrm{zs}i}(\mathrm{j}\omega)\right)_{i\in\mathcal{N}} \boldsymbol{L}\right),\ \forall \omega \in \mathbb{R} \\ \quad \text{does not encircle the origin of the complex plane.} & (6.7b) \end{cases}$$

If the amplification of all frequency response matrices $\boldsymbol{F}_{\mathrm{zs}i}(s)$, $(\forall i \in \mathcal{N})$ are small as claimed in Lemma 6.1, I/O stability of the overall closed-loop system can be guaranteed. Accordingly, the local stability condition $\mathcal{A}_i$ for the design of C_i consists of two claims which are:

$$\mathcal{A}_i: \begin{cases} 1.\ \text{I/O stability of the controlled subsystem } F_i & (6.8a) \\ 2.\ \text{the local amplification of the coupling output is limited according to} & (6.8b) \\ \qquad \|\boldsymbol{F}_{\mathrm{zs}i}(\mathrm{j}\omega)\| \sum_{j\in\mathcal{P}_i} \|\boldsymbol{L}_{ij}\| < 1, \quad \forall \omega \in \mathbb{R}. \end{cases}$$

The first claim (6.8a) is taken without changes from (6.7a), whereas the second claim (6.8b) grounds on Lemma 6.1. The relation between the local stability condition $\mathcal{A}_i$ and I/O stability of the overall closed-loop system is stated in the following theorem.

Theorem 6.1 (Local design conditions to guarantee global I/O stability) *If all controlled subsystems F_i, $(\forall i \in \mathcal{N})$ satisfy the corresponding local stability condition $\mathcal{A}_i$ defined in (6.8), then the overall closed-loop system F is I/O stable.*

Proof. Let the controlled subsystems F_i, $(\forall i \in \mathcal{N})$ be I/O stable by design. Based on the equality

$$\det\left(\mathbf{I} - \operatorname{diag}(\boldsymbol{F}_{\mathrm{zs}i}(\mathrm{j}\omega))_{i\in\mathcal{N}}\boldsymbol{L}\right) = \prod_{k=1}^{\sum_{i=1}^{N} r_{\mathrm{z}i}} \left(1 - \lambda_k\left(\operatorname{diag}(\boldsymbol{F}_{\mathrm{zs}i}(\mathrm{j}\omega))_{i\in\mathcal{N}}\boldsymbol{L}\right)\right),$$

where $r_{\mathrm{z}i}$ denotes the dimension of the coupling output $\boldsymbol{z}_i$, the following implications hold:

$$(6.7\mathrm{b})$$

$$\Uparrow$$

$$\mathrm{Re}\left(\prod_{k=1}^{\sum_{i=1}^{N} r_{\mathrm{z}i}} \left(1 - \lambda_k(\operatorname{diag}\left(\boldsymbol{F}_{\mathrm{zs}i}(\mathrm{j}\omega))_{i\in\mathcal{N}}\boldsymbol{L}\right)\right)\right) > 0 \tag{6.9}$$

$$\Uparrow$$

$$\max_k \left|\lambda_k\left(\operatorname{diag}(\boldsymbol{F}_{\mathrm{zs}i}(\mathrm{j}\omega))_{i\in\mathcal{N}}\boldsymbol{L}\right)\right| = \rho\left(\operatorname{diag}(\boldsymbol{F}_{\mathrm{zs}i}(\mathrm{j}\omega))_{i\in\mathcal{N}}\boldsymbol{L}\right) < 1 \tag{6.10}$$

$$\Uparrow$$

$$\lambda_{\mathrm{P}}\left(\left|\operatorname{diag}(\boldsymbol{F}_{\mathrm{zs}i}(\mathrm{j}\omega))_{i\in\mathcal{N}}\boldsymbol{L}\right|\right) < 1 \tag{6.11}$$

$$\Updownarrow$$

$$\left(\mathbf{I} - \left|\operatorname{diag}(\boldsymbol{F}_{\mathrm{zs}i}(\mathrm{j}\omega))_{i\in\mathcal{N}}\boldsymbol{L}\right|\right) \text{ is an M-matrix} \tag{6.12}$$

$$\Uparrow$$

$$\sum_{j\in\mathcal{P}_i} \|\boldsymbol{F}_{\mathrm{zs}i}(\mathrm{j}\omega)\boldsymbol{L}_{ij}\| < 1, \quad \forall i \in \mathcal{N} \tag{6.13}$$

$$\Uparrow$$

$$(6.8\mathrm{b}), \quad \forall i \in \mathcal{N}$$

Condition (6.9) requires the Nyquist plot of $\det\left(\mathbf{I} - \operatorname{diag}\left(\boldsymbol{F}_{\mathrm{zs}i}(\mathrm{j}\omega)\right)_{i\in\mathcal{N}}\boldsymbol{L}\right)$ always to stay in the right-half plane, which is implied by (6.10) forcing the maximal absolute eigenvalue (or equivalently the spectral radius ρ) to be less than one as claimed by (6.10).

In the Theorem A1.2 in [116], it has been shown that for an arbitrary square matrix $\boldsymbol{A}$ the relation

$$\rho(\boldsymbol{A}) \leq \lambda_{\mathrm{P}}(|\boldsymbol{A}|) \tag{6.14}$$

holds, where λ_{P} denotes the Perron root of a matrix. Due to the relation (6.14), the implication from (6.11) to (6.10) results.

The equivalence between condition (6.11) and (6.12) has been stated in Theorem A1.6 of

[116]. The square matrix

$$\mathbf{I} - \left|\mathrm{diag}(\boldsymbol{F}_{\mathrm{zs}i}(\mathrm{j}\omega))_{i\in\mathcal{N}}\boldsymbol{L}\right| = \begin{pmatrix} \mathbf{I} & -|\boldsymbol{F}_{\mathrm{zs}1}(\mathrm{j}\omega)\boldsymbol{L}_{12}| & \cdots & -|\boldsymbol{F}_{\mathrm{zs}1}(\mathrm{j}\omega)\boldsymbol{L}_{1N}| \\ -|\boldsymbol{F}_{\mathrm{zs}2}(\mathrm{j}\omega)\boldsymbol{L}_{21}| & \mathbf{I} & \cdots & -|\boldsymbol{F}_{\mathrm{zs}2}(\mathrm{j}\omega)\boldsymbol{L}_{2N}| \\ \vdots & & \ddots & \vdots \\ -|\boldsymbol{F}_{\mathrm{zs}N}(\mathrm{j}\omega)\boldsymbol{L}_{N1}| & -|\boldsymbol{F}_{\mathrm{zs}N}(\mathrm{j}\omega)\boldsymbol{L}_{N2}| & \cdots & \mathbf{I} \end{pmatrix}, \tag{6.15}$$

where all nondiagonal elements are nonnegative, is called an M-matrix if all its eigenvalues have positive real part. Based on Geršgorin's theorem [78, 183], the eigenvalues of the matrix (6.15) have positive real part if the diagonal dominance condition (6.13) holds. This condition is satisfied if (6.8b) holds for all $i \in \mathcal{N}$. Accordingly, if the local condition $\mathcal{A}_i$ is satisfied by all controlled subsystems F_i, $(\forall i \in \mathcal{N})$, the global stability condition $\mathcal{A}$ is satisfied and, thus, the overall closed-loop system is I/O stable. The proof is completed. □

The implications from (6.8b) to (6.9) shown in the proof of Theorem 6.1 indicate the conservatism in the local design condition $\mathcal{A}_i$. The severity of the conservatism depends amongst others on two things which are oulined in the following:

1. The number and the strength of the physical couplings: The smaller the number of neighbours and their coupling strength to a particular subsystem is, the less restrictive becomes the local design condition (6.8b). Moreover, Geršgorin's theorem states that the eigenvalues of the matrix (6.15) are located within N circles with the centre 1 and the radii $\sum_{j\in\mathcal{P}_i} \|\boldsymbol{F}_{\mathrm{zs}i}(\mathrm{j}\omega)\boldsymbol{L}_{ij}\|$. Accordingly, the smaller the number of neighbours and their coupling strength to a subsystem, the smaller becomes the uncertainty about the location of the eigenvalues of the matrix (6.15). As a consequence, the evaluation of the statement (6.12) by means of (6.13) is less conservative.

2. The individual dynamics of the controlled subsystems (cf. Section 6.2): This aspect takes effect if the overall system behaviour is analysed by means of a scalar value such as in conditions (6.10) and (6.11). For instance, if the real parts of (6.9) are close together, than the maximal eigenvalue (6.10) is an acceptable representation. Moreover, if the plots $\|\boldsymbol{F}_{\mathrm{zs}i}(\mathrm{j}\omega)\| \sum_{j\in\mathcal{P}_i} \|\boldsymbol{L}_{ij}\|$, $(\forall i \in \mathcal{N})$ are close to 1 for all ω, then the diagonal dominance criterion (6.8b) is less restrictive. In addition, if single input single output systems are considered (i.e., $F_{\mathrm{zs}i}(s) \in \mathbb{C}$, $(\forall i \in \mathcal{N})$), condition (6.8b) claims the attenuation of the the amplitude plot $|F_{\mathrm{zs}i}(\mathrm{j}\omega)|$, in contrast to the limitation of the norm $\|\boldsymbol{F}_{\mathrm{zs}i}(\mathrm{j}\omega)\|$. In this case, the conditions (6.13) and (6.8b) are equivalent.

The following example attains a better understanding about the conservatism of the local stability conditions $\mathcal{A}_i$.

Example 6.2 *Conservatism of the local stability condition*

This example compares the local stability condition (6.8) with the global stability condition (6.7) with the aim to emphasise the conservatism within the local condition.

Consider the three offices from Example 6.1 whose room temperature shall be controlled. The solid lines in Fig. 6.6a that the local stability conditions are satisfied. This implies the satisfaction of the global stability criterion visualised by the solid lined Nyquist plot in Fig. 6.6b.

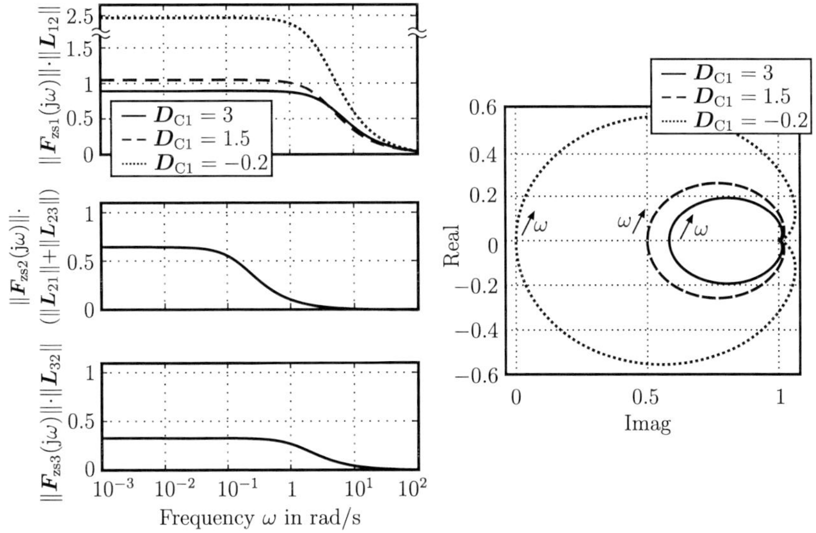

(a) Amplitude plot for the verification of the local stability condition

(b) Nyquist plot of $\det\left(\mathbf{I} - \operatorname{diag}\left(F_{zsi}(j\omega)\right)_{i\in\mathcal{N}} L\right)$

Figure 6.6: Room temperature control: Comparison of the global and the local stability condition.

Now, the thermostat valve is replaced by a new one which has the proportional gain $D_{C1} = 1.5$. This exchange has the effect that the local claim $\mathcal{A}_1$ is no longer satisfied as highlighted by the dashed lines in Fig. 6.6a. Thus, the stability of the overall closed-loop system can no longer be guaranteed if it is verified by D_1 based on the local conditions. However, the global claim is still satisfied as shown in Fig. 6.6b by the dashed lined Nyquist plot.

Moreover, Figure 6.6a highlights that the local amplification can be increased up to $\max_\omega(\|F_{zs1}(j\omega)\| \cdot \|L_{12}\|) \approx 2.5$ (corresponding to the feedback gain $D_{C1} = -0.2$) until the overall closed-loop system becomes unstable depicted by the dotted lined Nyquist plot in Fig. 6.6b. □

Remark 6.1 *The condition* (6.8b) *can be interpreted as a small-gain condition for physically interconnected systems. Similar small-gain conditions have been derived in [117] and [65]. [117] states a sufficient condition for I/O stability of continuous-time systems whereas [65] presents a sufficient condition for asymptotic stability of discrete-time systems. Both conditions* *are similar to the claim* (6.11) *with the consequence that some piece of model information from each controlled subsystem is required. Based on the condition from [65], a sufficient design condition which can be checked locally has been derived in [145] by means of a diagonal dominance criterion. This condition is similar to the condition* (6.8b) *presented in this thesis.*

6.3.2 I/O performance conditions

First, the performance condition of the overall closed-loop system is presented followed by the locally checkable performance condition.

The overall closed-loop performance is characterised by a desired individual performance of each controlled subsystem despite the dependence among each other through the physical couplings. In particular, the dynamics of a controlled subsystem F_i under consideration of the physical interaction $P_i^{\mathcal{N}}$, that are modelled by

$$F_i^{\mathcal{N}}: \; \boldsymbol{y}_i(s) = \underbrace{\left(\boldsymbol{F}_{\mathrm{yw}i}(s) + \boldsymbol{F}_{\Delta i}^{\mathcal{N}}(s)\right)}_{= \boldsymbol{F}_i^{\mathcal{N}}(s)} \, \boldsymbol{w}_i(s),$$

are separated into the isolated controlled subsystem

$$F_i|_{\boldsymbol{s}_i=\boldsymbol{0}}: \quad \hat{\boldsymbol{y}}_i(s) = \boldsymbol{F}_{\mathrm{yw}i}(s)\, \boldsymbol{w}_i(s)$$

and the error model

$$F_{\Delta i}^{\mathcal{N}}: \quad \boldsymbol{y}_{\Delta i}(s) = \boldsymbol{F}_{\Delta i}^{\mathcal{N}}(s)\, \boldsymbol{w}_i(s)$$

with

$$\boldsymbol{F}_{\Delta i}^{\mathcal{N}}(s) = \boldsymbol{F}_{\mathrm{ys}i}(s)\, \boldsymbol{P}_i^{\mathcal{N}}(s)\left(\mathbf{I} - \boldsymbol{F}_{\mathrm{zs}i}(s)\, \boldsymbol{P}_i^{\mathcal{N}}(s)\right)^{-1} \boldsymbol{F}_{\mathrm{zw}i}(s).$$

The overall closed-loop performance is characterised by the adherence of given dynamical requirements on the isolated controlled subsystems $F_i|_{\boldsymbol{s}_i=\boldsymbol{0}}$, $(\forall i \in \mathcal{N})$, while the effects of the error system $F_{\Delta i}^{\mathcal{N}}$, $(\forall i \in \mathcal{N})$ need to be attenuated. These claims comprise the the global performance condition $\mathcal{A}_{\mathrm{D}}$ which reads as follows:

$\mathcal{A}_{\mathrm{D}}:$

1. I/O stability of the overall closed-loop system F (6.16a)
2. satisfaction of given claims on the I/O behaviour of all isolated controlled subsystems $F_i|_{\boldsymbol{s}_i=\boldsymbol{0}}$, $\quad \forall i \in \mathcal{N}$ (6.16b)
3. for N given parameters $\varepsilon_i \in \mathbb{R}_+$ the influence of the physical interaction is limited according to (6.16c)

$$\|\boldsymbol{F}_{\Delta i}^{\mathcal{N}}(\mathrm{j}\omega)\| \leq \varepsilon_i, \quad \forall \omega \in \mathbb{R}, \quad \forall i \in \mathcal{N}$$

The desired I/O behaviour of the isolated controlled subsystem is reflected by the condition (6.16b). The admissible deviation $\|\boldsymbol{F}_{\Delta i}^{\mathcal{N}}(\mathrm{j}\omega)\|$ to the desired I/O behaviour can be adjusted by the design parameter ε_i. However, the construction of the transfer function matrix $\boldsymbol{F}_{\Delta i}^{\mathcal{N}}(s)$ requires the knowledge of all model sets $\mathcal{M}_i = \{S_i, C_i, K_i\}$, $(\forall i \in \mathcal{N})$.

As presented in Theorem 6.1, the claim (6.16a) is fulfilled if the local condition $\mathcal{A}_i$ is satisfied by all controlled subsystems. As a consequence, the dynamics $\boldsymbol{P}_i^{\mathcal{N}}(s)$ are bounded from above by $\bar{P}_i^{\{i\}}(\omega)$ and so are the dynamics $\boldsymbol{F}_{\Delta i}^{\mathcal{N}}(s)$, where the upper bound results to

$$\|\boldsymbol{F}_{\Delta i}^{\mathcal{N}}(\mathrm{j}\omega)\| < \bar{F}_{\Delta i}^{\{i\}}(\omega) = \|\boldsymbol{F}_{\mathrm{ys}i}(\mathrm{j}\omega)\| \sum_{j\in\mathcal{P}_i} \|\boldsymbol{L}_{ij}\|\,\varphi(\omega)\cdot \left(\mathbf{I} - \|\boldsymbol{F}_{\mathrm{zs}i}(\mathrm{j}\omega)\| \sum_{j\in\mathcal{P}_i} \|\boldsymbol{L}_{ij}\|\,\varphi(\omega)\right)^{-1} \|\boldsymbol{F}_{\mathrm{zw}i}(\mathrm{j}\omega)\|. \tag{6.17}$$

Accordingly, if $\bar{F}_{\Delta i}^{\{i\}}(\omega) \leq \varepsilon_i$ holds for all ω, then the claim (6.16c) is satisfied. Based on this deliberations, the local performance condition $\mathcal{A}_{\mathrm{D}i}$ can be formulated by four requirements which are:

$\mathcal{A}_{\mathrm{D}i}$:

1. I/O stability of the controlled subsystem F_i (6.18a)
2. for the given function $\varphi(\omega)\colon \mathbb{R} \to \mathbb{R}_+$ with $\varphi(\omega) \leq 1$ the local amplification of the coupling output is limited according to (6.18b)
   $$\textstyle\sum_{j\in\mathcal{P}_i} \|\boldsymbol{F}_{\mathrm{zs}i}(\mathrm{j}\omega)\boldsymbol{L}_{ij}\| < \varphi(\omega), \quad \forall\omega \in \mathbb{R}$$
3. satisfaction of given claims on the I/O behaviour of the isolated controlled subsystem $F_i|_{\boldsymbol{s}_i=\boldsymbol{0}}$ (6.18c)
4. for a given parameter $\varepsilon_i \in \mathbb{R}_+$ the influence of the physical interaction is limited according to (6.18d)
   $$\bar{F}_{\Delta i}^{\{i\}}(\omega) \leq \varepsilon_i, \quad \forall\omega \in \mathbb{R}$$

In conclusion, the relation between the local performance conditions $\mathcal{A}_{\mathrm{D}i}$ and the global performance $\mathcal{A}_{\mathrm{D}}$ is stated in the following theorem.

Theorem 6.2 (Local design condition for global I/O performance) *If all controlled subsystems F_i, $(\forall i \in \mathcal{N})$ satisfy the corresponding local performance condition $\mathcal{A}_{\mathrm{D}i}$ defined in* (6.18)*, then the overall closed-loop system F has a performance described by $\mathcal{A}_{\mathrm{D}}$ defined in* (6.16).

Based on the locally known upper bound $\bar{F}_{\Delta i}^{\{i\}}(\mathrm{j}\omega)$ a tube can be determined which encloses the amplitude plot $\|\boldsymbol{F}_i^{\mathcal{N}}(\mathrm{j}\omega)\|$. The tube results to

$$\mathbb{T}_i^{\{i\}}(\omega) := \left(\|\boldsymbol{F}_{\mathrm{yw}i}(\mathrm{j}\omega)\| - \bar{F}_{\Delta i}^{\{i\}}(\omega), \quad \|\boldsymbol{F}_{\mathrm{yw}i}(\mathrm{j}\omega)\| + \bar{F}_{\Delta i}^{\{i\}}(\omega) \right), \tag{6.19}$$

where $\|\boldsymbol{F}_i^{\mathcal{N}}(\mathrm{j}\omega)\| = \|\boldsymbol{F}_{\mathrm{yw}i}(\mathrm{j}\omega) + \boldsymbol{F}_{\Delta i}^{\mathcal{N}}(\mathrm{j}\omega)\| \in \mathbb{T}_i^{\{i\}}(\omega)$. That is,

$$\|\boldsymbol{F}_{\mathrm{yw}i}(\mathrm{j}\omega)\| - \bar{F}_{\Delta i}^{\{i\}}(\omega) < \|\boldsymbol{F}_i^{\mathcal{N}}(\mathrm{j}\omega)\| < \|\boldsymbol{F}_{\mathrm{yw}i}(\mathrm{j}\omega)\| + \bar{F}_{\Delta i}^{\{i\}}(\omega).$$

According to the design condition (6.18d), the tuning parameter ε_i enables the adjustment of the tube width. The tube can be calculated with local models only.

Similar to the local stability condition $\mathcal{A}_i$, the local performance condition $\mathcal{A}_{\mathrm{D}i}$ is also conservative due to the rough description of the interacting dynamics $\boldsymbol{P}_i^{\mathcal{N}}(s)$ by the upper bound $\bar{P}_i^{\{i\}}(\omega)$. The following example illustrates this conclusion by means of the size of the tube $\mathbb{T}_i^{\{i\}}(\omega)$.

Example 6.3 *Conservatism of the local performance condition*

Consider once more the three offices from Example 6.1. As it has been shown in Example 6.1, the upper bound $\bar{P}_1^{\{1\}}$ results to (6.5). In accordance with equation (6.17), the upper bound $\bar{F}_1^{\{1\}}(\omega)$ can be determined. Based on this upper bound, the tube $\mathbb{T}_1^{\{1\}}(\omega)$ can be derived as highlighted in Fig. 6.7. Due to the rough description of the interaction dynamics, the actual behaviour $\|\boldsymbol{F}_1^{\{1,2,3\}}(\mathrm{j}\omega)\|$ is only known with a great uncertainty. of the controlled subsystem under the consideration of the physical influence arises.

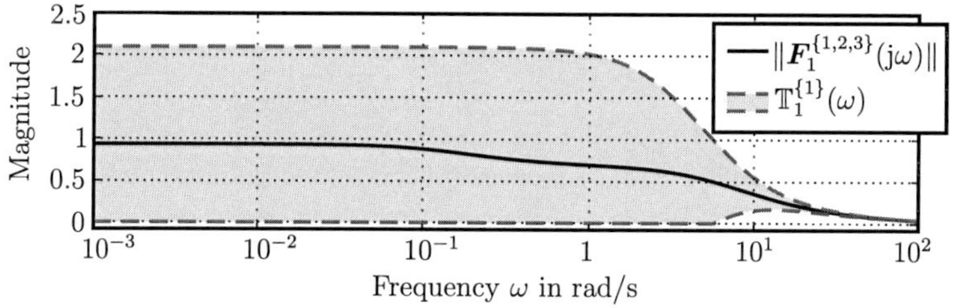

Figure 6.7: Room temperature control: Locally determined tube $\mathbb{T}_1^{\{1\}}(\omega)$.

Although the tube width is adjustable by the parameter ε_i, it is shown in Section 7.3.3 that the consideration of further model information will significantly narrow the tube. □

6.4 H_∞-controller design

This section proposes an H_∞-controller design regarding the local performance condition $\mathcal{A}_{\mathrm{D}i}$. LMIs are derived to efficiently solve the H_∞-optimal problem. First, the local performance conditions are transformed into H_∞-norm requirements on the transfer functions of the controlled subsystem F_i. These requirements are finally transformed into LMIs summarised in Lemma 2.2.

The subsystem to be controlled is represented by the state-space model

$$S_i : \begin{cases} \dot{\boldsymbol{x}}_i(t) &= \boldsymbol{A}_i \boldsymbol{x}_i(t) + \boldsymbol{b}_i \boldsymbol{u}_i(t) + \boldsymbol{e}_i s_i(t), \quad \boldsymbol{x}_i(0) = \boldsymbol{0} \\ \boldsymbol{y}_i(t) &= \boldsymbol{C}_i \boldsymbol{x}_i(t) \\ z_i(t) &= \boldsymbol{c}_{\mathrm{z}i}^\top \boldsymbol{x}_i(t), \end{cases} \tag{6.20}$$

where $s_i, z_i \in \mathbb{R}$. The equivalent frequency-domain representation is

$$S_i : \begin{cases} \boldsymbol{y}_i(t) &= \boldsymbol{S}_{\mathrm{yu}i}(s) \boldsymbol{u}_i(s) + \boldsymbol{S}_{\mathrm{ys}i}(s) s_i(s) \\ z_i(t) &= \boldsymbol{S}_{\mathrm{zu}i}(s) \boldsymbol{u}_i(s) + S_{\mathrm{zs}i}(s) s_i(s), \end{cases} \tag{6.21}$$

where

$$\boldsymbol{S}_{\mathrm{yu}i}(s) = \boldsymbol{C}_i \big(s\mathbf{I} - \boldsymbol{A}_i\big)^{-1} \boldsymbol{B}_i \in \mathbb{C}^{r_i \times m_i}, \qquad \boldsymbol{S}_{\mathrm{ys}i}(s) = \boldsymbol{C}_i \big(s\mathbf{I} - \boldsymbol{A}_i\big)^{-1} \boldsymbol{e}_i \in \mathbb{C}^{r_{\mathrm{z}i} \times 1}, \tag{6.22}$$

$$\boldsymbol{S}_{\mathrm{zu}i}(s) = \boldsymbol{c}_{\mathrm{z}i}^\top \big(s\mathbf{I} - \boldsymbol{A}_i\big)^{-1} \boldsymbol{B}_i \in \mathbb{C}^{1 \times m_i}, \qquad S_{\mathrm{zs}i}(s) = \boldsymbol{c}_{\mathrm{z}i}^\top \big(s\mathbf{I} - \boldsymbol{A}_i\big)^{-1} \boldsymbol{e}_i \in \mathbb{C}. \tag{6.23}$$

The model of the control station to be designed is represented in frequency domain by

$$C_i : \boldsymbol{u}_i(s) = \boldsymbol{C}_i(s) \left(\boldsymbol{w}_i(s) - \boldsymbol{y}_i(s)\right) \tag{6.24}$$

where $\boldsymbol{C}_i(s) = \boldsymbol{D}_{\mathrm{C}i} + \boldsymbol{C}_{\mathrm{C}i}\big(s\mathbf{I} - \boldsymbol{A}_{\mathrm{C}i}\big)^{-1} \boldsymbol{B}_{\mathrm{C}i}$. Hence, for the case of scalar coupling signals, the model of the controlled subsystem reads

$$F_i : \begin{cases} \boldsymbol{y}_i(t) &= \boldsymbol{F}_{\mathrm{yw}i}(s) \boldsymbol{w}_i(s) + \boldsymbol{F}_{\mathrm{ys}i}(s) s_i(s) \\ z_i(t) &= \boldsymbol{F}_{\mathrm{zw}i}(s) \boldsymbol{w}_i(s) + F_{\mathrm{zs}i}(s) s_i(s). \end{cases}$$

where, $\boldsymbol{F}_{\mathrm{yw}i}(s) \in \mathbb{C}^{r_i \times r_i}$, $\boldsymbol{F}_{\mathrm{ys}i}(s) \in \mathbb{C}^{r_{\mathrm{z}i} \times 1}$, $\boldsymbol{F}_{\mathrm{zw}i}(s) \in \mathbb{C}^{1 \times r_i}$, $F_{\mathrm{zs}i}(s) \in \mathbb{C}$.

Reformulation of the local performance conditions. In this paragraph H_∞-norm conditions are derived, which imply the satisfaction of the local performance condition $\mathcal{A}_{\mathrm{D}i}$. For the conditions (6.18b) and (6.18c) equivalent H_∞-norm conditions are formulated, while for conditions (6.18a) and (6.18d) sufficient H_∞-norm conditions are proposed.

It is started with the second condition (6.18b) of $\mathcal{A}_{\mathrm{D}i}$, which can be reformulated into the equivalent expression

$$\|W_{\mathrm{z}}(\mathrm{j}\omega)F_{\mathrm{zs}i}(\mathrm{j}\omega)\|_{\mathrm{H}_\infty} < \left(\sum_{j\in\mathcal{P}_i}\|\boldsymbol{L}_{ij}\|\right)^{-1}, \tag{6.25}$$

where the scalar transfer function $W_{\mathrm{z}}(s)$ is derived from the state-space model

$$W_{\mathrm{z}i}: \begin{cases} \dot{\boldsymbol{x}}_{\mathrm{Wz}i}(t) &= \boldsymbol{A}_{\mathrm{Wz}}\boldsymbol{x}_{\mathrm{Wz}i}(t) + \boldsymbol{b}_{\mathrm{Wz}}\boldsymbol{z}_i(t), \quad \boldsymbol{x}_{\mathrm{Wz}i}(0) = \boldsymbol{0} \\ v_{\mathrm{z}i}(t) &= \boldsymbol{c}_{\mathrm{Wz}}^{\top}\boldsymbol{x}_{\mathrm{Wz}i}(t) + d_{\mathrm{Wz}}\boldsymbol{z}_i(t), \end{cases} \tag{6.26}$$

with $\boldsymbol{x}_{\mathrm{Wz}i} \in \mathbb{R}^{n_{\mathrm{Wz}}}$. Accordingly, $W_{\mathrm{z}}(s) = d_{\mathrm{Wz}} + \boldsymbol{c}_{\mathrm{Wz}}^{\top}(s\mathbf{I} - \boldsymbol{A}_{\mathrm{z}})^{-1}\boldsymbol{b}_{\mathrm{Wz}}$, where the parameters of the model (6.26) has to be chosen such that the equality

$$|W_{\mathrm{z}}(\mathrm{j}\omega)| = (\varphi(\omega))^{-1}, \quad \forall\omega \in \mathbb{R}$$

holds for the given function $\varphi(\omega)$. The condition (6.25) represents a claim on the H_∞-norm of the extended subsystem with the scalar input $s_i(s)$ and the scalar output $v_{\mathrm{z}i}(s)$, shown in Fig. 6.8. Note that the equivalence between the claims (6.18b) and (6.25) hold for scalar coupling signals.

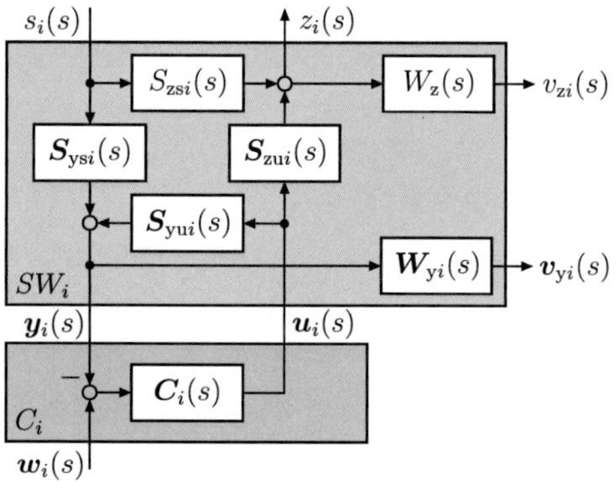

Figure 6.8: Subsystem extended by weight functions.

A desired I/O behaviour of the isolated controlled subsystem (cf. condition (6.18c)) can be specified by the introduction of further weight matrices as mentioned in Section 2.3.2. Accordingly, the I/O dynamics $\boldsymbol{F}_{\mathrm{yw}i}(s)$ are shaped by

$$\|\boldsymbol{W}_{\mathrm{y}i}(s)\boldsymbol{F}_{\mathrm{yw}i}(s)\|_{\mathrm{H}_\infty} < \gamma_{\mathrm{yw}i}, \tag{6.27}$$

where the weight function $\boldsymbol{W}_{\mathrm{y}i}(s)$ is derived from the state-space model

$$W_{\mathrm{y}i}: \begin{cases} \dot{\boldsymbol{x}}_{\mathrm{Wy}i}(t) = \boldsymbol{A}_{\mathrm{Wy}i}\boldsymbol{x}_{\mathrm{Wy}i}(t) + \boldsymbol{B}_{\mathrm{Wy}i}\boldsymbol{y}_i(t), \quad \boldsymbol{x}_{\mathrm{Wy}i}(0) = \boldsymbol{0} \\ \boldsymbol{v}_{\mathrm{y}i}(t) = \boldsymbol{C}_{\mathrm{Wy}i}\boldsymbol{x}_{\mathrm{Wy}i}(t) + \boldsymbol{D}_{\mathrm{Wy}i}\boldsymbol{y}_i(t), \end{cases} \tag{6.28}$$

with $\boldsymbol{x}_{\mathrm{Wy}i} \in \mathbb{R}^{n_{\mathrm{Wy}i}}$, $\boldsymbol{v}_{\mathrm{y}i} \in \mathbb{R}^{r_{\mathrm{Wy}i}}$ and $\boldsymbol{W}_{\mathrm{y}i}(s) = \boldsymbol{D}_{\mathrm{Wy}i} + \boldsymbol{C}_{\mathrm{Wy}i}\left(s\mathbf{I} - \boldsymbol{A}_{\mathrm{Wy}i}\right)^{-1}\boldsymbol{B}_{\mathrm{Wy}i}$.

To simultaneously fulfil the H_∞-norm requirements (6.25) and (6.27), the state-space models (6.26) and (6.28) of the weight functions are connected with the subsystem model (6.20) to result to the extended subsystem

$$SW_i = \mathrm{con}\left(\{S_i, W_{\mathrm{y}i}, W_{\mathrm{z}i}\}, \begin{pmatrix} \boldsymbol{u}_i \\ s_i \end{pmatrix}, \begin{pmatrix} \boldsymbol{y}_i \\ z_i \\ v_{\mathrm{z}i} \\ \boldsymbol{v}_{\mathrm{y}i} \end{pmatrix}\right)$$

with the respective state-space model

$$SW_i: \begin{cases} \begin{pmatrix} \dot{\boldsymbol{x}}_i(t) \\ \dot{\boldsymbol{x}}_{\mathrm{Wz}i}(t) \\ \dot{\boldsymbol{x}}_{\mathrm{Wy}i}(t) \end{pmatrix} = \underbrace{\begin{pmatrix} \boldsymbol{A}_i & \mathrm{O} & \mathrm{O} \\ \boldsymbol{b}_{\mathrm{Wz}i}\boldsymbol{c}_{\mathrm{z}i}^\top & \boldsymbol{A}_{\mathrm{Wz}} & \mathrm{O} \\ \boldsymbol{B}_{\mathrm{Wy}i}\boldsymbol{C}_i & \mathrm{O} & \boldsymbol{A}_{\mathrm{Wy}i} \end{pmatrix}}_{\boldsymbol{A}_{\mathrm{SW}i}} \underbrace{\begin{pmatrix} \boldsymbol{x}_i(t) \\ \boldsymbol{x}_{\mathrm{Wz}i}(t) \\ \boldsymbol{x}_{\mathrm{Wy}i}(t) \end{pmatrix}}_{\boldsymbol{x}_{\mathrm{SW}i}(t)} + \underbrace{\begin{pmatrix} \boldsymbol{B}_i \\ \mathrm{O} \\ \boldsymbol{B}_{\mathrm{Wy}i} \end{pmatrix}}_{\boldsymbol{B}_{\mathrm{SW}i}} \boldsymbol{u}_i(t) + \\ \qquad \underbrace{\begin{pmatrix} \boldsymbol{e}_i \\ \boldsymbol{0} \\ \boldsymbol{0} \end{pmatrix}}_{\boldsymbol{e}_{\mathrm{SW}i}} s_i(t), \quad \boldsymbol{x}_{\mathrm{SW}i}(0) = \boldsymbol{0} \\ \boldsymbol{y}_i(t) = \underbrace{\begin{pmatrix} \boldsymbol{C}_i & \mathrm{O} & \mathrm{O} \end{pmatrix}}_{\boldsymbol{C}_{\mathrm{SW}i}} \boldsymbol{x}_{\mathrm{SW}i}(t) \\ z_i(t) = \underbrace{\begin{pmatrix} \boldsymbol{c}_{\mathrm{z}i}^\top & \boldsymbol{0}^\top & \boldsymbol{0}^\top \end{pmatrix}}_{\boldsymbol{c}_{\mathrm{zSW}i}^\top} \boldsymbol{x}_{\mathrm{SW}i}(t) \\ v_{\mathrm{z}i}(t) = \underbrace{\begin{pmatrix} d_{\mathrm{Wz}i}\boldsymbol{c}_{\mathrm{z}i}^\top & \boldsymbol{c}_{\mathrm{Wz}}^\top & \boldsymbol{0}^\top \end{pmatrix}}_{\boldsymbol{c}_{\mathrm{SWz}i}^\top} \boldsymbol{x}_{\mathrm{SW}i}(t) \\ \boldsymbol{v}_{\mathrm{y}i}(t) = \underbrace{\begin{pmatrix} \boldsymbol{D}_{\mathrm{Wy}i}\boldsymbol{C}_i & \mathrm{O} & \boldsymbol{C}_{\mathrm{Wy}i} \end{pmatrix}}_{\boldsymbol{C}_{\mathrm{SWy}i}} \boldsymbol{x}_{\mathrm{SW}i}(t), \end{cases} \tag{6.29}$$

where $\boldsymbol{x}_{\mathrm{SW}i} \in \mathbb{R}^{n_i+n_{\mathrm{Wz}}+n_{\mathrm{Wy}i}}$. The block diagram of the controlled extended subsystem is shown in Fig. 6.8.

Now, it is focussed on the first condition (6.18a) and fourth condition (6.18d) of $\mathcal{A}_{\mathrm{D}i}$. This claim can also be expressed by means of H_∞-norms on the transfer functions $\boldsymbol{F}_{\mathrm{ys}i}(s)$, $F_{\mathrm{zs}i}(s)$ and $\boldsymbol{F}_{\mathrm{zw}i}(s)$ as follows:

$$\begin{aligned}
\bar{F}_{\Delta i}(\omega) &= \|\boldsymbol{F}_{\mathrm{ys}i}(\mathrm{j}\omega)\|_2 \sum_{i\in\mathcal{P}_i} \|\boldsymbol{L}_{ij}\|\varphi(\omega)\left(1 - \|F_{\mathrm{zs}i}(\mathrm{j}\omega)\| \sum_{i\in\mathcal{P}_i}\|\boldsymbol{L}_{ij}\|\,\varphi(\omega)\right)^{-1} \|\boldsymbol{F}_{\mathrm{zw}i}(\mathrm{j}\omega)\|_2 \\
&\leq \|\boldsymbol{F}_{\mathrm{ys}i}(\mathrm{j}\omega)\|_{\mathrm{H}_\infty} \sum_{i\in\mathcal{P}_i} \|\boldsymbol{L}_{ij}\|\varphi(\omega)\left(1 - \|F_{\mathrm{zs}i}(\mathrm{j}\omega)\|_{\mathrm{H}_\infty} \sum_{i\in\mathcal{P}_i}\|\boldsymbol{L}_{ij}\|\,\varphi(\omega)\right)^{-1} \|\boldsymbol{F}_{\mathrm{zw}i}(\mathrm{j}\omega)\|_{\mathrm{H}_\infty} \\
&\leq \gamma_{\mathrm{ys}i} \sum_{i\in\mathcal{P}_i} \|\boldsymbol{L}_{ij}\|\varphi(\omega)\left(1 - \gamma_{\mathrm{zs}i} \sum_{i\in\mathcal{P}_i}\|\boldsymbol{L}_{ij}\|\,\varphi(\omega)\right)^{-1} \gamma_{\mathrm{zw}i},
\end{aligned}$$

where

$$\|\boldsymbol{F}_{\mathrm{ys}i}(\mathrm{j}\omega)\|_{\mathrm{H}_\infty} < \gamma_{\mathrm{ys}i}, \tag{6.30}$$

$$\|F_{\mathrm{zs}i}(\mathrm{j}\omega)\|_{\mathrm{H}_\infty} < \gamma_{\mathrm{zs}i}, \tag{6.31}$$

$$\|\boldsymbol{F}_{\mathrm{zw}i}(\mathrm{j}\omega)\|_{\mathrm{H}_\infty} < \gamma_{\mathrm{zw}i}. \tag{6.32}$$

Accordingly, if the parameters $\gamma_{\mathrm{ys}i}$, $\gamma_{\mathrm{zs}i}$ and $\gamma_{\mathrm{zw}i}$ are chosen such that

$$\gamma_{\mathrm{ys}i} \sum_{i\in\mathcal{P}_i} \|\boldsymbol{L}_{ij}\|\,\varphi(\omega)\left(1 - \gamma_{\mathrm{zs}i} \sum_{i\in\mathcal{P}_i}\|\boldsymbol{L}_{ij}\|\,\varphi(\omega)\right)^{-1} \gamma_{\mathrm{zw}i} \leq \varepsilon_i, \tag{6.33}$$

then the original requirement (6.18d) is satisfied. Moreover, the H_∞-norm conditions (6.27) and (6.30)–(6.32) are sufficient for I/O stability of the controlled subsystem F_i.

LMI-based conditions for the H_∞-design problem. The H_∞-norm claims (6.25), (6.40) as well as (6.30)–(6.32) are reformulated into LMI-based conditions by means Lemma 2.2 to derive a design procedure for the control station C_i.

The following theorem proposes an LMI-based design of a dynamical control algorithm subject to the local performance condition $\mathcal{A}_{\mathrm{D}i}$.

Theorem 6.3 (LMI-based controller design regarding the local performance condition $\mathcal{A}_{\mathrm{D}i}$) *Let the weight functions* (6.26) *and* (6.28) *and the design parameter* $\varepsilon_i, \gamma_{\mathrm{yw}i} \in \mathbb{R}_+$ *be given. The subsystem* (6.20) *controlled by* (2.11) *satisfies the local performance condition* $\mathcal{A}_{\mathrm{D}i}$ *defined in* (6.18) *with* (6.27) *if there exists feasible solutions* $\boldsymbol{X} = \boldsymbol{X}^\top$, $\boldsymbol{Y} = \boldsymbol{Y}^\top$ *and* $\widetilde{\boldsymbol{A}}$, $\widetilde{\boldsymbol{B}}$, $\widetilde{\boldsymbol{C}}$, $\widetilde{\boldsymbol{D}}$ *to the LMIs*

$$\begin{cases}
\begin{pmatrix} \boldsymbol{X} & \mathbf{I} \\ \star & \boldsymbol{Y} \end{pmatrix} \succ 0 & \text{(6.34a)} \\
\begin{pmatrix} \boldsymbol{A}_{\mathrm{LMI}}^\top + \boldsymbol{A}_{\mathrm{LMI}} & \boldsymbol{e}_{\mathrm{LMI}} & \boldsymbol{c}_{\varphi\mathrm{LMI}} \\ \star & -\gamma_{\mathrm{zs}i} & 0 \\ \star & \star & -\gamma_{\mathrm{zs}i} \end{pmatrix} \prec 0, & \text{(6.34b)} \\
\begin{pmatrix} \boldsymbol{A}_{\mathrm{LMI}}^\top + \boldsymbol{A}_{\mathrm{LMI}} & \boldsymbol{B}_{\mathrm{LMI}} & \boldsymbol{C}_{\mathrm{yLMI}}^\top \\ \star & -\gamma_{\mathrm{yw}i}\mathbf{I} & \mathbf{O} \\ \star & \star & -\gamma_{\mathrm{yw}i}\mathbf{I} \end{pmatrix} \prec 0, & \text{(6.34c)} \\
\begin{pmatrix} \boldsymbol{A}_{\mathrm{LMI}}^\top + \boldsymbol{A}_{\mathrm{LMI}} & \boldsymbol{e}_{\mathrm{LMI}} & \boldsymbol{C}_{\mathrm{LMI}}^\top \\ \star & -\gamma_{\mathrm{ys}i} & \mathbf{0}^\top \\ \star & \star & -\gamma_{\mathrm{ys}i}\mathbf{I} \end{pmatrix} \prec 0, & \text{(6.34d)} \\
\begin{pmatrix} \boldsymbol{A}_{\mathrm{LMI}}^\top + \boldsymbol{A}_{\mathrm{LMI}} & \boldsymbol{B}_{\mathrm{LMI}} & \boldsymbol{c}_{\mathrm{zLMI}} \\ \star & -\gamma_{\mathrm{zw}i}\mathbf{I} & \mathbf{0} \\ \star & \star & -\gamma_{\mathrm{zw}i} \end{pmatrix} \prec 0, & \text{(6.34e)}
\end{cases}$$

for arbitrary $\gamma_{\mathrm{ys}i} > 0$, $\gamma_{\mathrm{zw}i} > 0$ *and* $\left(\sum_{j\in\mathcal{P}_i} ||\boldsymbol{L}_{ij}||\right)^{-1} \geq \gamma_{\mathrm{zs}i} > 0$ *which fulfil relation* (6.33), *where the matrices are constructed according to* (B.14)–(B.16) *in Appendix B.7. The parameters of the control station* (2.11) *result to*

$$\boldsymbol{D}_{\mathrm{C}i} = \widetilde{\boldsymbol{D}}, \quad \boldsymbol{C}_{\mathrm{C}i} = (\widetilde{\boldsymbol{C}} + \boldsymbol{D}_{\mathrm{C}i}\boldsymbol{C}_{\mathrm{SW}i}\boldsymbol{S})(\boldsymbol{N}^\top)^{-1}, \quad \boldsymbol{B}_{\mathrm{C}i} = \boldsymbol{M}^{-1}(\widetilde{\boldsymbol{B}} - \boldsymbol{R}\boldsymbol{B}_{\mathrm{SW}i}\boldsymbol{D}_{\mathrm{C}i}) \quad (6.35)$$

$$\begin{aligned}\boldsymbol{A}_{\mathrm{C}i} = \boldsymbol{M}^{-1}(\widetilde{\boldsymbol{A}} - \boldsymbol{R}(\boldsymbol{A}_{\mathrm{SW}i} - \boldsymbol{B}_{\mathrm{SW}i}\boldsymbol{D}_{\mathrm{C}i}\boldsymbol{C}_{\mathrm{SW}i})\boldsymbol{S} - \boldsymbol{R}\boldsymbol{B}_{\mathrm{SW}i}\boldsymbol{C}_{\mathrm{C}i}\boldsymbol{N}^\top + \\ \boldsymbol{M}\boldsymbol{B}_{\mathrm{C}i}\boldsymbol{C}_{\mathrm{SW}i}\boldsymbol{S})(\boldsymbol{N}^\top)^{-1},\end{aligned} \quad (6.36)$$

where

$$\boldsymbol{M}\boldsymbol{N}^\top = \boldsymbol{I} - \boldsymbol{X}\boldsymbol{Y} \quad (6.37)$$

with $\mathrm{rank}(\boldsymbol{M}) = \mathrm{rank}(\boldsymbol{N}) = n_i + n_{\mathrm{Wz}} + n_{\mathrm{Wy}i}$. *Feasible solutions of the LMIs exist only if the pair* $(\boldsymbol{A}_{\mathrm{SW}i}, \boldsymbol{B}_{\mathrm{SW}i})$ *is stabilisable and the pair* $(\boldsymbol{A}_{\mathrm{SW}i}, \boldsymbol{C}_{\mathrm{SW}i})$ *is detectable.*

Proof. See Appendix B.7. □

The LMI (6.34b) reflects the two requirements (6.25) and (6.31) due to the fact that

$$\|W_{zi}(j\omega)F_{zsi}(j\omega)\|_{H_\infty} < \gamma_{zsi} \quad \Rightarrow \quad \|F_{zsi}(j\omega)\|_{H_\infty} < \gamma_{zsi}.$$

This implication holds since $\|W_{zi}(j\omega)\| = (\varphi(\omega))^{-1}$ and $0 < \varphi(\omega) \leq 1$. The desired I/O behaviour of the controlled isolated subsystem $F_i|_{\boldsymbol{s}_i=\boldsymbol{0}}$ that is represented by the H_∞-norm condition (6.27) is regarded by the LMI (6.34c). Furthermore, the claims (6.30) and (6.32) are converted into the equivalent LMIs (6.34d) and (6.34e).

Note that the LMIs (6.34a) and (6.34b) are necessary and sufficient conditions for the local performance claim (6.18b) and for asymptotic stability of the controlled subsystem F_i in accordance with Lemma 2.2 (i.e., the claim (6.18a) of the local performance condition $\mathcal{A}_{Di}$ is fulfilled). Moreover, the isolated controlled subsystem $F_i|_{\boldsymbol{s}_i=\boldsymbol{0}}$ has a performance described by (6.27) if and only if the LMIs (6.34a) and (6.34c) hold (see the proof in Appendix B.7 for more detail). The LMIs (6.34a)–(6.34d) with γ_{ysi}, γ_{zsi} and γ_{zwi} satisfying the relation (6.33) are only sufficient for the satisfaction of the local performance condition (6.18d).

To use the design of Theorem 6.3 regarding the local stability condition $\mathcal{A}_i$, only the LMIs (6.34a) and (6.34b) are needed with $\gamma_{zsi} = \left(\sum_{j\in\mathcal{P}_i} \|\boldsymbol{L}_{ij}\|\right)^{-1}$.

Remarks on the controller design. There are some remarks concerning the design method as well as the resulting control algorithm which are the following:

- Choice of γ_{ysi}, γ_{zsi} and γ_{zwi}: There exist infinitely many combinations of the parameters γ_{ysi}, γ_{zsi} and γ_{zwi} which satisfy the relation (6.33). For example, an admissible tuple $(\gamma_{ysi}, \gamma_{zsi}, \gamma_{zwi})$ can be found randomly or systematically by the gradually decrease of the parameters (cf. γ-iteration [164]).

- Worst-case design: The H_∞-design aims at attenuating the maximal magnitude of the amplitude plot (cf. eqn (2.23)). Hence, the scalars γ_{ywi}, γ_{ysi}, γ_{zwi} and γ_{zsi} represent upper bounds of the whole amplitude plot for all frequencies. Moreover, the satisfaction of the performance claim (6.18d) by means of the H_∞ condition (6.33) leads to a conservative result, in addition to the already conservative local performance condition $\mathcal{A}_{Di}$

- High order controller: The resulting control station C_i has the same order than the extended subsystem SW_i, i.e., $n_{Ci} = n_i + n_{Wz} + n_{Wui}$. Some remarks are stated in [81, 153] to reduce the order of the resulting controller.

6.5 Application scenario: Integration of new subsystems

The scenario is considered in which a new subsystem S_{N+1} is added to the interconnected system and shall be integrated within the overall closed-loop performance. The integration of the new subsystem encompasses the design of control station C_{N+1} as well as the adjustment of neighbouring control stations due to newly appearing interactions between the existing subsystems and the new one.

For the considered scenario it is assumed that an initial controller exists.

A 6.2 *There exist N control stations C_i, $(\forall i \in \mathcal{N})$ such that each controlled subsystem F_i satisfies the corresponding local performance condition $\mathcal{A}_{\mathrm{D}i}$ defined in* (6.18).

Section 6.5.1 proposes Algorithm 6.1 that summarises the steps to be processed by the design agents so that S_{N+1} can be successfully integrated. This algorithm is applied to the thermofuid process in Section 6.5.2. An experiment shows that a new tank can successfully be integrated to the thermofluid process.

6.5.1 Plug-and-play integration algorithm

The connection of the new subsystem entails new physical couplings $\boldsymbol{L}_{N+1i}$ from existing subsystems S_i to the new subsystem S_{N+1} as well as couplings $\boldsymbol{L}_{iN+1}$ from the new subsystem to other subsystems S_i. Hence, after the physical connection of the new subsystem has been finished, design agent D_{N+1} has to register the new subsystem to its neighbouring design agents D_i. This registration triggers

- D_i to include $N{+}1$ into the set $\mathcal{P}_i$ (henceforth denoted by $\mathcal{P}_{i+}$) and
- D_i to update the model K_i by the new coupling gain $\boldsymbol{L}_{iN+1}$ (the updated model is henceforth denoted by K_{i+}).

It is assumed that the coupling gains $\boldsymbol{L}_{iN+1}$ are known by design agent D_i either from the new design agent D_{N+1} or as a result of an identification. Concerning the registration, the following assumption is made:

A 6.3 *The registration is completed at the same time instant when the new subsystem is connected.*

Let the set $\mathcal{P}^+$ collect all indices of those subsystems that are affected by the new subsystem S_{N+1} through the physical couplings, i.e.,

$$\mathcal{P}^+ := \{i\colon N+1 \in \mathcal{P}_{i+}\}.$$

In accordance with assumption A 6.3, after the connection of the new subsystem, the design agents D_i, $(\forall i \in \mathcal{P}^+)$ know the corresponding model set

$$\mathcal{M}_i = (C_i, S_i, K_{i+})$$

and design agent D_{N+1} knows

$$\mathcal{M}_{N+1} = (S_{N+1}, K_{N+1})\,.$$

Immediately after the registration, the integration of the new subsystem is processed as described by Algorithm 6.1.

Algorithm 6.1: Plug-and-play integration of subsystem S_{N+1}

Given: Control stations C_i, $(i \in \mathcal{N})$ exist such that F_i satisfy $\mathcal{A}_{\mathrm{D}i}$ for all $i \in \mathcal{N}$,
model set $\mathcal{M}_{N+1} = (S_{N+1}, K_{N+1})$ available to D_{N+1},
model set $\mathcal{M}_i = \{F_i, K_{i+}\}$ available to D_i, $(\forall i \in \mathcal{P}^+)$ and
local conditions $\mathcal{A}_{\mathrm{D}i}$ with the parameter ε_i and the models $W_{\mathrm{z}i}$ and $W_{\mathrm{y}i}$ of $\varphi(\omega)$ and $\mathbf{W}_{\mathrm{y}i}(s)$, respectively available to D_i, $(\forall i \in \mathcal{P}^+ \cup \{N+1\})$

Processing on D_{N+1}:

1. Determine $\bar{P}_{N+1}^{\{N+1\}}$ in accordance with (6.4)
2. Design control station C_{N+1} by means of Theorem 6.3
3. Implement control station C_{N+1} into the control equipment

Processing on D_i, $(\forall i \in \mathcal{P}_+)$:

4. Determine $\bar{P}_{i+}^{\{i\}}$ in accordance with (6.38)
5. **if** $\mathcal{A}_{\mathrm{D}i}$ is violated **then** redesign control station C_i by means of Theorem 6.3 **else** STOP (no reconfiguration necessary)
6. Implement the redesigned control station into the control equipment

Result: The overall closed-loop system that is composed of the controlled subsystems F_i, $(i = 1, ..., N+1)$ performs as claimed by $\mathcal{A}_{\mathrm{D}}$

In parallel, the design of control station C_{N+1} is processed (Steps 1–3) as well as the redesigns of the control stations C_i, $(\forall i \in \mathcal{P}_+)$ are initiated (Steps 4–6). In both cases the comparison system of the physical interaction is determined (Step 1 and Step 4), where

$$\bar{P}_{i+}^{\{i\}}: \quad \bar{s}_i(\omega) = \sum_{j \in \mathcal{P}_{i+}} \|\boldsymbol{L}_{ij}\| \, \varphi(\omega) \cdot \|\boldsymbol{z}_i(\mathrm{j}\omega)\|. \tag{6.38}$$

Thereafter, the new control station C_{N+1} is designed and implemented (Steps 2–3), while the design agents D_i, $(\forall i \in \mathcal{P}_+)$ check simultaneously whether the local condition $\mathcal{A}_{\mathrm{D}i}$ is still fulfilled and perform the redesign of the existing control station if required (Steps 5–6).

It should be emphasised that the design agents do not interact during the integration. However, if the design for at least one control station fails, the integration of the new subsystem is not possible.

Note that due to assumption A 3.4 the processing of Algorithm 6.1 does not consume time. Algorithm 6.1 can also be used if only I/O stability of the overall closed-loop system is desired. Therefore, the control stations have to be (re)designed with respect to the local stability condition $\mathcal{A}_i$ (Lines 2 and 5) and the design agents have to verify the respective local condition $\mathcal{A}_i$ in Line 5.

Removing subsystems. Consider the case that subsystem S_N shall be removed from the interconnected system. For this situation, the adherence of the global performance according to $\mathcal{A}_{\mathrm{D}}$ is still guaranteed without adapting any of the control stations. This is due to the fact that the comparison systems $\bar{P}_i^{\{i\}}$, $(\forall i \in \mathcal{N} \setminus N)$ are still comparison systems of the physical interaction without the subsystem S_N and the corresponding coupling gains $\boldsymbol{L}_{iN}$.

In particular, consider subsystem S_1 is influenced by S_2 and S_N. Then the upper bound of the physical interaction is given by

$$\bar{P}_1^{\{1\}}(\omega) = (\|\boldsymbol{L}_{12}\| + \|\boldsymbol{L}_{1N}\|) \, \varphi(\omega).$$

If S_N is removed, then the new upper bound denoted by $\bar{P}_{1-}^{\{1\}}(\omega)$ becomes

$$\bar{P}_{1-}^{\{1\}}(\omega) = \|\boldsymbol{L}_{12}\| \, \varphi(\omega) < \bar{P}_1^{\{1\}}(\omega).$$

Since the existing control station C_1 is designed based on $\bar{P}_1^{\{1\}}(\omega)$, the local condition $\mathcal{A}_{\mathrm{D}1}$ remain satisfied. A redesign might be desirable, if the overall closed-loop performance shall be improved. Accordingly, an arbitrary number of subsystems can be removed in arbitrary order without adapting the algorithm. That is, the initial controller is robust against the removing of subsystems.

6.5.2 Experiment: Interconnected thermofluid process

Plug-and-play integration (Algorithm 6.1) is applied to the thermofluid process introduced in Section 2.5.1 to integrate reactor TS. The effect of the connection of reactor TS to the behaviour of the level and temperature of reactor TB and tank T1 is analysed by simulations and experiments. The experimental results highlight the applicability of the presented concept to a real world process and reveals the conservatism of the local conditions. The amplitude plot of the experiment and simulations are identified from measurements of the corresponding time domain signals using the MATLAB System Identification Toolbox [111].

Initially, the thermofluid process consists only of the reactor TB and the tank T1, where the level dynamics of the fluid in the reactor TB and tank T1 are modelled by S_1 and S_3, respectively, and the temperature dynamics of the fluid are modelled by S_2. These models S_i, $(i = 1, 2, 3)$ are represented by (2.6) with the parameters from (A.8a)–(A.8c). The equivalent frequency domain representation (6.21) results from the relation (6.22) and (6.23). The physical couplings K_i, $(i = 1, 2, 3)$ among the tanks read

$$\begin{aligned} K_1: \quad & \boldsymbol{s}_1(t) = \boldsymbol{L}_{13}\boldsymbol{z}_3(t) \\ K_2: \quad & \boldsymbol{s}_2(t) = \boldsymbol{L}_{21}\boldsymbol{z}_1(t) + \boldsymbol{L}_{23}\boldsymbol{z}_3(t) \\ K_3: \quad & \boldsymbol{s}_3(t) = \boldsymbol{0} \end{aligned}$$

with the coupling gains from (A.9a)–(A.9c). Accordingly, the sets of predecessors are

$$\mathcal{P}_1 = \{3\}, \qquad \mathcal{P}_2 = \{1, 3\}, \qquad \mathcal{P}_3 = \emptyset. \tag{6.39}$$

The interconnection graph of this initial constellation is depicted in Fig. 6.9a and the local interconnection graphs which are known by the respective design agent D_i are shown in Figs. 6.9b–6.9d.

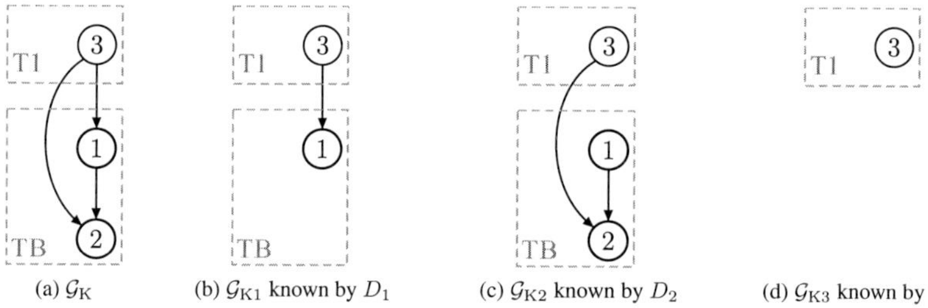

(a) $\mathcal{G}_{\mathrm{K}}$ (b) $\mathcal{G}_{\mathrm{K1}}$ known by D_1 (c) $\mathcal{G}_{\mathrm{K2}}$ known by D_2 (d) $\mathcal{G}_{\mathrm{K3}}$ known by D_3

Figure 6.9: Application: Interconnection graph $\mathcal{G}_{\mathrm{K}}$ and local interconnection graphs $\mathcal{G}_{\mathrm{K}i}$, $(i = 1, 2, 3)$ before the reactor TS is connected.

The subsystems are initially controlled by the control stations C_i, $(i = 1, 2, 3)$, modelled by (6.24). The initial control stations are design by means of Theorem 6.3 with respect to the local performance aim $\mathcal{A}_{\mathrm{D}i}$, where the design parameters are listed in Table 6.2. The dynamics of the control stations result to

$$\boldsymbol{C}_1(s) = -\frac{17.3s + 5.06 \cdot 10^4}{s + 40.29}, \quad \boldsymbol{C}_2(s) = \frac{2.9s^2 + 7.34s + 0.08}{s^2 + 2.2s + 0.02},$$
$$\boldsymbol{C}_3(s) = -\frac{8.42s^2 + 175.6s + 14.97}{s^2 + 1.78s + 0.18}.$$

Since the model W_{zi} that corresponds to the transfer function $W_{\mathrm{z}}(s)$ is static, the dimension of the control stations C_i, $(i = 1, 2, 3)$ result to $n_{\mathrm{C}1} = 1$ and $n_{\mathrm{C}2} = n_{\mathrm{C}3} = 2$.

Table 6.2: Application: Design parameters of the local performance condition $\mathcal{A}_{\mathrm{D}i}$

$\mathcal{A}_{\mathrm{D}i}$	Design parameters and functions		
	ε_i	$\boldsymbol{W}_{\mathrm{y}i}(s)$	$W_{\mathrm{z}}(s)$
$\mathcal{A}_{\mathrm{D}1}$	$\varepsilon_1 = 0.15$	not specified	$W_{\mathrm{z}}(s) = 1$
$\mathcal{A}_{\mathrm{D}2}$	$\varepsilon_2 = 0.2$	$\boldsymbol{W}_{\mathrm{y}2}(s) = 6s + 1$	$W_{\mathrm{z}}(s) = 1$
$\mathcal{A}_{\mathrm{D}3}$	not specified	$\boldsymbol{W}_{\mathrm{y}3}(s) = s + 1$	$W_{\mathrm{z}}(s) = 1$
$\mathcal{A}_{\mathrm{D}4}$	$\varepsilon_6 = 0.15$	not specified	$W_{\mathrm{z}}(s) = 1$
$\mathcal{A}_{\mathrm{D}5}$	$\varepsilon_5 = 0.2$	$\boldsymbol{W}_{\mathrm{y}5}(s) = 3s + 1$	$W_{\mathrm{z}}(s) = 1$

The first design parameter ε_i renders the admissible impact of the physical interaction on the output $\boldsymbol{y}_i(s)$. As tank T1 (i.e., subsystem S_3) is not strongly connected with the other tanks (see interconnection graph $\mathcal{G}_{\mathrm{K}}$ in Fig. 6.10a), no physical interaction with the other subsystems exists so that ε_3 left unspecified. The second column of Table 6.2 represents the desired behaviour of the isolated controlled subsystem $F_i|_{\boldsymbol{s}_i=\boldsymbol{0}}$ by means of the H_∞-norm claim (6.27). For the considered process, the temperatures $\vartheta_{\mathrm{TB}}(t)$ and $\vartheta_{\mathrm{TS}}(t)$ as well as the level $l_{\mathrm{T}1}(t)$ shall have PT1 characteristics (represented by $\boldsymbol{W}_{\mathrm{y}i}(s)$). The function $\varphi(\omega)$, represented by the filter functions $W_{\mathrm{z}}(s)$, $(i = 1, .., 5)$, are assumed to be known by all design agents D_i, $(i = 1, ..., 5)$.

Integration of reactor TS. At runtime, reactor TS is connected to the process. That means that simultaneously, subsystem S_4 which represents the level dynamics and subsystem S_5 which models the temperature dynamics are added to the interconnected system. Both subsystems S_4 and S_5 are modelled by (6.20) with (A.8d) and (A.8e). The reactor TB and the tank T1 feed tank TS with their heated liquids. Accordingly, the sets of predecessors results to

$$\mathcal{P}_4 = \{1, 2, 3\}, \qquad \mathcal{P}_5 = \{1, 3, 4\}$$

and the physical couplings K_4 and K_5 are modelled by

$$\begin{aligned} K_4: &\quad \boldsymbol{s}_4(t) = \boldsymbol{L}_{41}\boldsymbol{z}_1(t) + \boldsymbol{L}_{42}\boldsymbol{z}_2(t) + \boldsymbol{L}_{43}\boldsymbol{z}_3(t), \\ K_5: &\quad \boldsymbol{s}_5(t) = \boldsymbol{L}_{51}\boldsymbol{z}_1(t) + \boldsymbol{L}_{53}\boldsymbol{z}_3(t) + \boldsymbol{L}_{54}\boldsymbol{z}_4(t). \end{aligned}$$

with the coupling gains from (A.9d) and (A.9e). Conversely, liquid from reactor TS also flows in reactor TB such that the sets of predecessors (6.39) are updated

$$\mathcal{P}_{1+} = \{3, 4\}, \qquad \mathcal{P}_{2+} = \{1, 3, 4, 5\}$$

and, hence, the models K_1 and K_2 of the physical couplings too, which become

$$\begin{aligned} K_{1+}: &\quad \boldsymbol{s}_1(t) = \boldsymbol{L}_{13}\boldsymbol{z}_3(t) + \boldsymbol{L}_{14}\boldsymbol{z}_4(t), \\ K_{2+}: &\quad \boldsymbol{s}_2(t) = \boldsymbol{L}_{21}\boldsymbol{z}_1(t) + \boldsymbol{L}_{23}\boldsymbol{z}_3(t) + \boldsymbol{L}_{24}\boldsymbol{z}_4(t) + \boldsymbol{L}_{25}\boldsymbol{z}_5(t). \end{aligned}$$

The effect of the connection of reactor TS is highlighted by the interconnection graph depicted in Fig. 6.10a as well as by the local interconnection graphs shown in Figs. 6.10b–6.10f. It is to emphasise that the set $\mathcal{P}_3$ and the model K_3 do not change, since reactor TS does not have a direct influence on the level $l_{\mathrm{T1}}(t)$ as it can be seen by $\mathcal{G}_{\mathrm{K3}}$ in Fig. 6.10d.

Subsequently, the Algorithm 6.1 is performed to integrate subsystem S_4 and S_5 into the controlled thermofluid process. In parallel, the design agents D_4 and D_5 design the corresponding control stations and the design agents D_1 and D_2 check whether a redesign of the respective control stations C_1 and C_2 is necessary.

Step 1: The design agents D_4 and D_5 determine the comparison systems $\bar{P}_4^{\{5\}}$ and $\bar{P}_5^{\{5\}}$ according to (6.4) using the available models K_4 and K_5 as well as the locally known function $\varphi(\omega) = |W_{\mathrm{z}}(\mathrm{j}\omega)|^{-1} = 1$.

Step 2: With the locally available model sets

$$\mathcal{M}_4 = \left(S_4, K_4, \bar{P}_4^{\{4\}}\right), \qquad \mathcal{M}_5 = \left(S_5, K_5, \bar{P}_5^{\{5\}}\right)$$

and the available design conditions $\mathcal{A}_{\mathrm{D4}}$ and $\mathcal{A}_{\mathrm{D5}}$ with the parameters from Table 6.2, the design agents D_4 and D_5 independently perform the controller design by means of Theorem 6.3. The control stations C_4 and C_5 with

$$\boldsymbol{C}_4(s) = -\frac{64.78s + 10.38 \cdot 10^4}{s + 43.07}, \quad \boldsymbol{C}_5(s) = \frac{6.15s^2 + 10.57s + 0.12}{s^2 + 1.56s + 5.22 \cdot 10^{-3}}$$

result.

Step 3: The design agents D_4 and D_5 implement the control stations into the hardware and

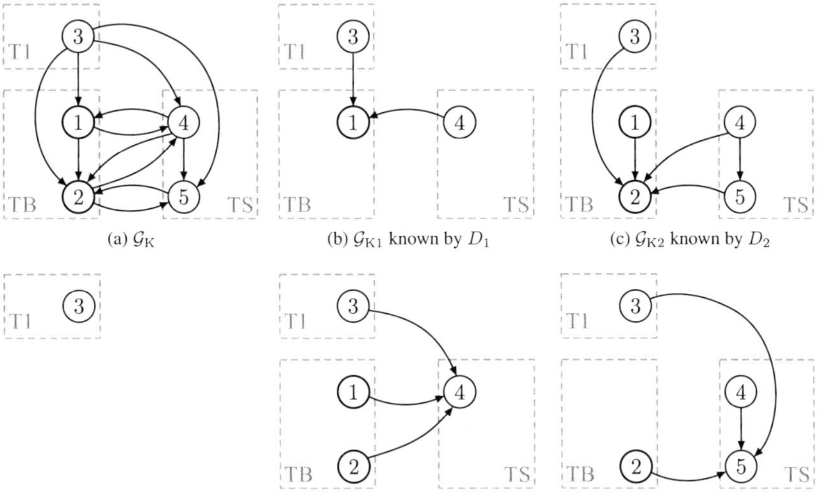

Figure 6.10: Application: Interconnection graph $\mathcal{G}_{\mathrm{K}}$ and local interconnection graphs $\mathcal{G}_{\mathrm{K}i}$, $(i = 1, ..., 5)$ after the connection and the registration of reactor TS.

(a) $\mathcal{G}_{\mathrm{K}}$ (b) $\mathcal{G}_{\mathrm{K1}}$ known by D_1 (c) $\mathcal{G}_{\mathrm{K2}}$ known by D_2 (d) $\mathcal{G}_{\mathrm{K3}}$ known by D_3 (e) $\mathcal{G}_{\mathrm{K4}}$ known by D_4 (f) $\mathcal{G}_{\mathrm{K5}}$ known by D_5

activate them.

Step 4: The design agents D_1 and D_2 construct their upper bounds $\bar{P}_{1+}^{\{1\}}$ and $\bar{P}_{2+}^{\{2\}}$ based on K_{1+} and K_{2+}, respectively, in order that the design agents have access to the respective models

$$\mathcal{M}_1 = \left(C_1, S_1, K_{1+}, \bar{P}_{1+}^{\{1\}}\right), \qquad \mathcal{M}_2 = \left(C_2, S_2, K_{2+}, \bar{P}_{2+}^{\{2\}}\right).$$

Step 4: With the model information at hand, both design agents check the local conditions $\mathcal{A}_{\mathrm{D1}}$ and $\mathcal{A}_{\mathrm{D2}}$. That is, due to the additional coupling influence, essentially the conditions (6.18b) and (6.18d). As it can be seen in Fig. 6.11a, the design conditions of $\mathcal{A}_{\mathrm{D1}}$ are still satisfied. Hence, D_1 stops running Step 4 of Algorithm 6.1. In contrast to this, the local performance condition $\mathcal{A}_{\mathrm{D2}}$ is violated since $\bar{F}_{\Delta 2+}^{\{2\}}(\omega) > \varepsilon_2$ for $\omega < 0.3\,\mathrm{rad/s}$ as highlighted in Fig. 6.11b. Accordingly, design agent D_2 performs the redesign of C_2 by means of the H_∞-design. The redesigned control station is denoted by $\mathring{C}_2$ with

$$\mathring{C}_2(s) = \frac{4.17s^2 + 13.77s + 0.14}{s^2 + 2.96s + 0.03}.$$

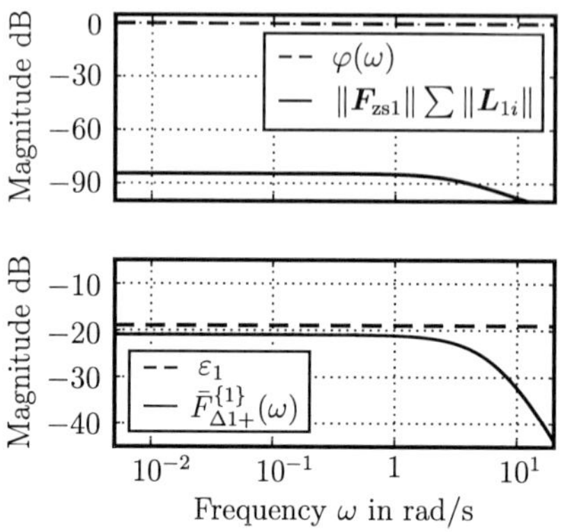

(a) Local conditions (6.18b) and (6.18d) from $\mathcal{A}_{\mathrm{D1}}$

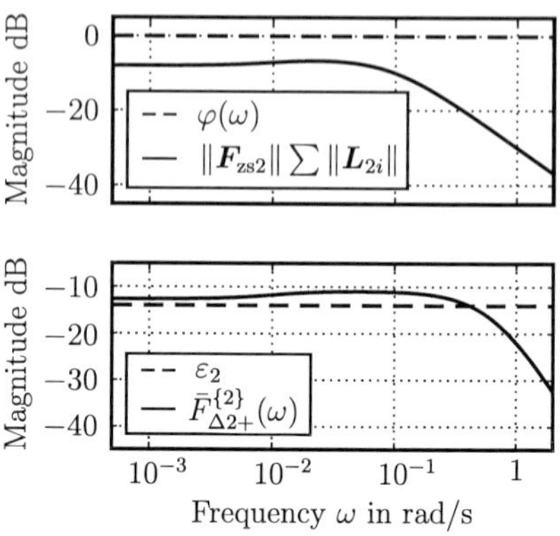

(b) Local conditions (6.18b) and (6.18d) from $\mathcal{A}_{\mathrm{D2}}$

Figure 6.11: Application: Verification of the local performance conditions $\mathcal{A}_{\mathrm{D1}}$ and $\mathcal{A}_{\mathrm{D2}}$.

Step 5: D_2 implements control station $\mathring{C}_2$ into the control equipment and starts it up. For the redesigned controlled subsystem

$$\mathring{F}_2 = \mathrm{con}\left(\left\{S_2,\, \mathring{C}_2\right\},\, \begin{pmatrix} \boldsymbol{w}_2 \\ \boldsymbol{s}_2 \end{pmatrix},\, \begin{pmatrix} \mathring{\boldsymbol{y}}_2 \\ \boldsymbol{z}_2 \end{pmatrix}\right).$$

the local condition (6.18d) is satisfied, i.e.,

$$\bar{\mathring{F}}^{\{2\}}_{\Delta 2}(\omega) = \|\mathring{\boldsymbol{F}}_{\mathrm{ys2}}(\mathrm{j}\omega)\|_2 \sum_{i\in\mathcal{P}_{i+}} \|\boldsymbol{L}_{2i}\| \left(1 - \|\mathring{\boldsymbol{F}}_{\mathrm{zs2}}(\mathrm{j}\omega)\|_2 \sum_{i\in\mathcal{P}_{i+}} \|\boldsymbol{L}_{2i}\|\right)^{-1} \|\mathring{\boldsymbol{F}}_{\mathrm{zw2}}(\mathrm{j}\omega)\|_2 \leq \varepsilon_2,$$

for all $\omega \in \mathbb{R}$ as highlighted in Fig. 6.12. Furthermore, Fig. 6.12 shows the actual dynamics $\|\mathring{\boldsymbol{F}}^{\{2\}}_{\Delta 2}(\mathrm{j}\omega)\|$ in order to highlight the relation

$$\|\mathring{\boldsymbol{F}}^{\{2\}}_{\Delta 2}(\mathrm{j}\omega)\| < \bar{\mathring{F}}^{\{2\}}_{\Delta 2}(\omega) \leq \varepsilon_2.$$

and to point out the conservatism of the local performance conditions by means of the difference between the upper bound $\bar{\mathring{F}}^{\{2\}}_{\Delta 2}(\omega)$ and the actual magnitude $\|\mathring{\boldsymbol{F}}^{\{2\}}_{\Delta 2}(\mathrm{j}\omega)\|$. In detail, the difference for the specific frequency $\omega = 10^{-3}\,\mathrm{rad/s}$ yields

$$\bar{\mathring{F}}^{\{2\}}_{\Delta 2}(\omega = 10^{-3}\,\mathrm{rad/s}) - \|\mathring{\boldsymbol{F}}^{\{2\}}_{\Delta 2}(\mathrm{j}\omega = \mathrm{j}10^{-3}\,\mathrm{rad/s})\| \approx 15\,\mathrm{dB} = 31.62.$$

Nevertheless, the simulation shows that although the local performance condition is conser-

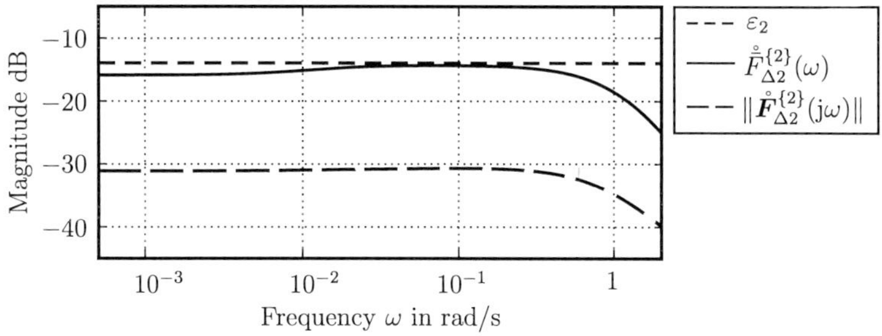

Figure 6.12: Application: Satisfaction of the conditions (6.18d) from $\mathcal{A}_{\mathrm{D}2}$ after the redesign of control station C_2.

vative, the integration was successful even for integrating two subsystems in parallel. Nevertheless, the success is owned to the small couplings among the subsystems (cf. parameters (A.9)).

Experimental results. The experiment aims at illustrating the actual behaviour of $\|\mathring{\boldsymbol{F}}^{\{2\}}_{\Delta 2}(\mathrm{j}\omega)\|$. The results of the experiment are shown in Fig. 6.13.

The amplitude plot $\|\mathring{\boldsymbol{F}}^{\{2\}}_{\Delta 2,\mathrm{exp}}(\mathrm{j}\omega)\|$ (solid line in Fig. 6.13) is identified from measurements on the thermofluid process using the MATLAB System Identification Toolbox. The identified linear model transfer function matches the nonlinear behaviour of the process by approximately $80\,\%$. This bad matching is attributable to noisy measurements and the neglect of nonlinear effects such as the actuation of the valves by pulse-width modulation. However, the local condition claim $\|\mathring{\boldsymbol{F}}^{\{2\}}_{\Delta 2,\mathrm{exp}}(\mathrm{j}\omega)\| \leq \varepsilon_2$ is not fulfilled. The violation of this condition is mainly attributable to actuator saturations of the valves and heating rods (cf. operating regions (A.1) in Appendix A.1). To substantiate this effect, the amplitude plot $\|\mathring{\boldsymbol{F}}^{\{2\}}_{\Delta 2,\mathrm{sat}}(\mathrm{j}\omega)\|$ is identified by means of simulations on the linear model under consideration of the actuator saturations (with a correlation of approximately $95\,\%$). Qualitatively, both plots are similar to each other. For comparison reasons, the amplitude plot $\|\mathring{\boldsymbol{F}}^{\{2\}}_{\Delta 2}(\mathrm{j}\omega)\|$ of the simulation of the linear model (without saturations) is shown.

Figure 6.13 highlights that the actuators saturate for a sinusoidal stimulations with $\omega \approx 0.1\,\mathrm{rad/s}$. To consider these actuator limitations within the H_∞-design, an additional weight function can be introduced to attenuate this frequency from the input $\boldsymbol{w}_2$ to the output $\boldsymbol{u}_2$. This claim is reflected by

$$\|\boldsymbol{W}_{\mathrm{u}2}(s)\boldsymbol{F}_{\mathrm{uw}2}(s)\|_{\mathrm{H}_\infty} < \gamma_{\mathrm{uw}2}. \tag{6.40}$$

The extended subsystem (6.29) need to be extended by a model that corresponds to the transfer

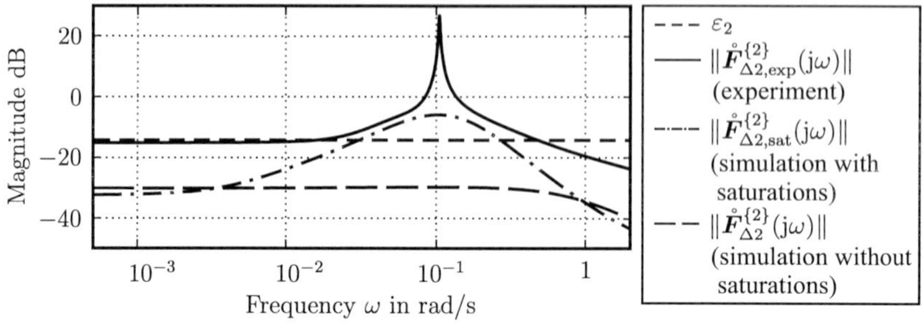

Figure 6.13: Application: Comparison of the simulation and the experiment.

function matrix $\boldsymbol{W}_{\mathrm{u}2}(s)$ and the LMIs (6.34) need to be extended by a further LMI reflecting the condition (6.40).

6.5.3 Further application scenarios

Due to the locally checkable condition $\mathcal{A}_{\mathrm{D}i}$, the control stations can be designed independently from each other. Accordingly, the proposed design can be used for the ordinary design without any changes, as it has been done for the design of the control stations C_1, C_2 and C_3.

Moreover, the redesign of a control station C_i can also be relevant to achieve fault-tolerance, as presented in detail in Section 5.5. To apply the proposed design method for the application scenario of fault-tolerant control, only slight changes of the plug-and-play reconfiguration algorithm (Algorithm 5.3) need to be applied. In particular, no model information need to be gathered over the network as all relevant model information are initially available to the design agent of the faulty subsystem (i.e., no procurement in Step 1 of Alg. 5.3). After the redesign of the control station of the faulty subsystem by Theorem 6.3 (i.e., Step 3 of Alg. 5.3), the overall closed-loop performance described by (6.16) is retrieved in accordance with assumption A 6.2. A particular H_∞-design of a virtual actuator is proposed in Section 7.4.1.

7 Plug-and-play control using an approximate model of the physical interaction

This chapter concentrates on the design of control station C_1 using an approximate model of subsystem S_1 under consideration of the physical interaction. Based on the initially stored models S_1 and K_1 of the subsystem and the local couplings, design agent D_1 has to set up an approximate model and has to design C_1 in a way that the global control aim is fulfilled. The main results of this chapter is a local decision rule that enables D_1 to set up an appropriate approximate model. The decision is made by the evaluation of the physical interaction strength of the other controlled subsystems with subsystem S_1. Furthermore, local design conditions are proposed that guarantee I/O stability and a certain I/O performance of the overall closed-loop system. It is shown that these conditions are less restrictive than the local conditions presented in Chapter 6. Finally, plug-and-play control is applied on the multizone furnace to guarantee fault-tolerance by reconfiguring control station C_1 using a virtual actuator.

7.1 Problem formulation

The situation is considered in which the control station C_1 shall be designed based on a model that approximately represents the behaviour of subsystem S_1 together with the physical interaction $P_1^{\mathcal{N}}$. The design process is organised by D_1 that solely knows the model of subsystem S_1 and the local coupling model K_1. The aim of this chapter is to enable D_1 to accomplish the design of C_1. Therefore, first, D_1 has to be able to set up the approximate model of appropriate accuracy. Second, local design conditions have to be found so as to guarantee I/O stability and a certain I/O performance of the overall closed-loop system. The design using an approximate model constitutes the middle course between the set up of an exact model (Chapter 4) and the stick on the local model (Chapters 5 and 6). On the one hand, the conservatism of the local design conditions from Chapter 6 is significantly relaxed and, on the other hand, the required amount of model information compared to Chapter 4 is reduced. Hence, this approach is preferable whenever computational limitations are not of prime importance but shall be considered.

This chapter concentrates on subsystems that are locally interconnected in the sense that S_i has direct influence on S_{i+1} and on S_{i-1} and vice versa as delineated in Fig. 7.1. A characteristic of locally interconnected systems is that the behaviour of subsystem S_1 is essentially represented by the dynamics of S_1 together with some neighbouring controlled subsystems F_i, $(i = 2, ..., v-1)$. In contrast to this, the interaction with the other controlled subsystems F_i, $(i = v, ..., N)$ is negligible. Thus, the dynamics of S_1 together with the physical interaction is appropriately approximated by the approximate model

$$\hat{S}_1^{\{1,..,v-1\}} = \text{con}\left(\{S_1\} \cup \left\{ F_i|_{\boldsymbol{w}_i=\boldsymbol{0}},\, i = 2, ..., v-1\right\} \cup \{K_i,\, i = 1, ..., v-1\},\, \boldsymbol{u}_1,\, \boldsymbol{y}_1\right)$$

as highlighted by the yellow box in Fig. 7.1, which is used to design control station C_1. All other controlled subsystems, which are accumulated to

$$E_1^{\{v,..,N\}} = \text{con}\left(\left\{ F_i|_{\boldsymbol{w}_i=\boldsymbol{0}},\, i = v, ..., N\right\} \cup \{K_i,\, i = v-1, ..., N\},\, \boldsymbol{p}_1,\, \boldsymbol{q}_1\right)$$

represent an admissible model uncertainty (grey box in Fig. 7.1)

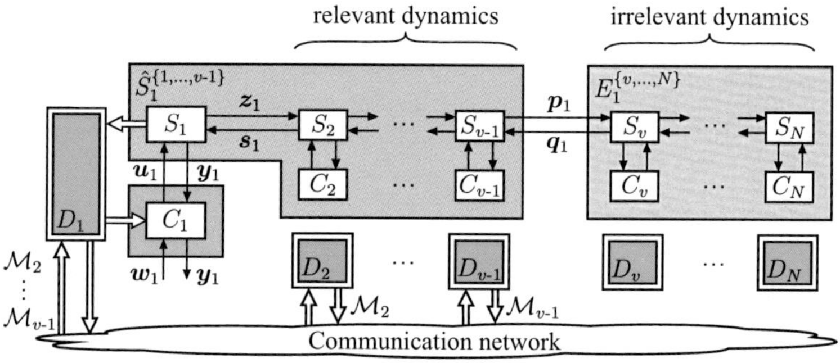

Figure 7.1: Structure of plug-and-play control using an approximate model of the interaction interaction: The approximation comprises only some controlled subsystem models (yellow box), while the remaining subsystems (grey box) are considered as model uncertainty.

However, design agent D_1 initially knows the subsystem model S_1 and has no information about the global interconnection structure. Based on this a priory knowledge, D_1 has to decide whether or not the controlled subsystem F_i has to be part of the approximate model. Consequently, the model set $\mathcal{M}_i = \left\{ F_i|_{\boldsymbol{w}_i=\boldsymbol{0}},\, K_i\right\}$ is either procured from D_i through the network (double arrows in Fig. 7.1) or not. This modelling issue is formalised as follows:

Problem 7.1 (Approximating the dynamics of the physical interaction)

Problem:	*An appropriate approximation of subsystem S_1 under consideration of the physical interaction is composed of subsystem S_1 together with some controlled subsystems only. This approximate model has to be set up by design agent D_1 that initially only knows the models S_1 and K_1 of the subsystem and the local couplings.*
Restrictions:	*Subsystems are local interconnected*
Given:	• *D_1 stores the model set $\mathcal{M}_1 = \{S_1, K_1\}$* • *$D_i$, $(\forall i \in \mathcal{N} \setminus \{1\})$ store the corresponding model set $\mathcal{M}_i = \{F_i, K_i\}$*
Find:	*Local conditions that enable D_1 to decide whether or not the model of a controlled subsystem is relevant for the approximate model.*
Approach:	*Evaluate the relevance of the model set $\mathcal{M}_i$ by means of the interaction strength of F_i with S_1 through the physical couplings.*

Although only the approximate model $\hat{S}_1^{\{1,..,v-1\}}$ is available to D_1 (and $E_1^{\{v,..,N\}}$ is unknown), the design of control station C_1 must guarantee I/O stability and a desired I/O performance of the overall closed-loop system. To enable this design, the method proposed in Chapter 6 is used. In short summary: The effect of the unknown dynamics is bounded if the controlled subsystems F_i, $(i = v, ..., N)$ attenuate the amplification from the coupling input signal to the coupling output signal. Then this bound can be determined without information from D_i, $(i = v, ..., N)$. Based on the bound of the unknown dynamics, design conditions are derived. These design conditions can be satisfied by the *cooperative design* of C_1 together with C_i, $(i = 2, ..., v{-}1)$. These design conditions together with a solution to Problem 7.1 represents a solution to the plug-and-play control problem (Problem 1.2). Moreover, an H_∞-design of the control station and a virtual actuator (for controller reconfiguration) is proposed. That is, a solution to Problem 1.1.

The first part of the chapter is devoted to the set-up of an approximate model that appropriately represents the dynamics of subsystem S_1 under the influence of the physical interaction. Section 7.2 introduces a locally verifiable condition (Theorem 7.1) to separate the controlled subsystems F_i, $(i = 2, ..., N)$ into strongly coupled and weakly coupled to subsystem S_1. While the approximation comprises the controlled subsystems that are strongly coupled to S_1, the weakly coupled controlled subsystems are considered as model uncertainty. The second part of the chapter focusses on the design conditions. In Section 7.3.1 conditions are stated to enable the design using the approximate model only. Based on these conditions, cooperative local stability conditions and cooperative performance conditions are proposed in Section 7.3 that guarantee I/O stability and a certain I/O performance of the overall closed-loop system. It will be shown that the proposed design conditions are less restrictive than the local design conditions from Chapter 6. The chapter ends with Section 7.4, where the proposed method is applied to achieve fault-tolerance. A virtual actuator is proposed that exploits the approximate model $\hat{S}_1^{\{1,..,v-1\}}$ to recover overall closed-loop I/O stability by reconfiguring control station C_1.

7.2 A local algorithm to approximate the dynamics of the physical interaction

This section presents Algorithm 7.1 that enable D_1 to find the controlled subsystems F_i, $(i = 2, ..., v-1)$ that approximate the physical interaction $P_1^{\mathcal{N}}$ appropriately. A decision threshold is introduced which, clever chosen, represents the admissible model uncertainty for the controller design. Therefore, first, the basic idea of separating the controlled subsystems into a relevant part and an irrelevant part for the controller design is recapitulated. Second, the property of a controlled subsystem to be strongly coupled or weakly coupled with another subsystem is introduced followed by a locally checkable threshold to categorise a controlled subsystem correspondingly. Moreover, the accuracy of the approximation in dependence on the decision threshold is studied.

7.2.1 Separation of relevant and irrelevant dynamics

This section reaffirms the partition of the physical interaction into a relevant part and an irrelevant part for the controller design (see Section 3.2.2).

As presented in Chapter 6, with the locally available models a comparison system $\bar{P}_1^{\{1\}}$ can be derived that roughly describes the physical interaction $P_1^{\mathcal{N}}$. Based on this comparison system local design conditions have been derived, which, thus, inherent a high degree of conservatism. Accordingly, this conservatism can be relaxed if a more accurate model of the physical interaction is used. Such a precise model is obtained by considering dynamics of neighbouring controlled subsystems. Before going into details, Example 6.1 is continued to emphasise this idea.

Example 6.1 (cont.) *Accuracy of the physical interaction model*

The first part of this example has shown that on the basis of local model information only an imprecise model of the physical interaction can be determined (see page 117).

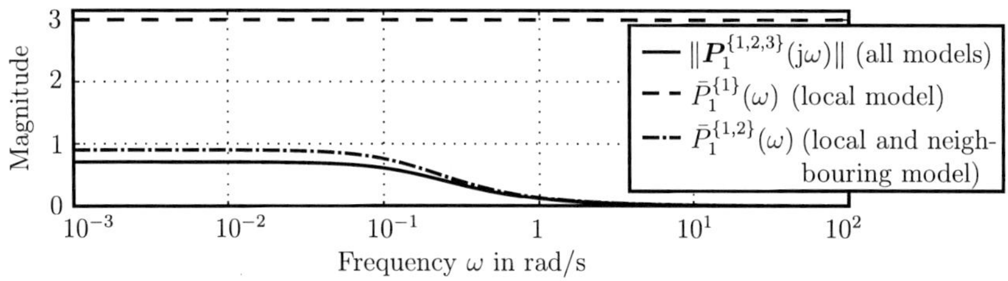

Figure 7.2: Example room temperature control: Improving of the accuracy of the comparison system of the physical interaction by taking the neighbouring room into account.

Figure 7.2 illustrates that the accuracy of the interaction model can significantly be improved by taking the model of F_2 of the neighbouring office into account. Moreover, the additional consideration of the model F_3 from office 3 does not entail a substantial improvement on the model accuracy. That means conversely that the neglect of the model F_3 is reasonable. □

Formally speaking, locally interconnected subsystems are considered according to the model (2.8) and the interconnection graph shown in Fig. 3.7a on page 60. The dynamics of the controlled subsystems F_i, $(\forall i \in \mathcal{N})$ are represented by the model

$$F_i : \begin{cases} \boldsymbol{y}_i(s) = \boldsymbol{F}_{\mathrm{yw}i}(s)\,\boldsymbol{w}_i(s) + \boldsymbol{F}_{\mathrm{ys}i}(s)\,\boldsymbol{s}_i(s) \\ \boldsymbol{z}_i(s) = \boldsymbol{F}_{\mathrm{zw}i}(s)\,\boldsymbol{w}_i(s) + \boldsymbol{F}_{\mathrm{zs}i}(s)\,\boldsymbol{s}_i(s). \end{cases}$$

From the view of subsystem S_1, the physical interaction

$$P_1^{\mathcal{N}} : \quad \boldsymbol{s}_1(s) = \boldsymbol{P}_1^{\mathcal{N}}(s)\,\boldsymbol{z}_1(s) \tag{7.1}$$

consists of all controlled subsystems F_i, $(i = 2, ..., N)$ as highlighted in Fig. 7.3. These controlled subsystems are partitioned into a *relevant part*, which is represented by

$$P_1^{\mathcal{N}_{\mathrm{S1}}} : \begin{cases} \boldsymbol{s}_1(s) = \hat{\boldsymbol{P}}_1^{\mathcal{N}_{\mathrm{S1}}}(s)\,\boldsymbol{z}_1(s) + \boldsymbol{P}_{\mathrm{sq1}}^{\mathcal{N}_{\mathrm{S1}}}(s)\,\boldsymbol{q}_1(s) \\ \boldsymbol{p}_1(s) = \boldsymbol{P}_{\mathrm{pz1}}^{\mathcal{N}_{\mathrm{S1}}}(s)\,\boldsymbol{z}_1(s) + \boldsymbol{P}_{\mathrm{pq1}}^{\mathcal{N}_{\mathrm{S1}}}(s)\,\boldsymbol{q}_1(s), \end{cases}$$

where $\mathcal{N}_{\mathrm{S1}} = \{1, ..., v-1\}$ and an *irrelevant part* which is modelled by

$$E_1^{\mathcal{N}_{\mathrm{W1}}} : \quad \boldsymbol{q}_1(s) = \boldsymbol{E}_1^{\mathcal{N}_{\mathrm{W1}}}(s)\,\boldsymbol{p}_1(s), \tag{7.2}$$

where $\mathcal{N}_{\mathrm{W1}} = \{v, ..., N\}$ (cf. Section 3.2.2).

This partition aims at the set up of an appropriate approximation of the physical interaction.

This approximation is described by the approximate model

$$\hat{P}_1^{\mathcal{N}_{S1}}: \quad \hat{\boldsymbol{s}}_1(s) = \hat{\boldsymbol{P}}_1^{\mathcal{N}_{S1}}(s)\boldsymbol{z}_1(s), \tag{7.3}$$

which is used for the controller design, while the irrelevant part $\boldsymbol{E}_1^{\mathcal{N}_{W1}}$ is considered as acceptable model uncertainty.

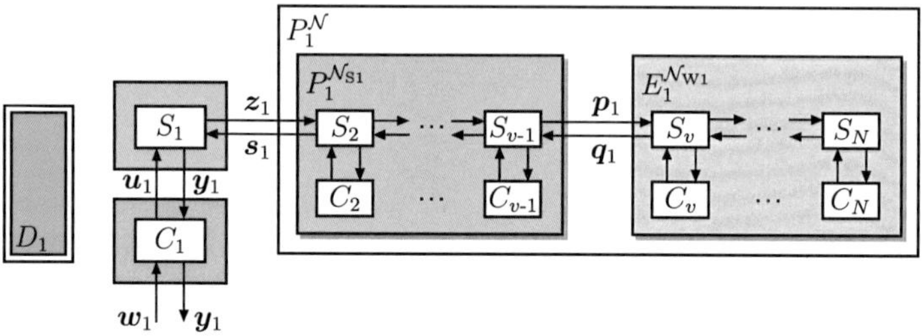

Figure 7.3: Structure of the partitioned physical interaction $P_1^{\mathcal{N}}$: The relevant part $P_1^{\mathcal{N}_{S1}}$ (yellow box) and irrelevant part $E_1^{\mathcal{N}_{W1}}$ (grey box).

In the next section it is shown that this separation is obtained by evaluating the interaction strength of the controlled subsystems F_i, $(i = 2, ..., N)$ with subsystem S_1. Moreover, it is shown that the more controlled subsystems are sorted as relevant for the controller design, the higher becomes the accuracy of the approximate model. However, a good approximation comprises only neighbouring controlled subsystems.

I/O stability of $P_1^{\mathcal{N}_{S1}}$ and $E_1^{\mathcal{N}_{W1}}$. For the remainder of this chapter it is essential to guarantee I/O stability of the relevant part $P_1^{\mathcal{N}_{S1}}$ and the irrelevant part $E_1^{\mathcal{N}_{W1}}$ independently of the separation boarder v and of the number of subsystems N. The following lemma proposes a sufficient condition:

Lemma 7.1 (Condition for I/O stability of the relevant part and the irrelevant part) *Let the transfer function matrices $\boldsymbol{F}_{zsi}(s)$, $(\forall i \in \mathcal{N})$ be I/O stable. The relevant part $P_1^{\mathcal{N}_{S1}}$ and the irrelevant part $E_1^{\mathcal{N}_{W1}}$ are I/O stable if*

$$\|\boldsymbol{F}_{zsi-1}(j\omega)\| \cdot \|\boldsymbol{L}_{i-1i}\| \cdot \|\boldsymbol{F}_{zsi}(j\omega)\| \cdot \|\boldsymbol{L}_{ii-1}\| < \tfrac{1}{4}, \quad \forall \omega \in \mathbb{R}, \quad \forall i = 2, .., N \tag{7.4}$$

holds.

Proof. See Appendix B.8 □

For the remainder of the chapter it is assumed that this condition is satisfied.

A 7.1 *There exist N control stations C_i, $(\forall i \in \mathcal{N})$ such that the condition* (7.4) *holds.*

Note that to satisfy the relation (7.4), the control stations has to be designed in a coordinated way. For instance, it is started with the design of control station C_1 followed by the design of C_2 and, subsequently, of C_3, etc. To design the control station C_i, $(i \geq 2)$, the model information F_{i-1} and K_{i-1} from the respective design agent D_{i-1} is required.

7.2.2 A measure of the coupling strength

This section introduces a measure to categorise the controlled subsystems as relevant or irrelevant for the controller design. This measure grounds on the evaluations of the interaction strength among the subsystems.

Strong connectivity between subsystems is a structural property to indicate feedback structures among the subsystems within the physical interconnection (cf. Def. 2.12). Due to this feedback structure, the local dynamics of subsystem S_i are subject to changes. To give a quantitative measure to these changes the terms *strong coupling* and *weak coupling* are introduced.

Definition 7.1 (Strong coupling and weak coupling) *Subsystem S_j is said to be strongly coupled with subsystem S_i, $(i \neq j)$ with respect to a given threshold $\xi(\omega)\colon \mathbb{R} \to \mathbb{R}_{+0}$ if*

1. *the subsystems S_j and S_i are strongly connected and*
2. *the interaction of S_j with S_i through the physical couplings exceeds the threshold $\xi(\omega)$.*

Subsystem S_j is said to be weakly coupled with subsystem S_i if at least one of the two conditions is not satisfied.

The first condition claims that an interaction between the subsystems S_i and S_j exists, whereas the second condition evaluates the strength of this interaction. In Definition 7.2 a more precise condition for a locally interconnected system is given.

Note that Definition 7.1 considers only the coupling strength of subsystem S_j on the subsystem S_i and not vice versa. That means that if S_j is strongly coupled to S_i, it is not implied that S_i is strongly coupled to S_j.

Remark 7.1 *It should be emphasised that in this thesis strong coupling is with respect to the strength of the physical interaction, whereas strong connectivity is a structural property as defined in Definition 2.12. In literature both term are usually used synonymously (e.g., [38, 40, 115, 117, 122]).*

Focussing on the particular situation, where locally interconnected controlled subsystems are considered, a more specific definition for strong coupling and weak coupling to subsystem S_1 is proposed based on Definition 7.1.

Definition 7.2 (Strong and weak coupling to S_1 for locally interconnected controlled subsystems) *Consider a local interconnection structure* (2.8) *and let* $\mathcal{N}_{\mathrm{S1}} = \{1, .., v-1\}$ *with* $v \geq 2$. *The controlled subsystem* F_v *is said to be weakly coupled to subsystem* S_1 *with respect to a given function* $\xi(\omega) \colon \mathbb{R} \to \mathbb{R}_{+0}$ *if the condition*

$$\|\hat{\boldsymbol{P}}_1^{\mathcal{N}_{\mathrm{S1}}\cup\{v\}}(\mathrm{j}\omega)\| - \|\hat{\boldsymbol{P}}_1^{\mathcal{N}_{\mathrm{S1}}}(\mathrm{j}\omega)\| \leq \xi(\omega), \quad \forall \omega \in \mathbb{R} \tag{7.5}$$

holds. The controlled subsystem F_v *is said to be strongly coupled to subsystem* S_1 *if*

$$\|\hat{\boldsymbol{P}}_1^{\mathcal{N}_{\mathrm{S1}}\cup\{v\}}(\mathrm{j}\omega)\| - \|\hat{\boldsymbol{P}}_1^{\mathcal{N}_{\mathrm{S1}}}(\mathrm{j}\omega)\| > \xi(\omega), \quad \exists \omega \in \mathbb{R} \tag{7.6}$$

holds.

The conditions (7.5) and (7.6) evaluate the difference between the dynamics of the approximate model (7.3) with and without considering the effect of the controlled subsystem F_v. That is, the *improvement of the approximation* $\hat{P}_1^{\{1,..,v-1\}}$ by considering F_v. Figure 7.4a depicts the structure of the approximate model $\hat{P}_1^{\{1,..,v-1\}}$ and Fig. 7.4b shows the structure of $\hat{P}_1^{\{1,..,v\}}$.

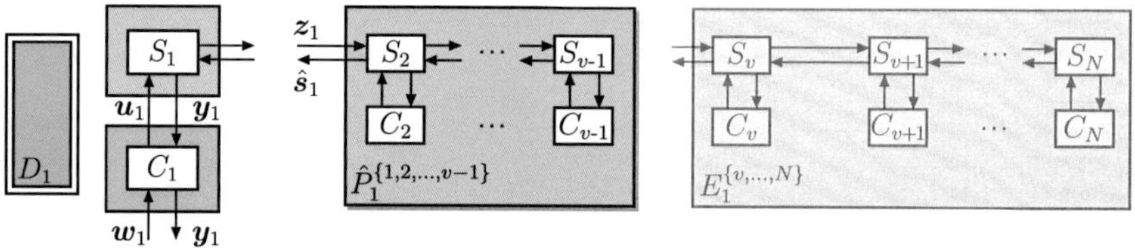

(a) Structure of the approximate model $\hat{P}_1^{\mathcal{N}_{\mathrm{S1}}}$

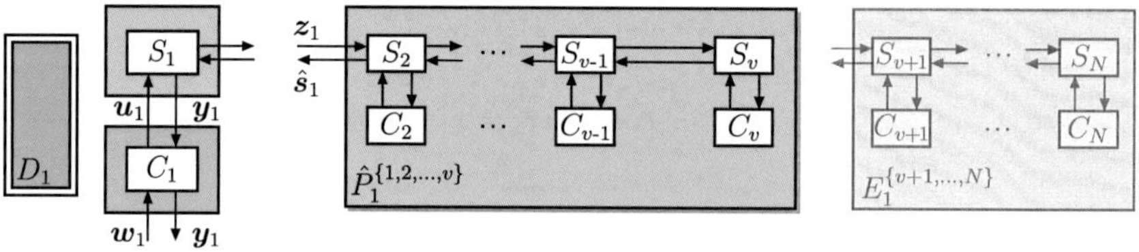

(b) Structure of the approximate model $\hat{P}_1^{\mathcal{N}_{\mathrm{S1}}\cup\{v\}}$

Figure 7.4: Structure of the approximate models $\hat{P}_1^{\mathcal{N}_{\mathrm{S1}}}$ and $\hat{P}_1^{\mathcal{N}_{\mathrm{S1}}\cup\{v\}}$ used to check whether controlled subsystem F_v is strongly coupled to subsystem S_1. The irrelevant part (grey box) is ignored.

The accuracy of the approximation is adjustable by the function $\xi(\omega)$. Accordingly, the smaller the threshold $\xi(\omega)$ is chosen, the more controlled subsystems are categorised as strongly coupled to S_1 and, thus, are considered for the approximate model. In other words, the function $\xi(\omega)$ adjusts the maximal model uncertainty of the physical interaction model it can be dealt with for the controller design. A detailed discussion about the choice of the function $\xi(\omega)$ is given at the end of Section 7.2.3.

7.2.3 A local condition to adjust the accuracy of the interaction model

This section proposes a local condition that enables D_1 to categorise a controlled subsystem into strongly coupled to subsystem S_1 or weakly coupled to subsystem S_1. The controlled subsystems that are strongly coupled to S_1 are assigned to the relevant part, which thus approximate the physical interaction. It is shown that the more controlled subsystems are considered to be relevant, the higher is the precision of the approximation. Moreover, it is evidenced that a good approximation is always composed of a small number of controlled subsystems.

To derive a locally checkable condition, the dynamics of the difference $\|\hat{\boldsymbol{P}}_1^{\mathcal{N}_{\mathrm{S1}}\cup\{v\}}(\mathrm{j}\omega)\| - \|\hat{\boldsymbol{P}}_1^{\mathcal{N}_{\mathrm{S1}}}(\mathrm{j}\omega)\|$ is analysed in detail. With eqn (C.9) from Appendix C.1 the following relation holds:

$$\begin{aligned}
&\|\hat{\boldsymbol{P}}_1^{\mathcal{N}_{\mathrm{S1}}\cup\{v\}}(\mathrm{j}\omega)\| - \|\hat{\boldsymbol{P}}_1^{\mathcal{N}_{\mathrm{S1}}}(\mathrm{j}\omega)\| \\
&\qquad = \Big\| \hat{\boldsymbol{P}}_1^{\mathcal{N}_{\mathrm{S1}}}(\mathrm{j}\omega) + \boldsymbol{P}_{\mathrm{sq1}}^{\mathcal{N}_{\mathrm{S1}}}(\mathrm{j}\omega)\left(\mathbf{I} - \boldsymbol{L}_{v-1v}\,\boldsymbol{F}_{\mathrm{zs}v}(\mathrm{j}\omega)\,\boldsymbol{L}_{vv-1}\,\boldsymbol{P}_{\mathrm{pq1}}^{\mathcal{N}_{\mathrm{S1}}}(\mathrm{j}\omega)\right)^{-1} \\
&\qquad\quad \boldsymbol{L}_{v-1v}\,\boldsymbol{F}_{\mathrm{zs}v}(\mathrm{j}\omega)\,\boldsymbol{L}_{vv-1}\,\boldsymbol{P}_{\mathrm{pz1}}^{\mathcal{N}_{\mathrm{S1}}}(\mathrm{j}\omega)1\Big\| - \|\hat{\boldsymbol{P}}_1^{\mathcal{N}_{\mathrm{S1}}}(\mathrm{j}\omega)\| \\
&\qquad \le \Big\| \boldsymbol{P}_{\mathrm{sq1}}^{\mathcal{N}_{\mathrm{S1}}}(\mathrm{j}\omega)\left(\mathbf{I} - \boldsymbol{L}_{v-1v}\,\boldsymbol{F}_{\mathrm{zs}v}(\mathrm{j}\omega)\,\boldsymbol{L}_{vv-1}\,\boldsymbol{P}_{\mathrm{pq1}}^{\mathcal{N}_{\mathrm{S1}}}(\mathrm{j}\omega)\right)^{-1} \\
&\qquad\quad \boldsymbol{L}_{v-1v}\,\boldsymbol{F}_{\mathrm{zs}v}(\mathrm{j}\omega)\,\boldsymbol{L}_{vv-1}\,\boldsymbol{P}_{\mathrm{pz1}}^{\mathcal{N}_{\mathrm{S1}}}(\mathrm{j}\omega)\Big\| \\
&\qquad \le \Delta_1^v(\omega)
\end{aligned} \tag{7.7}$$

with

$$\begin{aligned}
\Delta_1^v(\omega) :=& \|\boldsymbol{P}_{\mathrm{sq1}}^{\mathcal{N}_{\mathrm{S1}}}(\mathrm{j}\omega)\| \left(1 - \|\boldsymbol{L}_{v-1v}\| \cdot \|\boldsymbol{F}_{\mathrm{zs}v}(\mathrm{j}\omega)\,\boldsymbol{L}_{vv-1}\| \cdot \|\boldsymbol{P}_{\mathrm{pq1}}^{\mathcal{N}_{\mathrm{S1}}}(\mathrm{j}\omega)\|\right)^{-1} \\
& \|\boldsymbol{L}_{v-1v}\| \cdot \|\boldsymbol{F}_{\mathrm{zs}v}(\mathrm{j}\omega)\,\boldsymbol{L}_{vv-v}\| \cdot \|\boldsymbol{P}_{\mathrm{pz1}}^{\mathcal{N}_{\mathrm{S1}}}(\mathrm{j}\omega)\|.
\end{aligned} \tag{7.8}$$

Note that the inverse in (7.8) exists due to assumption A 7.1. Accordingly, $\Delta_1^v(\omega)$ represents an upper bound on the improvement of the approximation $\hat{P}_1^{\mathcal{N}_{\mathrm{S1}}}$ due to the consideration of F_v.

On the basis of (7.8) a sufficient condition is derived that enables design agent D_v to check whether or not the controlled subsystem F_v is strongly coupled to subsystem S_1 according to condition (7.5).

Theorem 7.1 (Local condition to classify strongly and weakly coupled subsystems) *Consider a local interconnection structure (2.8) and let $\mathcal{N}_{\mathrm{S1}} = \{1, .., v-1\}$ with $v \geq 2$. Then the controlled subsystem F_v is weakly coupled to subsystem S_1 according to (7.5) if*

$$\|\boldsymbol{F}_{\mathrm{zs}v}(\mathrm{j}\omega)\boldsymbol{L}_{vv-1}\| \leq \xi_v(\omega), \quad \forall \omega \in \mathbb{R} \tag{7.9}$$

with

$$\xi_v(\omega) = \|\boldsymbol{L}_{v-1v}\|^{-1} \left(\|\boldsymbol{P}_{\mathrm{sq1}}^{\mathcal{N}_{\mathrm{S1}}}(\mathrm{j}\omega)\| \cdot \left(\xi(\omega)\right)^{-1} \cdot \|\boldsymbol{P}_{\mathrm{pz1}}^{\mathcal{N}_{\mathrm{S1}}}(\mathrm{j}\omega)\| + \|\boldsymbol{P}_{\mathrm{pq1}}^{\mathcal{N}_{\mathrm{S1}}}(\mathrm{j}\omega)\| \right)^{-1} \tag{7.10}$$

holds. Moreover, if the controlled subsystem F_v is strongly coupled to subsystem S_1 according to (7.6), then

$$\|\boldsymbol{F}_{\mathrm{zs}v}(\mathrm{j}\omega)\boldsymbol{L}_{vv-1}\| > \xi_v(\omega), \quad \exists \omega \in \mathbb{R} \tag{7.11}$$

is satisfied.

Proof. Inserting (7.10) into (7.9) yields the claim

$$\Delta_1^v(\omega) \leq \xi(\omega), \quad \forall \omega \in \mathbb{R}.$$

Due to the relation (7.7) it follows that if (7.9) is fulfilled, the relation (7.5) holds, i.e.,

$$\|\boldsymbol{F}_{\mathrm{zs}v}(\mathrm{j}\omega)\boldsymbol{L}_{vv-1}\| \leq \xi_v(\omega),\ \forall \omega \in \mathbb{R} \ \Rightarrow \ \|\hat{\boldsymbol{P}}_1^{\mathcal{N}_{\mathrm{S1}}\cup\{v\}}(\mathrm{j}\omega)\| - \|\hat{\boldsymbol{P}}_1^{\mathcal{N}_{\mathrm{S1}}}(\mathrm{j}\omega)\| \leq \xi(\omega),\ \forall \omega \in \mathbb{R}$$

holds. Accordingly, inserting (7.10) into (7.11) yields the claim

$$\Delta_1^v(\omega) > \xi(\omega), \quad \exists \omega \in \mathbb{R}.$$

Due to the relation (7.7) it follows that if the relation (7.6) holds, then (7.11) is fulfilled, i.e.,

$$\|\hat{\boldsymbol{P}}_1^{\mathcal{N}_{\mathrm{S1}}\cup\{v\}}(\mathrm{j}\omega)\| - \|\hat{\boldsymbol{P}}_1^{\mathcal{N}_{\mathrm{S1}}}(\mathrm{j}\omega)\| > \xi(\omega),\ \exists \omega \in \mathbb{R} \ \Rightarrow \ \|\boldsymbol{F}_{\mathrm{zs}v}(\mathrm{j}\omega)\boldsymbol{L}_{vv-1}\| > \xi_v(\omega),\ \exists \omega \in \mathbb{R}$$

holds. □

The categorisation of a subsystem based on the conditions (7.11) ensures that all subsystems that are strongly coupled according to the original condition (7.6) are categorised as strongly coupled too. Conversely, if a subsystem is weakly coupled according to the original condition (7.5), it can, however, be categorised as strongly coupled by means of condition (7.11). This fact has the following consequence for the accuracy of the approximate model $\hat{P}_1^{\mathcal{N}_{\mathrm{S1}}}$: As mentioned before, the approximation comprises the controlled subsystems that are strongly coupled to S_1. Hence, the precision of the approximation becomes higher the more controlled subsystems are categorised as strongly coupled to S_1 (see next paragraph). As a consequence, the resulting approximation is at least as precise as desired.

The proposed conditions (7.9) and (7.11) are locally checkable in the sense that design agent D_1 can calculate the threshold $\xi_v(\omega)$ with its available model information. This threshold is subsequently transmitted to D_v, which is then able to check condition (7.9) and (7.11), respectively. This procedure is summarised in Algorithm 7.1 in Section 7.2.4.

Due to the local interconnection structure the improvement of the approximation if F_v is considered is greater that the improvement if F_{v+1} is taken into account. Accordingly, if the controlled subsystem F_v is weakly coupled to subsystem S_1, then all controlled subsystem F_i, $(\forall i \in \mathcal{N},\ i > v)$ are also weakly coupled to S_1. This fact is summarised by means of the upper bound $\Delta_1^v(\omega)$ defined in (7.8) in the following theorem.

Theorem 7.2 (Relation between controlled subsystems weakly coupled to S_1) *Consider a local interconnection structure* (2.8) *and let* $\mathcal{N}_{\mathrm{S1}} = \{1, .., v-1\}$ *with* $v \geq 2$. *The relation*

$$\Delta_1^{v+1}(\omega) < \Delta_1^v(\omega), \quad \forall \omega \in \mathbb{R} \tag{7.12}$$

holds.

Proof. See Appendix B.9. □

Due to the fact that $\Delta_1^{v+1}(\omega)$ is *strictly* less than $\Delta_1^v(\omega)$, the improvement of the approximation decreases monotonically with the number of controlled subsystems taken into account. This fact is studied in the following paragraph in more detail.

Analysis of the improvement of the approximation. A qualitative relation between the improvement of the approximation due to the consideration of F_v and of F_{v+1} is given by (7.12). This paragraph analyses the improvement quantitatively.

To derive this quantitative measure, the calculation of $\Delta_1^v(\omega)$ is analysed in detail. In particular, based on

$$\Delta_1^2(\omega) = \|\boldsymbol{L}_{12}\| \cdot \|\boldsymbol{F}_{\mathrm{zs2}}(\mathrm{j}\omega)\boldsymbol{L}_{21}\|$$

the following upper bounds for $\Delta_1^v(\omega)$, $(v = 3, 4)$ result to

$$\begin{aligned}\Delta_1^3(\omega) =& \|\boldsymbol{L}_{12}\boldsymbol{F}_{\mathrm{zs2}}(\mathrm{j}\omega)\| \cdot \|\boldsymbol{L}_{23}\| \cdot \|\boldsymbol{F}_{\mathrm{zs3}}(\mathrm{j}\omega)\boldsymbol{L}_{32}\| \cdot \|\boldsymbol{F}_{\mathrm{zs2}}(\mathrm{j}\omega)\boldsymbol{L}_{21}\| \\ & \left(1 - \|\boldsymbol{L}_{23}\| \cdot \|\boldsymbol{F}_{\mathrm{zs3}}(\mathrm{j}\omega)\boldsymbol{L}_{32}\| \cdot \|\boldsymbol{F}_{\mathrm{zs2}}(\mathrm{j}\omega)\|\right)^{-1}\end{aligned}$$

$$\begin{aligned}
&\leq \Delta_1^2(\omega) \cdot \|\boldsymbol{L}_{23}\| \cdot \|\boldsymbol{F}_{\mathrm{zs3}}(\mathrm{j}\omega)\boldsymbol{L}_{32}\| \cdot \|\boldsymbol{F}_{\mathrm{zs2}}(\mathrm{j}\omega)\| \\
&\qquad \left(1 - \|\boldsymbol{L}_{23}\| \cdot \|\boldsymbol{F}_{\mathrm{zs3}}(\mathrm{j}\omega)\boldsymbol{L}_{32}\| \cdot \|\boldsymbol{F}_{\mathrm{zs2}}(\mathrm{j}\omega)\|\right)^{-1} \\
&< \Delta_1^2(\omega) \cdot {}^1\!/_3 \qquad (7.13) \\
\Delta_1^4(\omega) &= \|\boldsymbol{P}_{\mathrm{sq1}}^{\{1,2,3\}}(\mathrm{j}\omega)\| \cdot \|\boldsymbol{L}_{34}\| \cdot \|\boldsymbol{F}_{\mathrm{zs4}}(\mathrm{j}\omega)\boldsymbol{L}_{43}\| \cdot \|\boldsymbol{P}_{\mathrm{pz1}}^{\{1,2,3\}}(\mathrm{j}\omega)\| \\
&\qquad \left(1 - \|\boldsymbol{L}_{34}\| \cdot \|\boldsymbol{F}_{\mathrm{zs4}}(\mathrm{j}\omega)\boldsymbol{L}_{43}\| \cdot \|\boldsymbol{P}_{\mathrm{pq1}}^{\{1,2,3\}}(\mathrm{j}\omega)\|\right)^{-1} \\
&< \Delta_1^2(\omega) \cdot {}^1\!/_3 \cdot {}^1\!/_2,
\end{aligned}$$

where the inversions exist due to assumption A 7.1. The relation (7.13) states that the maximal improvement of the approximation due to F_3 is less than $33.33\,\%$ of the improvement that is achieved by the consideration of F_2. Accordingly, F_4 improves the approximation by maximal $16.67\,\%$ compared to F_2. Continuing these calculations, the following result can be concluded.

Proposition 7.1 (Limited improvement of the approximation) *Let $\mathcal{N}_{\mathrm{S1}} = \{1, ..., v-1\}$ with $v \geq 3$. The improvement of the approximation $\hat{P}_1^{\mathcal{N}_{\mathrm{S1}}}$ due to the consideration of the controlled subsystem F_v is limited according to*

$$\Delta_1^v(\omega) < \frac{2}{v(v-1)} \Delta_1^2(\omega), \quad \forall \omega \in \mathbb{R}. \qquad (7.14)$$

The upper bounds (7.14) for $v = 3, 4, ..., 10$ are highlighted in Fig. 7.5. This consideration reveals, on the one hand, that the approximation is always improved if F_v is taken into account and, on the other hand, that the improvement monotonically decreases.

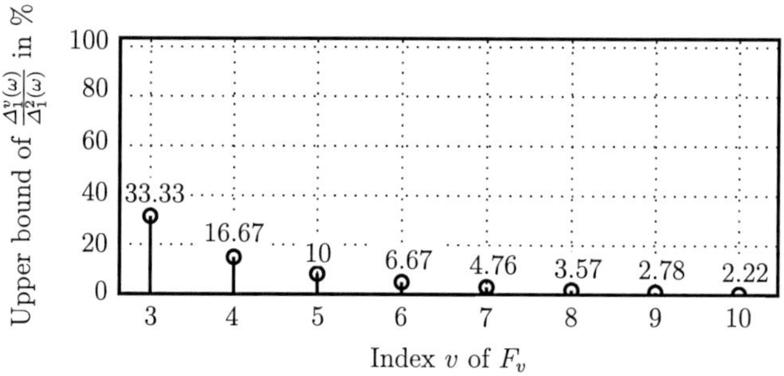

Figure 7.5: Characterisation of the upper bound of the improvement of the approximation $\hat{P}_1^{\mathcal{N}_{\mathrm{S1}}}$ due to F_v.

Based on Proposition 7.1, in the following it is shown that an upper bound of the physical interaction $P_1^{\mathcal{N}}$ exists which is independent of the number of subsystems. Moreover, it is shown that the dynamics of $P_1^{\mathcal{N}}$ are appropriately approximated by some controlled subsystems.

Consider an upper bound on the dynamics $\boldsymbol{P}_1^{\mathcal{N}}(s)$ that is expressed by $\Delta_1^v(\omega)$ as follows

$$\|\boldsymbol{P}_1^{\mathcal{N}}(\mathrm{j}\omega)\| \leq \sum_{i=2}^{N} \Delta_1^i(\omega) = \sum_{i=2}^{v-1} \Delta_1^i(\omega) + \sum_{i=v}^{N} \Delta_1^i(\omega). \tag{7.15}$$

The first addend represents an upper bound of the dynamics $\|\hat{\boldsymbol{P}}_1^{\mathcal{N}_{\mathrm{S1}}}(\mathrm{j}\omega)\|$ of the approximate model $\hat{P}_1^{\mathcal{N}_{\mathrm{S1}}}$, i.e.,

$$\|\hat{\boldsymbol{P}}_1^{\mathcal{N}_{\mathrm{S1}}}(\mathrm{j}\omega)\| \leq \sum_{i=2}^{v-1} \Delta_1^i(\omega). \tag{7.16}$$

The second addend represents an upper bound on the error dynamics

$$\|\boldsymbol{P}_{\Delta 1}^{\mathcal{N}_{\mathrm{S1}}}(\mathrm{j}\omega)\| \leq \sum_{i=v}^{N} \Delta_1^i(\omega). \tag{7.17}$$

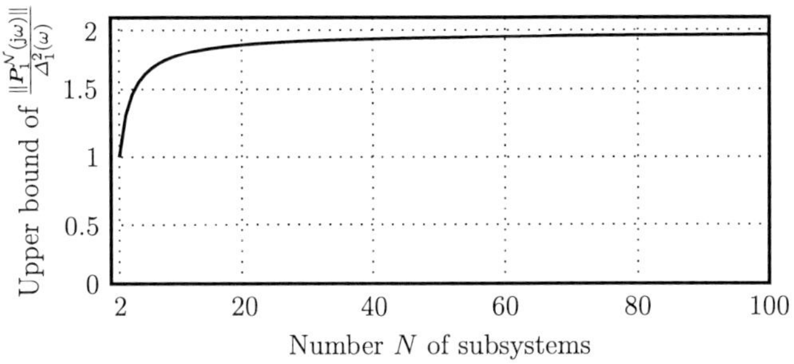

Figure 7.6: Resulting upper bound of the physical interaction dynamics $\|\boldsymbol{P}_1^{\mathcal{N}}(\mathrm{j}\omega)\|$ for an increasing number of subsystems.

Based on (7.14), the inequality (7.15) can be rewritten into

$$\|\boldsymbol{P}_1^{\mathcal{N}}(\mathrm{j}\omega)\| \leq \Delta_1^2(\omega) \sum_{i=2}^{N} \frac{\Delta_1^i(\omega)}{\Delta_1^2(\omega)} \leq \Delta_1^2(\omega) \sum_{i=2}^{N} \frac{2}{i(i-1)}. \tag{7.18}$$

For $N \to \infty$, the series (7.18) converges to

$$\sum_{v=2}^{\infty} \frac{2}{v(v-1)} = 2 \sum_{v=2}^{\infty} \frac{1}{v-1} - \frac{1}{v} = 2 \cdot \left(\frac{1}{1} - \frac{1}{\infty}\right) = 2$$

in order that the maximal impact of all other controlled subsystems on subsystem S_1 through the physical interaction is limited by

$$\|\boldsymbol{P}_1^{\mathcal{N}}(\mathrm{j}\omega)\| \leq 2 \cdot \Delta_1^2(\omega), \tag{7.19}$$

despite an unlimited number of subsystems (see also the proof of Lemma 7.1). Figure 7.6 visualises the evaluation of the series (7.18) for $N = 100$ subsystems. It can be seen that the upper bound (7.19) will be asymptotically reached.

Now, it is focused on the accuracy of the approximate model $\hat{P}_1^{\mathcal{N}_{\mathrm{S1}}}$. From the relation (7.19) can be concluded that $50\,\%$ of the upper bound (7.19) is attributable to the effect of the controlled subsystem F_2 (i.e., $\Delta_1^2(\omega)$). Hence, the consideration of the controlled subsystem F_2 as approximation seems reasonable. In particular, an upper bound on the approximated dynamics $\|\hat{\boldsymbol{P}}_1^{\mathcal{N}_{\mathrm{S1}}}(\mathrm{j}\omega)\|$ is derived from (7.16) using (7.14). This upper bound results to

$$\|\hat{\boldsymbol{P}}_1^{\mathcal{N}_{\mathrm{S1}}}(\mathrm{j}\omega)\| \leq \sum_{i=2}^{v-1} \Delta_1^i(\omega) \leq \sum_{i=2}^{v-1} \frac{2}{i(i-1)} \Delta_1^2(\omega).$$

For example, consider the situation in which the approximation comprises the controlled subsystems F_2 and F_3 (i.e., $\mathcal{N}_{\mathrm{S1}} = \{1,2,3\}$). Then, the plot $\|\hat{\boldsymbol{P}}_1^{\{1,2,3\}}(\mathrm{j}\omega)\|$ is bounded by

$$\|\hat{\boldsymbol{P}}_1^{\{1,2,3\}}(\mathrm{j}\omega)\| < 1.33\,\Delta_1^2(\omega) = 0.67\,\|\boldsymbol{P}_1^{\mathcal{N}}(\mathrm{j}\omega)\|$$

if $N \to \infty$. That is, the approximation $\|\hat{\boldsymbol{P}}_1^{\{1,2,3\}}(\mathrm{j}\omega)\|$ constitutes up to $66.67\,\%$ of the actual dynamics $\|\boldsymbol{P}_1^{\mathcal{N}}(\mathrm{j}\omega)\|$. Furthermore, the smaller the number of subsystems is, the bigger becomes this percentage indication. For instance, for $N = 10$ the same approximation $\|\hat{\boldsymbol{P}}_1^{\{1,2,3\}}(\mathrm{j}\omega)\|$ already constitutes up to $74.07\,\%$ of the actual dynamics (Fig. 7.7) and for $N = 3$ it yields $\|\hat{\boldsymbol{P}}_1^{\{1,2,3\}}(\mathrm{j}\omega)\| = \|\boldsymbol{P}_1^{\mathcal{N}}(\mathrm{j}\omega)\|$.

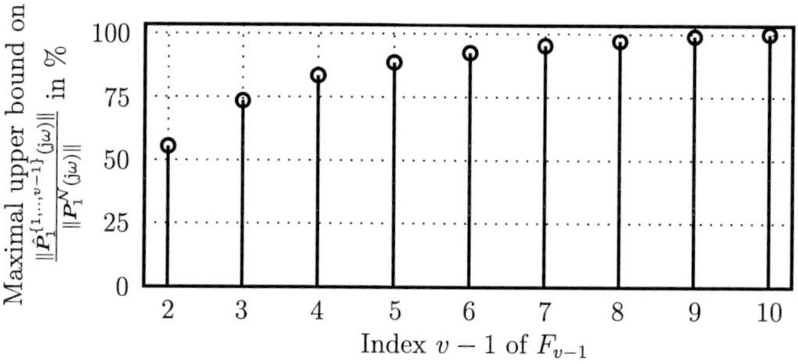

Figure 7.7: Maximal percentage of the approximation $\|\hat{\boldsymbol{P}}_1^{\{1,\dots,v-1\}}(\mathrm{j}\omega)\|$ on the upper bound of $\|\boldsymbol{P}_1^{\mathcal{N}}(\mathrm{j}\omega)\|$ for $N = 10$.

Choice of the function $\xi(\omega)$. On the basis of the forgoing analysis, this paragraph focuses on the choice of the threshold $\xi(\omega)$ so that the admissible uncertainty of the physical interaction model is adjustable. That is the answer to the question how $\xi(\omega)$ shall be chosen such that

$$\|\boldsymbol{P}_{\Delta 1}^{\mathcal{N}_{\mathrm{S}1}}(\mathrm{j}\omega)\| \leq \eta(\omega) \tag{7.20}$$

for a given threshold $\eta(\omega)\colon \mathbb{R} \to \mathbb{R}_{+0}$.

Let the threshold $\xi(\omega)$ be given. Consider that the particular condition

$$\|\boldsymbol{F}_{\mathrm{zs}2}(\mathrm{j}\omega)\boldsymbol{L}_{21}\| \leq \xi_2(\omega) \tag{7.21}$$

is satisfied, i.e., $\Delta_1^2(\omega) \leq \xi(\omega)$. Then it can be concluded from the relations (7.17) and (7.19) that

$$\|\boldsymbol{P}_{\Delta 1}^{\{1\}}(\mathrm{j}\omega)\| \leq \Delta_1^2(\omega) \sum_{i=2}^{N} \frac{2}{i(i-1)} \leq 2 \cdot \xi(\omega). \tag{7.22}$$

The right hand side of (7.22) represents an upper bound of the error dynamics $\|\boldsymbol{P}_{\Delta 1}^{\{1\}}(\mathrm{j}\omega)\|$. Similarly, if the condition (7.21) is violated but

$$\|\boldsymbol{F}_{\mathrm{zs}3}(\mathrm{j}\omega)\boldsymbol{L}_{32}\| \leq \xi_3(\omega) \tag{7.23}$$

holds, i.e., $\Delta_1^3(\omega) \leq \xi(\omega)$, then the relation

$$\|\boldsymbol{P}_{\Delta 1}^{\{1,2\}}(\mathrm{j}\omega)\| < \Delta_1^3 + \Delta_1^2(\omega) \sum_{i=4}^{N} \frac{2}{i(i-1)} \leq \xi(\omega) + \Delta_1^2(\omega) \cdot 0.67$$

holds. Accordingly,

$$\|\boldsymbol{P}_{\Delta 1}^{\{1,2,3\}}(\mathrm{j}\omega)\| < \xi(\omega) + \Delta_1^2(\omega) \cdot 0.5.$$

results if (7.21) and (7.23) are violated but $\|\boldsymbol{F}_{\mathrm{zs}4}(\mathrm{j}\omega)\boldsymbol{L}_{43}\| \leq \xi_4(\omega)$ holds.

Proceeding this considerations, it can be concluded that an upper bound on the error dynamics generally results to

$$\|\boldsymbol{P}_{\Delta 1}^{\mathcal{N}_{\mathrm{S}1}}(\mathrm{j}\omega)\| < \begin{cases} 2 \cdot \xi(\omega) & \text{for } v = 2 \\ \xi(\omega) + \Delta_1^2(\omega) \sum_{i=v+1}^{\infty} \frac{2}{i(i-1)} & \text{for } v \geq 3. \end{cases}$$

In order to satisfy the claim (7.20), the threshold $\xi(\omega)$ has to be adapted according to

$$\xi(\omega) = \begin{cases} 0.5 \cdot \eta(\omega) & \text{for } v = 2 \\ \eta(\omega) - \Delta_1^2(\omega) \sum_{i=v+1}^{\infty} \frac{2}{i(i-1)} & \text{for } v \geq 3. \end{cases} \tag{7.24}$$

7.2.4 Model procurement algorithm

This section presents a local algorithm to enable D_1 to set up an appropriate approximate model $\hat{P}_1^{\mathcal{N}_{\mathrm{S1}}}$ of the physical interaction $P_1^{\mathcal{N}}$.

As proposed by assumption A 3.2, each design agent D_i has initially available the model set $\mathcal{M}_i = \{S_i, C_i, K_i\}$ that consists of the model of subsystem S_i, control station C_i and the local couplings K_i, respectively. Additionally, each design agent stores its local design condition $\mathcal{A}_{\mathrm{D}i}$ defined in (6.18) with the design parameters ε_i and $\varphi(\omega)$. Moreover, the design agent D_1 also knows its decision threshold $\xi(\omega)$.

The model procurement is processed iteratively. Starting from the initially available model information, design agent D_1 can calculate $\xi_2(\omega)$ according to (7.10) and transmits this threshold to D_2 (i.e., D_1 requests the model set $\mathcal{M}_2$). D_2 responds with the model set $\mathcal{M}_2$ if the condition (7.9) is violated. With the model sets $\mathcal{M}_1$ and $\mathcal{M}_2$ at hand, design agent D_1 is able to calculate the threshold $\xi_3(\omega)$ and can request the model set $\mathcal{M}_3$ from D_3. This processing is summarised in Algorithm 7.1.

Algorithm 7.1: Approximating the interaction dynamics by D_1

Given: Model set $\mathcal{M}_i = (C_i, S_i, K_i)$ available to D_i, $(\forall i \in \mathcal{N})$,
local design condition $\mathcal{A}_{\mathrm{D}i}$ available to D_i, $(\forall i \in \mathcal{N})$ and
threshold $\xi(\omega)$ or $\eta(\omega)$ available to D_1

Initialise: $\kappa = 0, \quad \mathcal{N}_{\mathrm{S1}} = \{1\}, \quad \mathcal{P}_1 = \{2\}$

Processing on D_1:

do
- Set $\kappa = \kappa + 1$
- Calculate $\xi_i(\omega)$ according to (7.10)
- **run** `res = getApproximation(`i`,` $\xi_i(\omega)$`)`
- **if** `res=` $\{\mathcal{M}_i,\ \mathcal{A}_{\mathrm{D}i}\}$ **then** set $\mathcal{N}_{\mathrm{S1}} = \mathcal{N}_{\mathrm{S1}} \cup \{i\}$, $\mathcal{P}_1 = \mathcal{P}_v \setminus \mathcal{N}_{\mathrm{S1}}$ and $\mathcal{M}_1 = \mathcal{M}_1 \cup \mathcal{M}_i$ **else** calculate $\hat{P}_1^{\mathcal{N}_{\mathrm{S1}}}$ and break.

while $\mathcal{N}_{\mathrm{S1}} \subseteq \mathcal{N}$

Processing on D_i:

function `getApproximation(`i`,` $\xi_i(\omega)$`)`
- **if** condition (7.9) is violated **then** return $\{\mathcal{M}_i,\ \mathcal{A}_{\mathrm{D}i}\}$ **else** return $\emptyset$

end

Result: Approximate model $\hat{P}_1^{\mathcal{N}_{\mathrm{S1}}}$ and local design condition $\mathcal{A}_{\mathrm{D}i}$, $(\forall i \in \mathcal{N}_{\mathrm{S1}})$ available to design agent D_1

The variable κ represents the number of iterations of the algorithm and accordingly counts the number of communications among the design agents as analysed in the next paragraph. The function `getApproximation` (Lines 7–9) is implemented on each design agent, whereas the loop from Line 1 to Line 6 is only processed by the design agent that wants to set up the approximate model. At the end of the Algorithm 7.1, the sets $\mathcal{N}_{\mathrm{S1}} = \{1, ..., v-1\}$ and $\mathcal{P}_1 = \{v\}$, the model $\hat{P}_1^{\mathcal{N}_{\mathrm{S1}}}$ as well as the design condition $\mathcal{A}_{\mathrm{D}i}$, $(\forall i \in \mathcal{N}_{\mathrm{S1}})$ are available to D_1. Based on these local design conditions $\mathcal{A}_{\mathrm{D}i}$, D_1 can derive the cooperative performance condition $\mathcal{A}_{\mathrm{D}}^{\mathcal{N}_{\mathrm{S1}}}$ defined in (7.40) in Section 7.3.3.

Note that Algorithm 7.1 can easily be adapted to be applicable for arbitrary design agents D_i. Merely, the loop (Lines 1 – 6) has to be executed by D_i in order to request the design agents D_j, $(\forall j \in \mathcal{N},\ j < i)$ and D_k, $(\forall k \in \mathcal{N},\ k > i)$. This request can be processed in parallel.

Communication among the design agents. During the processing of Algorithm 7.1, the design agents communicate models, control algorithms and design conditions among each other. The communication is visualised by the communication graph $\mathcal{G}_{\mathrm{D}}(\kappa)$ shown in Fig. 7.8. As it can be seen, merely $v-1$ communications are necessary to procure the required model sets $\mathcal{M}_i$, $(\forall i \in \mathcal{N}_{\mathrm{S1}})$ and design conditions $\mathcal{A}_{\mathrm{D}i}$, $(\forall i \in \mathcal{N}_{\mathrm{S1}})$ by design agent D_1. Thereafter, design agent D_1 knows the model set

$$\mathcal{M}_1 = \{S_1,\, S_2, \cdots, S_{v-1}\} \cup \{C_1,\, C_2, \cdots, C_{v-1}\} \cup \{K_1,\, K_2, \cdots, K_{v-1}\}\,.$$

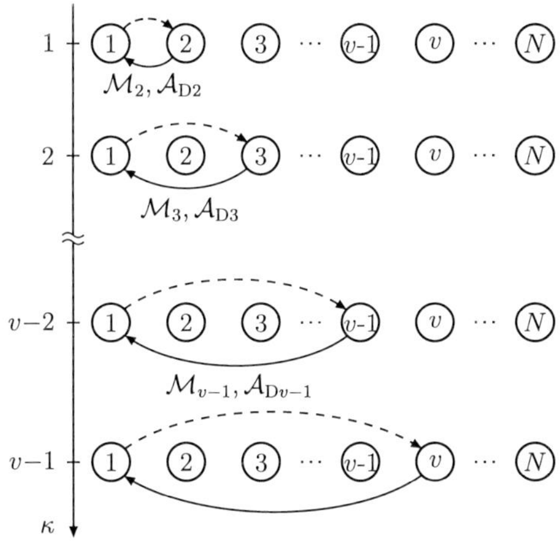

Figure 7.8: Communication graphs $\mathcal{G}_{\mathrm{D}}(\kappa)$ during the processing of the procurement Algorithm 7.1: Dashed edges denote requests and solid edges denote responses.

Note that D_v responds not with its models but informs the design agent D_1 only about the adherence of condition (7.9). That is, D_1 knows that F_v is weakly coupled to S_1 and, thus, all other controlled subsystems F_i, $(i > v)$.

7.3 Cooperative design conditions

This sections, first, proposes a condition that enables the controller design using the subsystem model S_1 together with the approximate model $\hat{P}_1^{\mathcal{N}_{S1}}$ only, while no model information from the design agents D_i, $(\forall i \in \mathcal{N}_{W1})$ is required. Based on this condition, cooperative design conditions are derived that aim at I/O stability and a certain performance of the overall closed-loop system. These design conditions are cooperative in the sense that amongst the design of C_1, the design of $C_2, ..., C_{v-1}$ is allowed so as cooperatively satisfy the design conditions. Moreover, it will be shown that the conservatism of the local design conditions $\mathcal{A}_i$ and $\mathcal{A}_{Di}$ from Section 6.3 will be relaxed.

7.3.1 Conditions to enable the design with an approximate model of the physical interaction

Section 7.2.1 has outlined the idea of subdividing the physical interaction into a relevant part and an irrelevant part. While the relevant part which comprises the strongly coupled controlled subsystems forms the approximate model of the physical interaction and are known by D_1 (after executing Alg. 7.1), the irrelevant part remains unknown. In direct analogy to Lemma 6.1, conditions on the controlled subsystems are proposed that enables D_1 to set up a comparison system of the irrelevant part without the usage of any model information from the design agents D_i, $(\forall i \in \mathcal{N}_{W1})$.

Consider the connection of all controlled subsystems F_i that has been categorised as strongly coupled to subsystem S_1 according to

$$F^{\mathcal{N}_{S1}} = \text{con}\left(\{F_i,\, i \in \mathcal{N}_{S1}\} \cup \{K_i,\, i \in \mathcal{N}_{S1}\},\, \begin{pmatrix} \boldsymbol{w}^{\mathcal{N}_{S1}}(s) \\ \boldsymbol{q}_1(s) \end{pmatrix},\, \begin{pmatrix} \boldsymbol{y}^{\mathcal{N}_{S1}}(s) \\ \boldsymbol{p}_1(s) \end{pmatrix}\right),$$

where $\boldsymbol{q}_1(s) = \boldsymbol{L}_{v-1v}\boldsymbol{z}_v(s)$ and $\boldsymbol{p}_1(s) = \boldsymbol{z}_{v-1}(s)$. The resulting model $F^{\mathcal{N}_{S1}}$ is called the *cooperative controlled subsystem* and is represented in frequency domain by

$$F^{\mathcal{N}_{S1}}: \begin{cases} \boldsymbol{y}^{\mathcal{N}_{S1}}(s) = \boldsymbol{F}_{\text{yw}}^{\mathcal{N}_{S1}}(s)\,\boldsymbol{w}^{\mathcal{N}_{S1}}(s) + \boldsymbol{F}_{\text{yq}}^{\mathcal{N}_{S1}}(s)\,\boldsymbol{q}_1(s) \\ \boldsymbol{p}_1(s) = \boldsymbol{F}_{\text{pw}}^{\mathcal{N}_{S1}}(s)\,\boldsymbol{w}^{\mathcal{N}_{S1}}(s) + \boldsymbol{F}_{\text{pq}}^{\mathcal{N}_{S1}}(s)\,\boldsymbol{q}_1(s) \end{cases}$$

with the cooperative input signal and cooperative output signal

$$\boldsymbol{w}^{\mathcal{N}_{\mathrm{S1}}}(s) = \begin{pmatrix} \boldsymbol{w}(s) \\ \vdots \\ \boldsymbol{w}_{v-1}(s) \end{pmatrix}, \quad \boldsymbol{y}^{\mathcal{N}_{\mathrm{S1}}}(s) = \begin{pmatrix} \boldsymbol{y}(s) \\ \vdots \\ \boldsymbol{y}_{v-1}(s) \end{pmatrix},$$

respectively, and with the transfer function matrices from Appendix C.2. The overall closed-loop system F can, thus, be recomposed by the controlled subsystems F_i, $(\forall i \in \mathcal{N}_{\mathrm{W1}})$ and by the cooperatively controlled subsystem $F^{\mathcal{N}_{\mathrm{S1}}}$ according to

$$F = \mathrm{con}\left(\left\{F^{\mathcal{N}_{\mathrm{S1}}},\, F_i,\, i \in \mathcal{N}_{\mathrm{W1}}\right\} \cup \left\{K_{v-1}, K_i,\, i \in \mathcal{N}_{\mathrm{W1}}\right\},\, \boldsymbol{w},\, \boldsymbol{y}\right) \tag{7.25}$$

as highlighted by the structure shown in Fig. 7.9. For this recomposed overall closed-loop system, the results from Chapter 6 can directly be transferred. In particular, this section proposes conditions to enable the design using only the approximate model (cf. Section 6.2). Moreover, Section 7.3 proposes design conditions that guarantee I/O stability and a certain I/O performance of the overall closed-loop system (cf. Section 6.3).

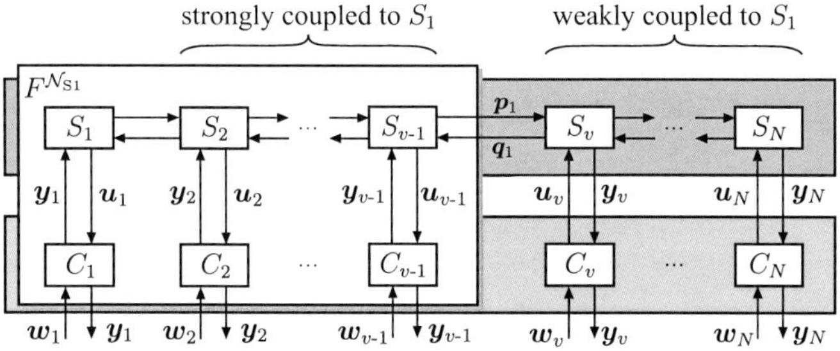

Figure 7.9: Structure of the overall closed-loop system composed of the cooperatively controlled subsystem $F^{\mathcal{N}_{\mathrm{S1}}}$ and the controlled subsystems F_i, $(\forall i \in \mathcal{N}_{\mathrm{W1}})$.

In accordance with Lemma 6.1, the following lemma proposes conditions on the cooperatively controlled subsystem $F^{\mathcal{N}_{\mathrm{S1}}}$ and on the controlled subsystems F_i, $(\forall i \in \mathcal{N}_{\mathrm{W1}})$ to enable the modelling of the physical interaction $P_1^{\mathcal{N}}$ with relevant models only.

Lemma 7.2 (Cooperative conditions to constrain the effect of the physical interaction)
Consider a local interconnection (2.8) *and let* $\mathcal{N}_{\mathrm{S1}} = \{1, ..., v-1\}$ *and* $\mathcal{N}_{\mathrm{W1}} = \{v, ..., N\}$

with $v \geq 2$. If the relations

$$\|\boldsymbol{F}_{\mathrm{pq}}^{\mathcal{N}_{\mathrm{S1}}}(\mathrm{j}\omega)\boldsymbol{L}_{v-1v}\| < \varphi(\omega), \quad \forall \omega \in \mathbb{R} \tag{7.26}$$

$$\|\boldsymbol{F}_{\mathrm{zs}i}(\mathrm{j}\omega)\| \sum_{j\in\mathcal{P}_i} \|\boldsymbol{L}_{ij}\| < \varphi(\omega), \quad \forall \omega \in \mathbb{R}, \quad \forall i \in \mathcal{N}_{\mathrm{W1}} \tag{7.27}$$

are satisfied for a given function $\varphi(\omega)\colon \mathbb{R} \to \mathbb{R}_+$ with $\varphi(\omega) \leq 1$, then

- *the system*

$$\bar{E}_i^{\{i\}}: \quad \bar{p}_i(\omega) = \underbrace{\|\boldsymbol{L}_{v-1v}\|\varphi(\omega)}_{\bar{E}_i^{\{i\}}(\omega)} \cdot \|\boldsymbol{q}_i(\mathrm{j}\omega)\|, \tag{7.28}$$

  is a comparison system of the irrelevant part $E_i^{\mathcal{N}_{\mathrm{W1}}}$ represented by (7.2) for all $i \in \mathcal{N}_{\mathrm{S1}}$ and

- *the system (6.4) is a comparison system of the physical interaction $P_i^{\mathcal{N}}$ represented by (7.1) for all $i \in \mathcal{N}_{\mathrm{W1}}$.*

Proof. See Appendix B.10. □

In Lemma 6.1 it is claimed that the magnitude of each transfer function $\boldsymbol{F}_{\mathrm{zs}i}(s)$, $(i \in \mathcal{N}_{\mathrm{S}i})$ has to be small. Due to the accumulation of the controlled subsystems F_i, $(i \in \mathcal{N}_{\mathrm{S}i})$ to the cooperatively controlled subsystem $F^{\mathcal{N}_{\mathrm{S1}}}$, this claim is relaxed in the sense that only the cooperative transfer function $\boldsymbol{F}_{\mathrm{pq}}^{\mathcal{N}_{\mathrm{S1}}}(s)$ has to be small. That is, high magnitudes of single transfer functions are compensated by neighbours with small magnitudes. This fact is considered in depth in Section 7.3.2.

Analysis from the local view of S_1. From the local view of the subsystems S_1, a comparison system of the physical interaction $P_1^{\mathcal{N}}$ can be determined by the combination of the relevant part $P_1^{\mathcal{N}_{\mathrm{S1}}}$ and the comparison system $\bar{E}_1^{\{1\}}$ of the irrelevant part according to

$$\bar{P}_1^{\mathcal{N}_{\mathrm{S1}}} = \mathrm{con}\left(\left\{\|P_1^{\mathcal{N}_{\mathrm{S1}}}\|,\ \bar{E}_1^{\{1\}}\right\},\ \|\boldsymbol{z}_1\|,\ \bar{s}_1\right).$$

The structure of the comparison system $\bar{P}_1^{\mathcal{N}_{\mathrm{S1}}}$ is illustrated in Fig. 7.10. This comparison system is represented by

$$\bar{P}_1^{\mathcal{N}_{\mathrm{S1}}}: \quad \bar{s}_1(\omega) = \underbrace{\left(\|\hat{\boldsymbol{P}}_1^{\mathcal{N}_{\mathrm{S1}}}(\mathrm{j}\omega)\| + \bar{P}_{\Delta 1}^{\mathcal{N}_{\mathrm{S1}}}(\omega)\right)}_{=\,\bar{P}_1^{\mathcal{N}_{\mathrm{S1}}}(\omega)} \|\boldsymbol{z}_1(\mathrm{j}\omega)\|, \tag{7.29}$$

where $\hat{\boldsymbol{P}}_1^{\mathcal{N}_{S1}}(s)$ approximates the dynamics of the physical interaction $\boldsymbol{P}_1^{\mathcal{N}}(s)$ (cf. approximate model (7.3)) and

$$\bar{P}_{\Delta 1}^{\mathcal{N}_{S1}}(\omega) = \|\boldsymbol{P}_{\mathrm{sq1}}^{\mathcal{N}_{S1}}(\mathrm{j}\omega)\| \cdot \|\boldsymbol{L}_{v-1v}\|\varphi(\omega) \left(1 - \|\boldsymbol{P}_{\mathrm{pq1}}^{\mathcal{N}_{S1}}(\mathrm{j}\omega)\| \cdot \|\boldsymbol{L}_{v-1v}\|\varphi(\omega)\right)^{-1} \|\boldsymbol{P}_{\mathrm{pz1}}^{\mathcal{N}_{S1}}(\mathrm{j}\omega)\|$$

represents an upper bound of the error between the approximated dynamics $\hat{\boldsymbol{P}}_1^{\mathcal{N}_{S1}}(s)$ and the actual dynamics $\boldsymbol{P}_1^{\mathcal{N}}(s)$. The detailed construction rule of the transfer functions is presented in Appendix C.1. Moreover, from the local view of the other subsystems S_i, $(\forall i \in \mathcal{N}_{S1})$ that are categorised as relevant, the irrelevant part $E_i^{\mathcal{N}_{W1}}$ becomes representable by the comparison system $\bar{E}_i^{\{i\}}$ according to (7.28). Hence, the comparison systems $\bar{P}_i^{\mathcal{N}_{S1}}$, $(\forall i \in \mathcal{N}_{S1})$ can be calculated in direct analogy to $\bar{P}_1^{\mathcal{N}_{S1}}$.

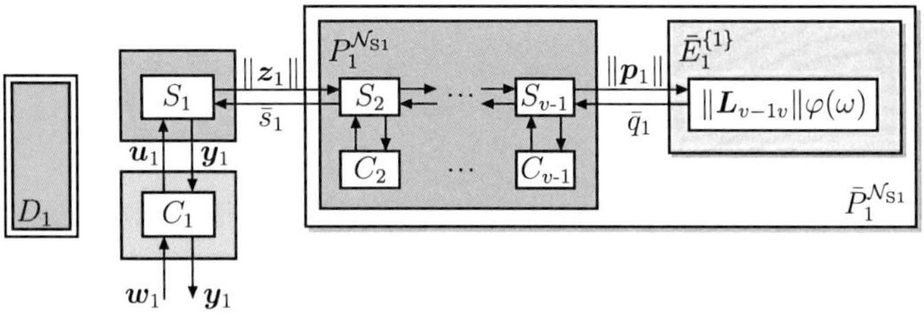

Figure 7.10: Structure of the comparison system of the physical interaction $P_1^{\mathcal{N}}$: The relevant part $P_1^{\mathcal{N}_{S1}}$ is exactly known by D_1 (after running Alg. 7.1), while the irrelevant part $E_1^{\mathcal{N}_{W1}}$ is known as comparison system $\bar{E}_1^{\{1\}}$. This comparison system can be determined with the information available to D_1.

As solely the irrelevant dynamics $\boldsymbol{E}_1^{\mathcal{N}_{W1}}(s)$ are treated as unknown-but-bounded model uncertainty, the comparison system $\bar{P}_1^{\mathcal{N}_{S1}}$ is more precise than the comparison system $\bar{P}_1^{\{1\}}$ described by (6.4), where the whole dynamics $\boldsymbol{P}_1^{\mathcal{N}}(s)$ are considered to be unknown. This relation is summarised in the following theorem:

Theorem 7.3 (Improvement of the precision of the upper bound) *Let* $\mathcal{N}_{S1} = \{1, ..., v-1\}$ *with* $v > 2$. *The relations*

$$\bar{P}_i^{\mathcal{N}_{S1}}(\omega) < \bar{P}_i^{\{i\}}(\omega), \quad \forall \omega \in \mathbb{R}, \quad \forall i \in \mathcal{N}_{S1} \tag{7.30}$$

hold.

Proof. See Appendix B.11. □

From Lemma 7.2 and Theorem 7.3 follows the relation

$$||\boldsymbol{P}_i^{\mathcal{N}}(\mathrm{j}\omega)|| < \bar{P}_i^{\mathcal{N}_{\mathrm{S1}}}(\omega) < \bar{P}_i^{\{i\}}(\omega), \quad \forall \omega \in \mathbb{R}, \quad \forall i \in \mathcal{N}_{\mathrm{S1}}, \tag{7.31}$$

which is essentially the reason why the design conditions proposed in Section 7.3 are less conservative than the local design conditions from Chapter 6.

Still, the upper bounds $\bar{P}_i^{\{i\}}(\omega)$, $(\forall i \in \mathcal{N}_{\mathrm{W1}})$ can be determined by the corresponding design agents D_i with the initially available model information $\mathcal{M}_i = (S_i, C_i, K_i)$, whereas the design agents D_i, $(\forall i \in \mathcal{N}_{\mathrm{S1}})$ have to gather the model sets from their neighbouring design agents to determine the corresponding upper bound $\bar{P}_i^{\mathcal{N}_{\mathrm{S1}}}(\omega)$. Accordingly, the design agent D_1 requires the models collected in the set

$$\mathcal{M}_1 = \left(S_1, C_1, K_1, \left\{ F_i|_{\boldsymbol{w}_i=\boldsymbol{0}}, K_i, i \in \mathcal{N}_{\mathrm{S1}} \setminus \{1\}\right\}\right)$$

to set up the comparison system $\bar{P}_1^{\mathcal{N}_{\mathrm{S1}}}$. Therefore, Algorithm 7.1 proposed in Section 7.2.4 is used.

As the comparison system $\bar{P}_i^{\mathcal{N}_{\mathrm{S1}}}$, which is used for the controller design, is only composed of the controlled subsystems F_i, $(\forall i \in \mathcal{N}_{\mathrm{S1}})$, the control stations C_i, $(\forall i \in \mathcal{N}_{\mathrm{S1}})$ can be designed independently from the control stations C_i, $(\forall i \in \mathcal{N}_{\mathrm{W1}})$, but can no longer be designed independently from each other. This issue is discussed in detail in Section 7.3.4.

Existence of the upper bound $\bar{P}_i^{\mathcal{N}_{\mathrm{S1}}}(\omega)$. The condition (7.4) of Lemma 7.1 also states a sufficient condition for the existence of the upper bound $\bar{P}_i^{\mathcal{N}_{\mathrm{S1}}}(\omega)$ from the comparison system $\bar{P}_i^{\mathcal{N}_{\mathrm{S1}}}$ as summarised in the following corollary:

Proposition 7.2 (Existence of an upper bound $\bar{P}_i^{\mathcal{N}_{\mathrm{S1}}}(\omega)$) *Let the transfer functions $\boldsymbol{F}_{\mathrm{zs}i}(s)$, $(\forall i \in \mathcal{N})$ be I/O stable and let the conditions* (7.26) *and* (7.27) *of Lemma 7.2 be satisfied. The upper bounds $\bar{P}_i^{\mathcal{N}_{\mathrm{S1}}}(\omega)$, $(\forall i \in \mathcal{N}_{\mathrm{S1}})$ exist if the claim* (7.4) *of Lemma 7.1 is satisfied.*

Proof. See Appendix B.12 □

7.3.2 I/O stability conditions

This section presents stability conditions that have to be satisfied by the cooperatively controlled subsystems $F^{\mathcal{N}_{\mathrm{S1}}}$ as well as by the controlled subsystems F_i, $(\forall i \in \mathcal{N}_{\mathrm{W1}})$ so that I/O stability of the overall closed-loop system is guaranteed. Moreover, it is shown that the proposed cooperative stability condition is less conservative than the local stability condition $\mathcal{A}_i$

from Chapter 6.

Consider the overall closed-loop system that is composed of the cooperatively controlled subsystem $F^{\mathcal{N}_{\mathrm{S1}}}$ and the remaining controller subsystems F_i, $(\forall i \in \mathcal{N}_{\mathrm{W1}})$ according to (7.25). The overall closed-loop system is modelled by

$$F: \quad \begin{aligned} \boldsymbol{y}(s) = {} & \mathrm{diag}\left(\boldsymbol{F}_{\mathrm{yw}}^{\mathcal{N}_{\mathrm{S1}}}(s),\, \boldsymbol{F}_{\mathrm{yw}i}(s)\right)_{i\in\mathcal{N}_{\mathrm{W1}}} \boldsymbol{w}(s) + \mathrm{diag}\left(\boldsymbol{F}_{\mathrm{yq}}^{\mathcal{N}_{\mathrm{S1}}}(s),\, \boldsymbol{F}_{\mathrm{ys}i}(s)\right)_{i\in\mathcal{N}_{\mathrm{W1}}} \boldsymbol{L} \\ & \left(\mathbf{I} - \mathrm{diag}\left(\boldsymbol{F}_{\mathrm{pq}}^{\mathcal{N}_{\mathrm{S1}}}(s),\, \boldsymbol{F}_{\mathrm{zs}i}(s)\right)_{i\in\mathcal{N}_{\mathrm{W1}}} \boldsymbol{L}\right)^{-1} \mathrm{diag}\left(\boldsymbol{F}_{\mathrm{pw}}^{\mathcal{N}_{\mathrm{S1}}}(s),\, \boldsymbol{F}_{\mathrm{zw}i}(s)\right)_{i\in\mathcal{N}_{\mathrm{W1}}} \boldsymbol{w}(s). \end{aligned} \tag{7.32}$$

For the system (7.32), the necessary and sufficient condition $\mathcal{A}$ defined in (6.7) is rewritten into the equivalent condition

$$\mathcal{A}: \begin{cases} 1.\ \text{I/O stability of the cooperatively controlled subsystem } F^{\mathcal{N}_{\mathrm{S1}}} \text{ and of the} \\ \quad \text{controlled subsystems } F_i, \quad \forall i \in \mathcal{N}_{\mathrm{W}i} \\ 2.\ \text{the Nyquist plot} \\ \qquad \det\left(\mathbf{I} - \mathrm{diag}\left(\boldsymbol{F}_{\mathrm{pq}}^{\mathcal{N}_{\mathrm{S1}}}(\mathrm{j}\omega),\, \boldsymbol{F}_{\mathrm{zs}i}(\mathrm{j}\omega)\right)_{i\in\mathcal{N}_{\mathrm{W1}}} \boldsymbol{L}\right), \quad \forall \omega \in \mathbb{R} \\ \quad \text{does not encircle the origin of the complex plane.} \end{cases}$$

From the local stability condition $\mathcal{A}_i$, the cooperative stability condition $\mathcal{A}^{\mathcal{N}_{\mathrm{S1}}}$ for the cooperatively controlled subsystem $F^{\mathcal{N}_{\mathrm{S1}}}$ can be derived. $\mathcal{A}^{\mathcal{N}_{\mathrm{S1}}}$ consists of the following two claim:

$$\mathcal{A}^{\mathcal{N}_{\mathrm{S1}}}: \begin{cases} 1.\ \text{I/O stability of the cooperatively controlled subsystem } F^{\mathcal{N}_{\mathrm{S1}}} & (7.34\mathrm{a}) \\ 2.\ \text{the cooperative amplification of the coupling output is limited} \\ \quad \text{according to} & (7.34\mathrm{b}) \\ \qquad \|\boldsymbol{F}_{\mathrm{pq}}^{\mathcal{N}_{\mathrm{S1}}}(\mathrm{j}\omega)\boldsymbol{L}_{v-1v}\| < 1, \quad \forall \omega \in \mathbb{R}. \end{cases}$$

In direct analogy to Theorem 6.1, the relation between I/O stability of the overall closed-loop system and the stability conditions $\mathcal{A}^{\mathcal{N}_{\mathrm{S1}}}$ and $\mathcal{A}_i$, $(\forall i \in \mathcal{N}_{\mathrm{W1}})$ is summarised as follows.

Corollary 7.1 (Cooperative design condition for global I/O stability)
Let $\mathcal{N}_{\mathrm{S1}} = \{1, ..., v-1\}$ *and* $\mathcal{N}_{\mathrm{W1}} = \{v, ..., N\}$ *with* $v \geq 2$. *If the cooperatively controlled subsystem* $F^{\mathcal{N}_{\mathrm{S1}}}$ *satisfies the cooperative stability condition* $\mathcal{A}^{\mathcal{N}_{\mathrm{S1}}}$ *defined in* (7.34) *and the controlled subsystems* F_i, $(\forall i \in \mathcal{N}_{\mathrm{W1}})$ *satisfy the corresponding local stability condition* $\mathcal{A}_i$ *defined in* (6.8), *then the overall closed-loop system* F *is I/O stable.*

Proof. The proof follows the same steps as the proof of Theorem 6.1 just with the cooperative controlled subsystem $F^{\mathcal{N}_{\mathrm{S1}}}$ and the cooperative stability condition $\mathcal{A}^{\mathcal{N}_{\mathrm{S1}}}$. □

As presented in Section 6.3.1 each controlled subsystem F_i has to satisfy its local objective $\mathcal{A}_i$ on its own. However, due to the accumulation of the relevant controlled subsystems, they are allowed to cooperate to satisfy the cooperative condition $\mathcal{A}^{\mathcal{N}_{S1}}$. As a consequence, the conservatism of the local stability condition $\mathcal{A}_i$ can be reduced as stated in the following.

Corollary 7.2 (Reduction of the conservatism of the local stability condition by cooperation) *Let* $\mathcal{N}_{S1} = \{1, ..., v-1\}$ *and* $\mathcal{N}_{W1} = \{v, ..., N\}$ *with* $v \geq 2$. *The cooperatively controlled subsystem* $F^{\mathcal{N}_{S1}}$ *satisfies* $\mathcal{A}^{\mathcal{N}_{S1}}$ *defined in* (7.34) *if the controlled subsystems* F_i, $(\forall i \in \mathcal{N}_{S1})$ *satisfy* $\mathcal{A}_i$ *defined in* (6.8).

Proof. The local claims

$$\|\boldsymbol{F}_{\mathrm{zs}i}(\mathrm{j}\omega)\| \sum_{j\in\mathcal{P}_i} \|\boldsymbol{L}_{ij}\| < 1, \quad \forall \omega \in \mathbb{R}, \quad \forall i \in \mathcal{N}_{S1} \tag{7.35}$$

can be compactly written into the equivalent condition

$$\left\|\mathrm{diag}(\boldsymbol{F}_{\mathrm{zs}i}(\mathrm{j}\omega))_{i\in\mathcal{N}_{S1}}\| \cdot \|\boldsymbol{L}_{\mathcal{N}_{S1},\{v\}}\right\| < \left(\mathbf{I} - \left\|\mathrm{diag}(\boldsymbol{F}_{\mathrm{zs}i}(\mathrm{j}\omega))_{i\in\mathcal{N}_{S1}}\| \cdot \|\boldsymbol{L}_{\mathcal{N}_{S1}}\right\|\right) \cdot \mathbf{1}, \tag{7.36}$$

as presented in the proof of Lemma 6.1. The norm applies block-element wise (cf. eqn (B.7)). The relation (7.36) can be rewritten into the equivalent claim

$$\left(\mathbf{I} - \left\|\mathrm{diag}(\boldsymbol{F}_{\mathrm{zs}i}(\mathrm{j}\omega))_{i\in\mathcal{N}_{S1}}\| \cdot \|\boldsymbol{L}_{\mathcal{N}_{S1}}\right\|\right)^{-1} \left\|\mathrm{diag}(\boldsymbol{F}_{\mathrm{zs}i}(\mathrm{j}\omega))_{i\in\mathcal{N}_{S1}}\| \cdot \|\boldsymbol{L}_{\mathcal{N}_{S1},\{v\}}\right\| < \mathbf{1}, \tag{7.37}$$

where the inverse exists due to the fulfilment of the condition (6.8b) for all $i \in \mathcal{N}_{S1}$ (cf. proof of Theorem 6.1). If the relation (7.37) is satisfied, then the relation

$$\left\|(\mathbf{I} - \mathrm{diag}(\boldsymbol{F}_{\mathrm{zs}i}(\mathrm{j}\omega))_{i\in\mathcal{N}_{S1}} \cdot \boldsymbol{L}_{\mathcal{N}_{S1}})^{-1} \,\mathrm{diag}(\boldsymbol{F}_{\mathrm{zs}i}(\mathrm{j}\omega))_{i\in\mathcal{N}_{S1}} \boldsymbol{L}_{\mathcal{N}_{S1},\{v\}}\right\| < \mathbf{1} \tag{7.38}$$

holds. In accordance with the construction rule of $\boldsymbol{F}_{\mathrm{pq}}^{\mathcal{N}_{S1}}(s)$ from (C.16), the last element of (7.38) represents the claim

$$\|\boldsymbol{F}_{\mathrm{pq}}^{\mathcal{N}_{S1}}(\mathrm{j}\omega)\boldsymbol{L}_{v-1v}\| < 1, \forall \omega \in \mathbb{R}.$$

Moreover, since all controlled subsystems are I/O stable (cf. (6.8a)) and

$$\left(\mathbf{I} - \left\|\mathrm{diag}(\boldsymbol{F}_{\mathrm{zs}i}(\mathrm{j}\omega))_{i\in\mathcal{N}_{S1}}\| \cdot \|\boldsymbol{L}_{\mathcal{N}_{S1}}\right\|\right)$$

is an M-Matrix (concluded from the proof of Theorem 6.1 and Theorem A1.3 from [117]), the transfer functions (C.13)–(C.16) from Appendix C.2 of the model $F^{\mathcal{N}_{S1}}$ are I/O stable.

Hence, if the local stability conditions (6.8) are satisfied for F_i, ($\forall i \in \mathcal{N}_{\mathrm{S1}}$), the cooperative condition (7.34) holds, which completes the proof. □

Although Corollary 7.2 does not quantify the relaxation of the conservatism of the local conditions, the proof gives a hint of this relaxation. In particular, I/O stability of the cooperatively controlled subsystem is claimed (cf. condition(7.34a)) in contrast to the I/O stability of each controlled subsystem as claimed by condition (6.8a). Moreover, the fulfilment of condition (7.35) implies the satisfaction of (7.38) and, hence, of (7.34b). The effect of this implication is highlighted by the following example.

Example 6.2 (cont.) *Conservatism of the local stability condition*

It has been shown in the first part of this example that although the local stability conditions are violated, the overall closed-loop system is I/O stable (cf. page 123). Now, the model set $\mathcal{M}_2$ of the second office is taken into account to improve the precision of the interaction model and, thus, to reduce the conservatism of the local stability condition. In particular, the amplitude plot $\|\boldsymbol{F}_{\mathrm{pq}}^{\{1,2\}}(\mathrm{j}\omega)\boldsymbol{L}_{23}\|$ for the cooperatively controlled subsystem is depicted in Fig. 7.11. The reduction of the conservatism is shown by the fact that although the local conditions are violated the stability condition (7.34b) is still satisfied.

Moreover, Fig. 7.11 highlights by the dotted plot that the cooperative condition is violated if the local magnitude $\|\boldsymbol{F}_1(\mathrm{j}\omega)\| \cdot \|\boldsymbol{L}_{12}\|$ is approximately twice as big as claimed in the local condition $\mathcal{A}_1$. □

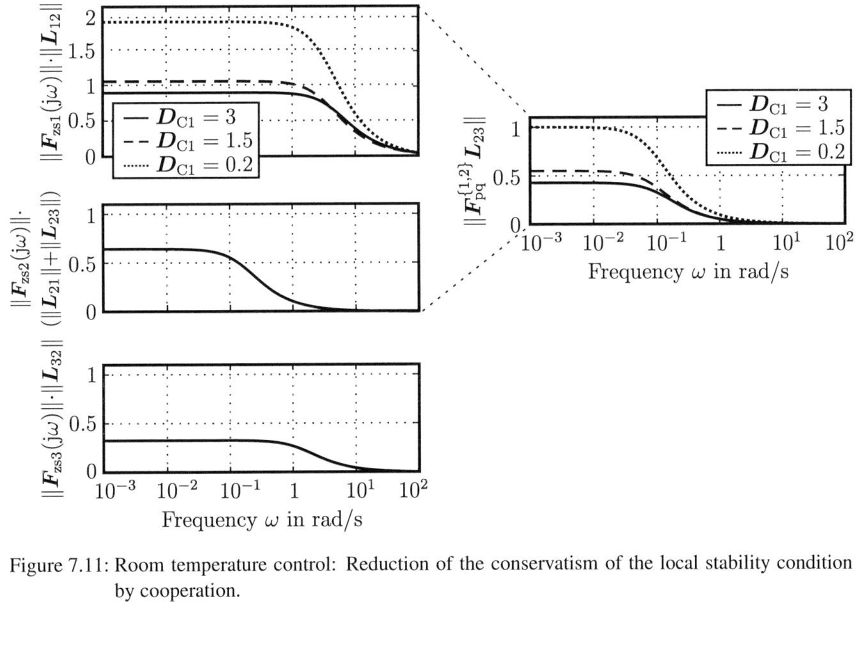

Figure 7.11: Room temperature control: Reduction of the conservatism of the local stability condition by cooperation.

7.3.3 I/O performance conditions

The section proposes the cooperative design condition $\mathcal{A}_{\mathrm{D}}^{\mathcal{N}_{\mathrm{S1}}}$ for the cooperatively controlled subsystem $F^{\mathcal{N}_{\mathrm{S1}}}$. It is shown that the conservatism of the local design condition $\mathcal{A}_{\mathrm{D1}}$ is reduced if the approximate model $\hat{P}_1^{\mathcal{N}_{\mathrm{S1}}}$ is used for the controller design.

As introduced in Section 6.3.2 on page 124, the desired overall closed-loop performance is reflected by the global control aim $\mathcal{A}_{\mathrm{D}}$. The result from Lemma 7.2 has particular effect on the design condition (6.16c)

$$\|\boldsymbol{F}_{\Delta i}^{\mathcal{N}}(\mathrm{j}\omega)\| < \varepsilon_i, \quad \forall \omega \in \mathbb{R}$$

as shown in the following. Due to the resulting upper bound $\bar{P}_1^{\mathcal{N}_{\mathrm{S1}}}(\omega)$ described by (7.29) the following relations hold:

$$\begin{aligned}
\|\boldsymbol{F}_{\Delta i}^{\mathcal{N}}(\mathrm{j}\omega)\| < \bar{F}_{\Delta i}^{\{i\}}(\omega) &= \|\boldsymbol{F}_{\mathrm{ys}i}(\mathrm{j}\omega)\|\, \bar{P}_i^{\{i\}}(\omega) \\
&\quad \left(1 - \|\boldsymbol{F}_{\mathrm{zs}i}(\mathrm{j}\omega)\|\, \bar{P}_i^{\{i\}}(\omega)\right)^{-1} \|\boldsymbol{F}_{\mathrm{zw}i}(\mathrm{j}\omega)\|, \ \forall \omega \in \mathbb{R},\ \forall i \in \mathcal{N}_{\mathrm{W1}}, \\
\|\boldsymbol{F}_{\Delta i}^{\mathcal{N}}(\mathrm{j}\omega)\| < \bar{F}_{\Delta i}^{\mathcal{N}_{\mathrm{S1}}}(\omega) &= \|\boldsymbol{F}_{\mathrm{ys}i}(\mathrm{j}\omega)\|\, \bar{P}_i^{\mathcal{N}_{\mathrm{S1}}}(\omega) \\
&\quad \left(1 - \|\boldsymbol{F}_{\mathrm{zs}i}(\mathrm{j}\omega)\|\, \bar{P}_i^{\mathcal{N}_{\mathrm{S1}}}(\omega)\right)^{-1} \|\boldsymbol{F}_{\mathrm{zw}i}(\mathrm{j}\omega)\|, \ \forall \omega \in \mathbb{R},\ \forall i \in \mathcal{N}_{\mathrm{S1}}. \quad (7.39)
\end{aligned}$$

Based on the upper bound (7.39), the cooperative performance condition $\mathcal{A}_{\mathrm{D}}^{\mathcal{N}_{\mathrm{S1}}}$ can be derived. $\mathcal{A}_{\mathrm{D}}^{\mathcal{N}_{\mathrm{S1}}}$ comprises the following four claims:

$\mathcal{A}_{\mathrm{D}}^{\mathcal{N}_{\mathrm{S1}}}$:

1. I/O stability of the cooperatively controlled subsystem $F^{\mathcal{N}_{\mathrm{S1}}}$ (7.40a)
2. for the given function $\varphi(\omega)\colon \mathbb{R} \to \mathbb{R}_+$ with $\varphi(\omega) \leq 1$ the cooperative amplification of the coupling output is limited according to (7.40b)
$$\|\boldsymbol{F}_{\mathrm{pq}}^{\mathcal{N}_{\mathrm{S1}}}(\mathrm{j}\omega)\boldsymbol{L}_{v-1v}\| < \varphi(\omega), \quad \forall \omega \in \mathbb{R}$$
3. satisfaction of given claims on the I/O behaviour of the isolated controlled subsystems $F_i|_{\boldsymbol{s}_i=\boldsymbol{0}}$, $\forall i \in \mathcal{N}_{\mathrm{S1}}$ (7.40c)
4. for a given parameter $\varepsilon_i \in \mathbb{R}_+$ the influence of the physical interaction is limited according to (7.40d)
$$\bar{F}_{\Delta i}^{\mathcal{N}_{\mathrm{S1}}}(\omega) \leq \varepsilon_i, \quad \forall \omega \in \mathbb{R}, \quad \forall i \in \mathcal{N}_{\mathrm{S1}}$$

The claims (7.40a) and (7.40b) reflect the cooperative condition $\mathcal{A}^{\mathcal{N}_{\mathrm{S1}}}$, while the conditions (7.40c) and (7.40d) claim a robust performance of the cooperatively controlled subsystem $F^{\mathcal{N}_{\mathrm{S1}}}$. The following corollary of Theorem 6.2 is concluded.

Corollary 7.3 (Cooperative design conditions for global I/O performance) *Let* $\mathcal{N}_{\rm S1} = \{1, ..., v-1\}$ *and* $\mathcal{N}_{\rm W1} = \{v, ..., N\}$ *with* $v \geq 2$. *If the cooperatively controlled subsystem* $F^{\mathcal{N}_{\rm S1}}$ *satisfies the cooperative performance condition* $\mathcal{A}_{\rm D}^{\mathcal{N}_{\rm S1}}$ *defined in* (7.40) *and the controlled subsystems* F_i, $(\forall i \in \mathcal{N}_{\rm W1})$ *satisfy the corresponding local performance condition* $\mathcal{A}_{{\rm D}i}$ *defined in* (6.18), *then the overall closed-loop system* F *has a performance described by* $\mathcal{A}_{\rm D}$ *defined in* (6.16).

The conservatism of the local performance conditions is reduced due to the more precise model of the physical interaction. In particular, from the proved relation (7.31) it can be concluded that

$$\|\boldsymbol{F}_{\Delta i}^{\mathcal{N}}({\rm j}\omega)\| < \bar{F}_{\Delta i}^{\mathcal{N}_{\rm S1}}(\omega) < \bar{F}_{\Delta i}^{\{i\}}(\omega), \quad \forall \omega \in \mathbb{R}, \quad \forall i \in \mathcal{N}_{\rm S1} \tag{7.41}$$

holds, which has a direct consequence on the conservatism of the local performance claim (7.40d). Based on the relation (7.41) and with the result from Corollary 7.2, the following statement can be derived:

Corollary 7.4 (Reduction of the conservatism of the local performance condition by cooperation) *The cooperatively controlled subsystem* $F^{\mathcal{N}_{\rm S1}}$ *satisfies the cooperative performance conditions* $\mathcal{A}_{\rm D}^{\mathcal{N}_{\rm S1}}$ *defined in* (7.34) *if the controlled subsystems* F_i, $(\forall i \in \mathcal{N}_{\rm S1})$ *satisfy the local performance conditions* $\mathcal{A}_{\rm D1}$ *defined in* (6.8).

In the Bode plot, the amplitude plots $\|\boldsymbol{F}_i^{\mathcal{N}}({\rm j}\omega)\| = \|\boldsymbol{F}_{{\rm yw}i}({\rm j}\omega) + \boldsymbol{F}_{\Delta i}^{\mathcal{N}}({\rm j}\omega)\|$, $(\forall i \in \mathcal{N}_{\rm S1})$ are enclosed in the corresponding tube

$$\|\boldsymbol{F}_{{\rm yw}i}({\rm j}\omega)\| - \bar{F}_{\Delta i}^{\mathcal{N}_{\rm S1}}(\omega) < \|\boldsymbol{F}_i^{\mathcal{N}}({\rm j}\omega)\| < \|\boldsymbol{F}_{{\rm yw}i}({\rm j}\omega)\| + \bar{F}_{\Delta i}^{\mathcal{N}_{\rm S1}}(\omega).$$

That is,

$$\|\boldsymbol{F}_i^{\mathcal{N}}({\rm j}\omega)\| \in \mathbb{T}_i^{\mathcal{N}_{\rm S1}}(\omega) := \left(\|\boldsymbol{F}_{{\rm yw}i}({\rm j}\omega)\| - \bar{F}_{\Delta i}^{\mathcal{N}_{\rm S1}}(\omega), \quad \|\boldsymbol{F}_{{\rm yw}i}({\rm j}\omega)\| + \bar{F}_{\Delta i}^{\mathcal{N}_{\rm S1}}(\omega)\right), \forall \omega \in \mathbb{R}$$

for all $i \in \mathcal{N}_{\rm S1}$. Due to relation (7.41), this tube is tighter than the tube $\mathbb{T}_i^{\{i\}}(\omega)$ from (6.19) according to

$$\mathbb{T}_i^{\mathcal{N}_{\rm S1}}(\omega) \subset \mathbb{T}_i^{\{i\}}(\omega), \quad \forall \omega \in \mathbb{R}, \quad i \in \mathcal{N}_{\rm S1}.$$

As a consequence, lower control actions are required to shrink the tube to the desired size (cf. (7.40d)). The following example illustrates the statement form Corollary 7.4.

Example 6.3 (cont.) *Conservatism of the local performance condition*

The first part of this example highlights that only wide tube can be determined with local model information (cf. page 126). Taking the model of the neighbouring room into account tightens the tube significantly. In particular, with the model set $\mathcal{M}_1$ and $\mathcal{M}_2$ at hand of design agent D_1 the tube $\mathbb{T}_1^{\{1,2\}}(\omega)$ can be calculated as depicted in Fig. 7.12. □

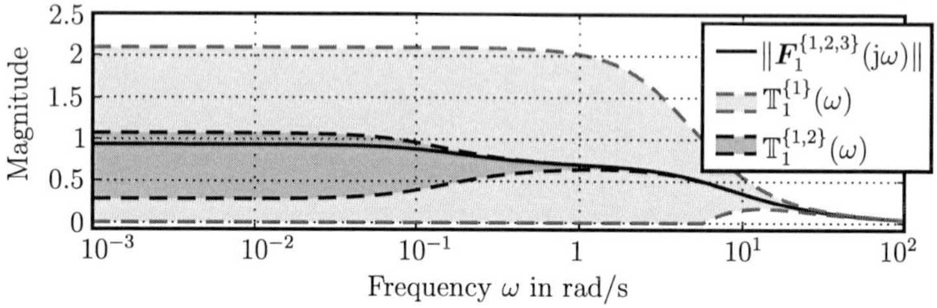

Figure 7.12: Example room temperature control: Reduction of the conservatism of the local performance condition.

7.3.4 H_∞-controller design

This section gives a brief remark on the controller design regarding the cooperative condition $\mathcal{A}^{\mathcal{N}_{\mathrm{S1}}}$ and $\mathcal{A}_{\mathrm{D}}^{\mathcal{N}_{\mathrm{S1}}}$ and proposes an iterative design procedure.

The burden to obtain less conservative design conditions is that control station C_1 can no longer be designed independently from the control stations $C_2, ..., C_{\upsilon-1}$. That is due to the fact that the upper bound $\bar{F}_{\Delta i}^{\mathcal{N}_{\mathrm{S1}}}(\omega)$ that is used for the design depends upon the control stations C_i, $(\forall i \in \mathcal{N}_{\mathrm{S1}})$. A possible design procedure with the help of the H_∞-design that has been proposed in Section 6.4 can be used as presented in Algorithm 7.2, where D_1 manages the design of all control stations C_i, $(\forall i \in \mathcal{N}_{\mathrm{S1}})$.

The models and design conditions available to D_1 can be gathered by means of Algorithm 7.1. The initial design according to the local aims gives a starting point of the iterative controller design (Step 1–6). After the initial design, initial H_∞-gains $\gamma_{\mathrm{ys}i}$, $\gamma_{\mathrm{zs}i}$ and $\gamma_{\mathrm{zw}i}$, $(\forall i \in \mathcal{N}_{\mathrm{S1}})$ exist which are iteratively relaxed (Step 2) during the loop, while D_1 continuously checks whether the cooperative aim is still satisfied (Step 5). The step size can be freely chosen and can be independent for each design agent and can differ for the increase step (Step 2) and decrease step (Step 5) as well as for each iteration. Hence, the final number of iterations of Algorithm 7.2 depends on the chosen step size of the change of the H_∞-gains as well as on the termination condition (Step 6).

To consider the stability aim $\mathcal{A}^{\mathcal{N}_{\mathrm{S1}}}$ two slight changes need to be made in Algorithm 7.2. First, only the LMI (6.34b) has to be considered within the initial design (Initialise) and within

Algorithm 7.2: Cooperative controller design procedure supervised by D_1

Given: Model sets $\mathcal{M}_i = (S_i, K_i)$, $(\forall i \in \mathcal{N}_{\mathrm{S1}})$ available to D_1 and
local condition $\mathcal{A}_{\mathrm{D}i}$, $(\forall i \in \mathcal{N}_{\mathrm{S1}})$ with ε_i and $\varphi(\omega)$ available to D_1

Initialise: $i = 1$,
D_1 designs C_i, $(\forall i \in \mathcal{N}_{\mathrm{S1}})$ subject to $\mathcal{A}_{\mathrm{D}i}$ using Theorem 6.3

Processing on D_1**:**

1. **do**
2. Increase the H_∞-norm bounds $\gamma_{\mathrm{ys}i}$, $\gamma_{\mathrm{zs}i}$ and $\gamma_{\mathrm{zw}i}$
3. Redesign C_i using Theorem 6.3
4. Recompose the other upper bound $\bar{P}_j^{\mathcal{N}_{\mathrm{S1}}}(\omega)$, $(\forall j \in \mathcal{N}_{\mathrm{S1}})$
5. **if** condition $\mathcal{A}_{\mathrm{D}}^{\mathcal{N}_{\mathrm{S1}}}$ is satisfied **then** set $i = i + 1$ and goto Step 2
   else decrease the H_∞-norm bounds $\gamma_{\mathrm{ys}i}$, $\gamma_{\mathrm{zs}i}$ and $\gamma_{\mathrm{zw}i}$ and goto Step 3
6. **while** modifications on $\gamma_{\mathrm{ys}i}$, γ_{zs} and $\gamma_{\mathrm{zw}i}$ are not too small

Result: Control stations C_i, $(\forall i \in \mathcal{N}_{\mathrm{S1}})$ so that $F^{\mathcal{N}_{\mathrm{S1}}}$ satisfies $\mathcal{A}_{\mathrm{D}}^{\mathcal{N}_{\mathrm{S1}}}$

the adaptation in Step 3. Thus, only the H_∞-gain $\gamma_{\mathrm{zs}i}$ has to be manipulated in Step 2 and Step 5. Second, the cooperatively controlled subsystem $F^{\mathcal{N}_{\mathrm{S1}}}$ has to be composed in Step 4 in order to check the adherence of the stability condition $\mathcal{A}^{\mathcal{N}_{\mathrm{S1}}}$ in Step 5.

7.4 Application scenario: Fault-tolerant control

The situation is considered in which an actuator in subsystem S_1 breaks and the control station C_1 shall be reconfigured. Due to the physical interconnections among the subsystems, the failure has effect to all other controlled subsystems and, thus, leads to instability of the overall closed-loop system. The overall closed-loop I/O stability must be recovered by means of a virtual actuator that is implemented on control station C_1 only. The proposed virtual actuator uses an approximate model of the faulty subsystem under the influence of the physical interaction. This approach differs to the reconfiguration approach presented in Section 4.4.1, where the virtual actuator makes use of the exact model.

To focus on the reconfiguration process, the following two assumptions are made:

A 7.2 *A diagnosis system detects the fault and uniquely identifies the model* S_{1f} *of the faulty subsystem. As a result, the model* S_{1f} *is known by design agent* D_1*. The diagnostic process does not consume any time.*

A 7.3 *There exist N control stations C_i, $(\forall i \in \mathcal{N})$ such that each controlled subsystem F_i satisfies the corresponding local stability condition $\mathcal{A}_i$ defined in* (6.8) *and satisfies condition* (7.4).

Concerning assumption A 7.2, a local diagnosis system is proposed in Chapter 8. Due to assumption A 7.3, I/O stability of the overall closed-loop system is retained if the virtual actuator is designed to fulfil the stability claim $\mathcal{A}^{\mathcal{N}_{\mathrm{S}1}}$ (cf. Corollary 7.1). This design is accomplished by an H_∞-design presented in Corollary 7.5. The complete reconfiguration procedure that is managed by design agent D_1 is summarised in Algorithm 7.3.

7.4.1 Local reconfiguration with a virtual actuator using an approximate model of the physical interaction

This section presents the reconfiguration of control station C_1 by a virtual actuator in order to recover overall closed-loop I/O stability after an actuator in subsystem S_1 has failed. An H_∞-design is proposed to design the virtual actuator.

At time instant $t_{\mathrm{f}} \geq 0$ an actuator failure in subsystem S_1 occurs. The model S_{1f} of the faulty subsystem is represented by (2.26), where the corresponding column in $\boldsymbol{B}_1$ that represents the failed actuator is set to zero. To compensate the loss of the broken actuator, the virtual actuator uses the approximate model

$$\hat{S}_{1f}^{\mathcal{N}_{\mathrm{S}1}} = \mathrm{con}\left(\left\{S_{1f},\, \hat{P}_1^{\mathcal{N}_{\mathrm{S}1}}\right\},\, \boldsymbol{u}_{1f},\, \boldsymbol{y}_{1f}\right)$$

as shown in Fig. 7.13. The model $\hat{S}_{1f}^{\mathcal{N}_{\mathrm{S}1}}$ is represented in state space by

$$\hat{S}_{1f}^{\mathcal{N}_{\mathrm{S}1}}: \begin{cases} \dot{\boldsymbol{x}}_{1f}^{\mathcal{N}_{\mathrm{S}1}}(t) &= \boldsymbol{A}_1^{\mathcal{N}_{\mathrm{S}1}} \boldsymbol{x}_{1f}^{\mathcal{N}_{\mathrm{S}1}}(t) + \boldsymbol{B}_{1f}^{\mathcal{N}_{\mathrm{S}1}} \boldsymbol{u}_{1f}(t), \quad \boldsymbol{x}_{1f}^{\mathcal{N}_{\mathrm{S}1}}(t_{\mathrm{f}}) = \boldsymbol{0} \\ \boldsymbol{y}_{1f}(t) &= \boldsymbol{C}_1^{\mathcal{N}_{\mathrm{S}1}} \boldsymbol{x}_{1f}^{\mathcal{N}_{\mathrm{S}1}}(t), \end{cases}$$

where $\boldsymbol{x}_{1f}^{\mathcal{N}_{\mathrm{S}1}}(t)$ accumulates the state $\boldsymbol{x}_{1f}(t)$ of the faulty subsystem and the states $\boldsymbol{x}_{\mathrm{F}i}(t)$ of the controlled subsystems F_i, $(\forall i \in \mathcal{N}_{\mathrm{S}1} \setminus \{1\})$ that are strongly coupled to S_1. Accordingly, the virtual actuator is modelled by

$$VA_1: \begin{cases} \dot{\boldsymbol{x}}_{\delta 1}(t) &= \boldsymbol{A}_{\delta 1} \boldsymbol{x}_{\delta 1}(t) + \boldsymbol{B}_1^{\mathcal{N}_{\mathrm{S}1}} \mathring{\boldsymbol{u}}_1(t), \quad \boldsymbol{x}_{\delta 1}(t_{\mathrm{f}}) = \boldsymbol{0} \\ \boldsymbol{u}_{1f}(t) &= \boldsymbol{M}_1 \boldsymbol{x}_{\delta 1}(t) \\ \mathring{\boldsymbol{y}}_1(t) &= \boldsymbol{y}_{1f}(t) + \boldsymbol{C}_1^{\mathcal{N}_{\mathrm{S}1}} \boldsymbol{x}_{\delta 1}(t), \end{cases} \tag{7.42}$$

where $\boldsymbol{x}_{\delta 1}(t) = \boldsymbol{x}_{1f}^{\mathcal{N}_{\mathrm{S}1}}(t) - \boldsymbol{x}_1^{\mathcal{N}_{\mathrm{S}1}}(t)$ is the difference state and $\boldsymbol{A}_{\delta 1} = \boldsymbol{A}_1^{\mathcal{N}_{\mathrm{S}1}} - \boldsymbol{B}_{1f}^{\mathcal{N}_{\mathrm{S}1}} \boldsymbol{M}_1$. The virtual actuator VA_1 together with the nominal control station C_1 represent the reconfigured control station $\mathring{C}_1$ as highlighted in Fig. 7.13.

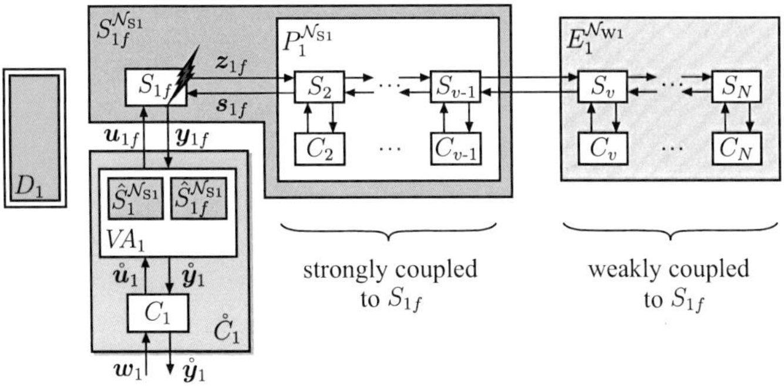

Figure 7.13: Structure of the reconfigured closed-loop system: Local reconfiguration of the control station C_1 using a virtual actuator VA_1. The result is the reconfigured control station $\mathring{C}_1$. For the design phase and the execution phase of the virtual actuator the approximate model $\hat{S}_1^{\mathcal{N}_{S1}}$ and $\hat{S}_{1f}^{\mathcal{N}_{S1}}$ are required which are obtained using Alg. 7.1.

In order to recover overall closed-loop system I/O stability, the feedback gain $\boldsymbol{M}_1$ has to be designed such that the cooperative stability condition $\mathcal{A}^{\mathcal{N}_{S1}}$ is fulfilled (cf. Corollary 7.1). This gain $\boldsymbol{M}_1$ is obtained by means of LMI-based H_∞-design as stated in Corollary 7.5.

Corollary 7.5 (Stabilising reconfiguration after an actuator failure in S_1 using a virtual actuator) *The virtual actuator* (7.42) *recovers I/O stability of the faulty overall closed-loop system F_f (model* (4.6)*) if there exit feasible solutions $\boldsymbol{X} = \boldsymbol{X}^\top \prec 0$ and $\boldsymbol{Y}$ to the LMI*

$$\begin{pmatrix} \boldsymbol{A}_1^{\mathcal{N}_{S1}}\boldsymbol{X} + \boldsymbol{X}\boldsymbol{A}_1^{\mathcal{N}_{S1}\top} + \boldsymbol{B}_{1f}^{\mathcal{N}_{S1}}\boldsymbol{Y} + \boldsymbol{Y}\boldsymbol{B}_{1f}^{\mathcal{N}_{S1}\top} & \boldsymbol{B}_1^{\mathcal{N}_{S1}} & \boldsymbol{X}\boldsymbol{C}_1^{\mathcal{N}_{S1}\top} \\ \star & -\gamma_{\mathrm{VA}}\mathbf{I} & \mathrm{O} \\ \star & \star & -\gamma_{\mathrm{VA}}\mathbf{I} \end{pmatrix} \prec 0 \tag{7.43}$$

with γ_{VA} from (B.38) *in Appendix B.13. The feedback gain results to $\boldsymbol{M}_1 = -\boldsymbol{Y}\boldsymbol{X}^{-1}$. Feasible solutions to the LMI exist only if the pair $\left(\boldsymbol{A}_1^{\mathcal{N}_{S1}}, \boldsymbol{B}_{1f}^{\mathcal{N}_{S1}}\right)$ is stabilisable.*

Proof. See Appendix B.13. □

The multi-objective design aims at recovering the I/O stability of the overall closed-loop system and establishes a weak amplification to the irrelevant part $E_1^{\mathcal{N}_{W1}}$ in accordance to the design condition $\mathcal{A}^{\mathcal{N}_{S1}}$. Due to the claim that the pair $\left(\boldsymbol{A}_1^{\mathcal{N}_{S1}}, \boldsymbol{B}_{1f}^{\mathcal{N}_{S1}}\right)$ is stabilisable, the cooperative stability condition $\mathcal{A}^{\mathcal{N}_{S1}}$ can be recovered by the redesign of C_1 only.

It has to be emphasised that for the special case, where no approximate model of the physical interaction is used (i.e., only S_1 is used for the design), the proposed design regards the local stability condition $\mathcal{A}_i$. That is plug-and-play reconfiguration with local model information (see Section 6.5.3).

In comparison with the virtual actuator (4.5) introduced in Section 4.4.1, the complexity of the virtual actuator (7.42) is reduced. This is due to the fact that only the models F_i, $(\forall i \in \mathcal{N}_{S1})$ which are strongly coupled to the faulty subsystem S_{1f} are used during the design phase as well as during the execution phase. In contrast to this, the virtual actuator (4.5) uses all controlled subsystems that are strongly connected to S_{1f}. That is for locally interconnected subsystems (according to (2.8)), the virtual actuator uses $\upsilon - 1$ models in contrast to N. However, Corollary 7.5 states only a sufficient condition for overall closed-loop I/O stability, while Corollary 4.1 is necessary and sufficient for asymptotic stability.

7.4.2 Plug-and-play reconfiguration algorithm

This section summarises the reconfiguration steps of control station C_1, supervised by D_1.

Initially, the design agents have available only their local model information. Based on the locally known threshold $\xi(\omega)$, D_1 procures the models F_i, which have a meaningful influence on the faulty subsystem S_{1f} and accumulates them to the approximate model $\hat{P}_1^{\mathcal{N}_{S1}}$. Based on the model S_{1f} together with the approximation $\hat{P}_1^{\mathcal{N}_{S1}}$ of the physical interaction, the virtual actuator (7.42) can be designed using Corollary 7.5. The reconfiguration steps are summarised in Algorithm 7.3.

Alternatively to the design of the virtual actuator (Step 2), the control station C_1 together with the control stations $C_2, ..., C_{\upsilon-1}$ can be redesigned using Algorithm 7.2. Therefore, the recovery of the overall closed-loop performance according to $\mathcal{A}_{\mathrm{D}}$ can be considered.

Note that the reconfiguration does not consume any time in accordance with assumption A 3.4. Moreover, note that the cooperative stability condition $\mathcal{A}^{\mathcal{N}_{S1}}$ is known by D_1 since it does not provide specific design parameters compared to the performance condition $\mathcal{A}_{\mathrm{D}}^{\mathcal{N}_{S1}}$.

7.4.3 Example: Multizone furnace

Plug-and-play reconfiguration (Algorithm 7.3) is applied to the multizone furnace from Section 2.5.2. The complete brake down of the heater in zone 1 requires the usage of an auxiliary fan as a back-up solution. It is shown that although the local condition $\mathcal{A}_1$ is violated, overall closed-loop I/O stability can be verified if the dynamics of the neighbouring zone are taken into account.

The four zone S_i, $(i = 1, .., 4)$ are modelled by (2.6) with (A.10a)–(A.10d) and are locally

Algorithm 7.3: Cooperative plug-and-play reconfiguration after actuator failures in S_1

Given: Control stations C_i, $(\forall i \in \mathcal{N})$ exist such that F_i satisfy $\mathcal{A}_i$ and (7.4) for all $i \in \mathcal{N}$,
model set $\mathcal{M}_1 = (S_1, S_{1f}, C_1, K_1)$ available to D_1,
model set $\mathcal{M}_i = (S_i, C_i, K_i)$ available to D_i, $(\forall i \in \mathcal{N} \setminus \{1\})$ and
cooperative condition $\mathcal{A}^{\mathcal{N}_{\mathrm{S1}}}$ and decision threshold $\xi(\omega)$ available to D_1

Processing on D_1:

1. Run Algorithm 7.1 to gather the model set
$\mathcal{M}_1 = \{S_1, S_{f1}, C_1, K_1\} \cup \{F_i, K_i,\ i \in \mathcal{N}_{\mathrm{S1}} \setminus \{1\}\}$ in order to set up the approximate models $\hat{S}_1^{\mathcal{N}_{\mathrm{S1}}}$ and $\hat{S}_{1f}^{\mathcal{N}_{\mathrm{S1}}}$ and the cooperatively controlled subsystem $F_1^{\mathcal{N}_{\mathrm{S1}}}$
2. **if** $\mathcal{A}^{\mathcal{N}_{\mathrm{S1}}}$ is violated **then** design the virtual actuator (7.42) according to Corollary 7.5, **else** STOP (no reconfiguration necessary)
3. Implement the virtual actuator into the control equipment

Result: I/O stable reconfigured overall closed-loop system $\mathring{F}$

interconnected according to the model (2.8) with (A.11a)–(A.11d). The zones are initially controlled by PI-controllers represented by (2.11), where

$$\begin{aligned}
\boldsymbol{A}_{\mathrm{C}i} &= 0, \quad \boldsymbol{B}_{\mathrm{C}i} = 1, \quad \boldsymbol{C}_{\mathrm{C}i} = 2, \quad \boldsymbol{D}_{\mathrm{C}i} = 1, \qquad i = 1,4\\
\boldsymbol{A}_{\mathrm{C}i} &= 0, \quad \boldsymbol{B}_{\mathrm{C}i} = 1, \quad \boldsymbol{C}_{\mathrm{C}i} = 4, \quad \boldsymbol{D}_{\mathrm{C}i} = 3, \qquad i = 2,3.
\end{aligned}$$

These control stations are designed such that the nominal controlled subsystems F_i, $(i = 1, ..., 4)$ satisfy the requirement (7.4) (cf. assumption A 7.1) as well as the corresponding local stability condition $\mathcal{A}_i$ defined in (6.8).

Fault occurrence in zone 1. At runtime, the primary heater breaks down and it is switched immediately to the heating fan. In particular,

$$\boldsymbol{B}_1 = \begin{pmatrix} 1.34 \\ 0.1 \end{pmatrix} \xrightarrow{\text{break down}} \boldsymbol{B}_{1f} = \begin{pmatrix} 0 \\ 0 \end{pmatrix} \xrightarrow{\text{switching}} \boldsymbol{B}_{1f} = 10^{-2} \begin{pmatrix} 4.5 \\ 5 \end{pmatrix}.$$

This fan is still actuated by the nominal control station C_1.

Although the faulty controlled subsystem F_{1f} is still I/O stable, the claim (6.8b)

$$\|\boldsymbol{F}_{\mathrm{zs}1f}(\mathrm{j}\omega)\| \cdot \|\boldsymbol{L}_{12}\| < 1$$

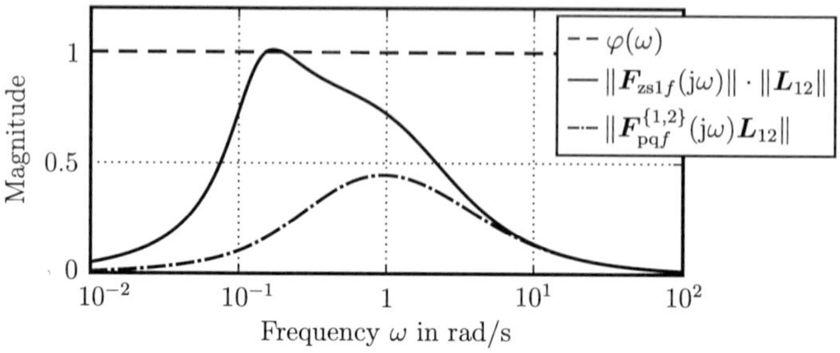

Figure 7.14: Application: Verification of the local stability condition $\mathcal{A}_1$ to be fulfilled by F_1 and cooperative stability condition $\mathcal{A}^{\{1,2\}}$ to be fulfilled by $F^{\{1,2\}}$.

is violated as shown in Fig. 7.14. Hence, design agent D_1 starts Algorithm 7.3:

Step 1: The decision threshold is chosen based on the admissible uncertainty of the physical interaction model $P_1^{\{1,2,3,4\}}$. In particular, the impact of the error dynamics

$$\|\boldsymbol{S}_{\Delta 1}^{\mathcal{N}_{S1}}(j\omega)\| = \|\boldsymbol{S}_{ys1}(j\omega)\| \cdot \|\boldsymbol{P}_{\Delta 1}^{\mathcal{N}_{S1}}(j\omega)\| \left(1-\|\boldsymbol{S}_{zs1}(j\omega)\| \cdot \|\boldsymbol{P}_{\Delta 1}^{\mathcal{N}_{S1}}(j\omega)\|\right)^{-1} \|\boldsymbol{S}_{zu1f}(j\omega)\|$$

has to be ten times smaller than the effect of the isolated subsystem dynamics $\boldsymbol{S}_{1f}(s)$, i.e.,

$$\|\boldsymbol{S}_{\Delta 1}^{\mathcal{N}_{S1}}(j\omega)\| \leq 0.1 \cdot \max_{\omega\in\mathbb{R}} \|\boldsymbol{S}_{1f}(j\omega)\|,$$

where the threshold $\eta(\omega)$ of the uncertainty $\|\boldsymbol{P}_{\Delta 1}^{\mathcal{N}_{S1}}(j\omega)\|$ results to

$$\eta(\omega) = \left(\|\boldsymbol{S}_{ys1}(j\omega)\| \cdot (0.1 \cdot \max_{\omega\in\mathbb{R}} \|\boldsymbol{S}_{1f}(j\omega)\|)^{-1} \cdot \|\boldsymbol{S}_{zu1}(j\omega)\| + \|\boldsymbol{S}_{zs1}(j\omega)\|\right)^{-1}.$$

According to the equations (7.10) and (7.24), design agent D_1 calculates the threshold

$$\xi_2(\omega) = \|\boldsymbol{L}_{12}\|^{-1}\, 0.5\, \eta(\omega)$$

and transmits this threshold to design agent D_2. Figure 7.15a shows that the relation

$$\|\boldsymbol{F}_{zs2}(j\omega)\boldsymbol{L}_{21}\| > \xi_2(\omega), \quad \exists \omega \in \mathbb{R}$$

holds. Accordingly, the controlled subsystem F_2 is categorised as strongly coupled to S_1. Thus, D_2 transmits its model set $\mathcal{M}_2 = \{F_2, K_2\}$ to design agent D_1. With this model information at

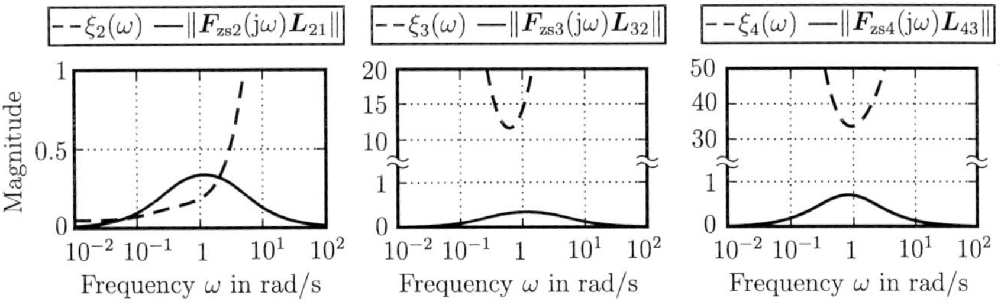

(a) Classification of F_2 (b) Classification of F_3 (c) Classification of F_4

Figure 7.15: Application: Local classification of neighbouring controlled subsystems into weakly and strongly coupled with the faulty subsystem S_{1f}.

hand, D_1 determines the threshold

$$\xi_3(\omega) = \|\boldsymbol{L}_{23}\|^{-1}\left(\|\boldsymbol{L}_{12}\boldsymbol{F}_{\text{zs2}}(\mathrm{j}\omega)\|\left(\eta(\omega) - \Delta_1^2(\omega)0.67\right)^{-1}\|\boldsymbol{F}_{\text{zs2}}(\mathrm{j}\omega)\boldsymbol{L}_{21}\| + \|\boldsymbol{F}_{\text{zs2}}(\mathrm{j}\omega)\|\right)^{-1}$$

where $\Delta_1^2(\omega) = \|\boldsymbol{L}_{12}\| \cdot \|\boldsymbol{F}_{\text{zs2}}(\mathrm{j}\omega)\boldsymbol{L}_{21}\|$ and sends it to design agent D_3. Figure 7.15b shows that the local claim

$$\|\boldsymbol{F}_{\text{zs3}}(\mathrm{j}\omega)\boldsymbol{L}_{32}\| \leq \xi_3(\omega), \quad \forall\omega \in \mathbb{R}$$

is satisfied in order that the controlled subsystem F_3 is categorised as weakly coupled to S_{1f}. Hence, D_3 responses this information to D_1, which subsequently stops the procurement phase. In accordance with Theorem 7.2, the controlled subsystem F_4 is also weakly coupled. Figure 7.15c highlights that the local claim $\|\boldsymbol{F}_{\text{zs4}}(\mathrm{j}\omega)\boldsymbol{L}_{43}\| \leq \xi_4(\omega)$ is satisfied.

At the end of the procurement phase, D_1 has interacted with D_2 and D_3 with the result that D_1 stored the models

$$\mathcal{M}_1 = \{S_{1f},\, S_1,\, S_2,\, C_1,\, C_2,\, K_1,\, K_2\}$$

and constructs the cooperatively controlled subsystem $F^{\{1,2\}}$ and the approximate model $S_1^{\{1,2\}}$ and $S_{1f}^{\{1,2\}}$. The communication graph in Fig. 7.16 depicts the interaction among the design agents. It can be seen that no communication link to design agent D_4 has been established.

Step 2: The adherence of the cooperative stability condition

$$\mathcal{A}_1^{\{1,2\}}: \begin{cases} \text{1. I/O stability of the cooperatively controlled subsystem } F^{\{1,2\}} \\ \text{2. the cooperative amplification of the coupling output is limited} \\ \quad \text{according to} \\ \qquad \|\boldsymbol{F}_{\text{pq}}^{\{1,2\}}(\mathrm{j}\omega)\boldsymbol{L}_{23}\| < 1, \quad \forall\omega \in \mathbb{R} \end{cases}$$

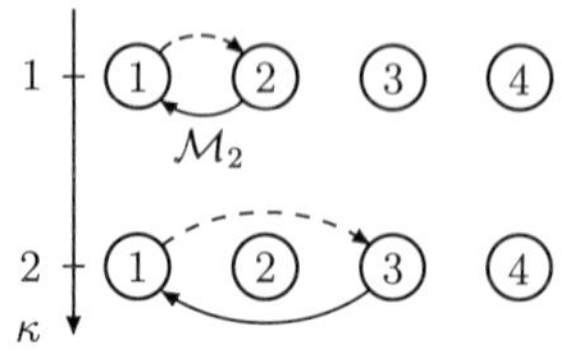

Figure 7.16: Application: Communication graph $\mathcal{G}_{\mathrm{D}}(\kappa)$ for the processing of Algorithm 7.1. Dashed edges denote requests and solid edges denote responses.

is verified. The first claim is checked by analysing the eigenvalues of the system matrix

$$\boldsymbol{A}_{\mathrm{F}}^{\{1,2\}} = \begin{pmatrix} \boldsymbol{A}_{\mathrm{F1}} & \boldsymbol{E}_{\mathrm{F1}} \boldsymbol{L}_{12} \boldsymbol{C}_{\mathrm{Fz2}} \\ \boldsymbol{E}_{\mathrm{F2}} \boldsymbol{L}_{21} \boldsymbol{C}_{\mathrm{Fz1}} & \boldsymbol{A}_{\mathrm{F2}} \end{pmatrix}$$

of the cooperatively controlled subsystem $F^{\{1,2\}}$, which all have negative real parts:

$$\lambda_1 = -4.41, \quad \lambda_{2,3} = -0.62 \pm 0.35\mathrm{i}, \quad \lambda_{4,5} = -0.78 \pm 0.1\mathrm{i}, \quad \lambda_6 = -0.1.$$

The second claim is satisfied too, as visualised by the dash-dot line in Fig. 7.14. That is due to the fact that the impermissible magnitude of $\|\boldsymbol{F}_{\mathrm{zs}1f}(\mathrm{j}\omega)\| \cdot \|\boldsymbol{L}_{12}\|$ is compensated by the neighbouring zone. Accordingly, Algorithm 7.3 terminates and no reconfiguration is necessary.

This example illustrates the value of model information for the reconfiguration process by means of the accuracy of the approximate model

$$\hat{S}_{1f}^{\mathcal{N}_{\mathrm{S1}}} : \quad \hat{\boldsymbol{y}}_{1f}(s) = \boldsymbol{S}_{1f}^{\mathcal{N}_{\mathrm{S1}}}(s)\, \boldsymbol{u}_{1f}(s)$$

of the faulty interacting subsystem $S_{1f}^{\mathcal{N}}$. Figure 7.17 highlights that the consideration of the model of zone 2 significantly improves the accuracy of the approximation.

In particular, Fig. 7.17 shows that the dynamics $\|\boldsymbol{S}_{1f}(\mathrm{j}\omega)\|$ of the faulty isolated subsystem $S_{1f}|_{\boldsymbol{s}_{1f}=\boldsymbol{0}}$ do not adequately represent the dynamics $\|\boldsymbol{S}_{1f}^{\{1,2,3,4\}}(\mathrm{j}\omega)\|$ of the faulty interacting subsystem. In contrast to this, the dynamics $\|\hat{\boldsymbol{S}}_{1f}^{\{1,2\}}(\mathrm{j}\omega)\|$ yields a good approximation, as shown by the solid line in Fig. 7.17. The additional inclusion of the local information from design agent D_3 does not significantly improve the approximation. Hence, it is reasonable to ignore this model information for the overall closed-loop system analysis and, thus, for the controller reconfiguration.

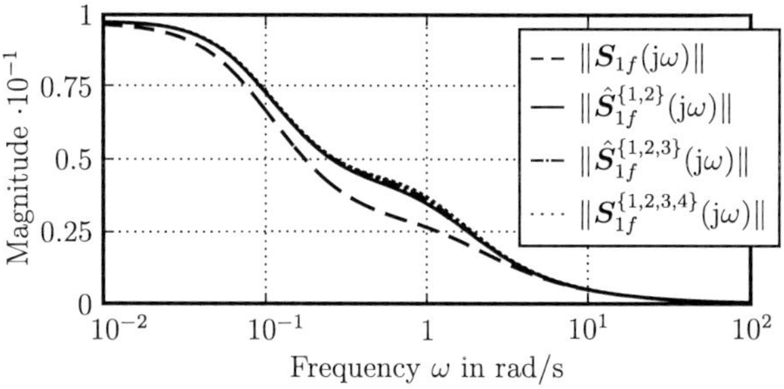

Figure 7.17: Application: Improvement of the accuracy of the approximate model $\hat{S}_{1f}^{\mathcal{N}_{S1}}$ by taking neighbouring models into account. The approximate model $\hat{S}_{1f}^{\{1,2\}}$ adequately approximates the faulty interacting subsystem $S_{1f}^{\{1,2,3,4\}}$.

7.4.4 Further application scenarios

The presented plug-and-play reconfiguration algorithm (Algorithm 7.3) can easily be adapted to the usage for the ordinary controller design as well as for the integration of a new subsystem.

For the ordinary design, Algorithm 7.2 has to be performed by design agent D_1 to obtain the control stations C_i, $(\forall i \in \mathcal{N}_{S1})$. The control stations C_i, $(\forall i \in \mathcal{N}_{S1} \setminus \{1\})$ are transmitted to the corresponding design agents to be implemented into the control hardware. Thereafter, Algorithm 7.2 is performed by design agent D_v.

To integrate subsystem S_{N+1}, first, the design agent D_{N+1} gathers relevant model information and, second, designs the control stations C_i, $(\forall i \in \mathcal{N}_{SN+1})$ with respect to the cooperative condition $\mathcal{A}^{\mathcal{N}_{S1}}$ or $\mathcal{A}_{D}^{\mathcal{N}_{S1}}$, respectively. A possible redesign has to be performed on design agent D_N, due to the local interconnection structure. Hence, control station C_N has to be adjusted if it has not already been redesigned by the cooperative design of C_{N+1}.

8 Plug-and-play diagnosis using an approximate model of the physical interaction

This chapter addresses the design of a diagnosis system to detect and isolate faults that can occur in subsystem S_1. A local diagnosis system is proposed that evaluates the local measurements of S_1 by means of an approximate model of S_1 under the influence of the physical interaction. The required accuracy of this approximate model has to be determined by design agent D_1, although only the model S_1 and some information about the physical couplings are initially available. The main results of this chapter are a robust detectability condition and a robust isolability condition that are used to quantify the necessary accuracy of the approximation. Furthermore, it is shown that a PI-based residual generator is well suited for residual generation in the presence of model uncertainties. The chapter closes with the demonstration of plug-and-play diagnosis on the multizone furnace.

8.1 Problem formulation

This chapter focuses on the design of a local diagnosis system by design agent D_1 that shall detect and isolate faults that can occur in subsystem S_1. Therefore, the proposed diagnosis system requires a model of the fault-free subsystem S_1 as well as of the faulty subsystem S_{1f} under consideration of the physical interaction to evaluate the local measurements $\boldsymbol{u}_1(t)$ and $\boldsymbol{y}_1(t)$ (Fig. 8.1). However, design agent D_1 initially knows the model S_1 of the healthy subsystem, the models S_{1f}, $(\forall f \in \mathcal{F})$ of the subsystem affected by the fault f and the model K_1 of the local couplings only. The aim of this chapter is to enable D_1 to set up the model for the diagnosis system that only has to be as precise as necessary to detect and isolate the fault. The purposive neglect of models that are irrelevant for diagnosis reduces the computational workload for both the design phase as well as the execution phase of the diagnosis system.

This chapter concentrates on subsystems that are locally interconnected in the sense that S_i directly interacts with S_{i+1} and S_{i-1} as indicated in Fig. 8.1. Although all controlled subsystems

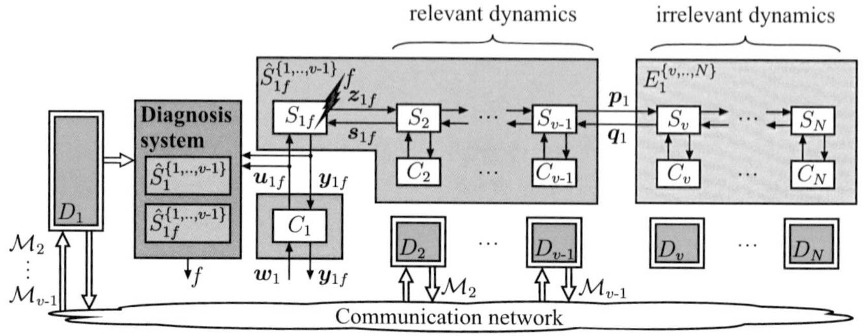

Figure 8.1: Structure of plug-and-play diagnosis using an approximation of the physical interaction: The approximation comprises only some controlled subsystem models (yellow box), while the remaining subsystems (grey box) are considered as model uncertainty.

F_i, $(i = 2, ..., N)$ interact with the faulty subsystem S_{1f}, the behaviour of S_{1f} is essentially characterised by the dynamics of subsystem S_{1f} together with some neighbouring controlled subsystems F_i, $(i = 2, .., v-1)$. This approximation of the

$$\hat{S}_{1f}^{\{1,..,v-1\}} = \text{con}\left(S_{1f} \cup \left\{ F_i|_{\boldsymbol{w}_i=\boldsymbol{0}}, \, i = 2, .., v{-}1 \right\} \cup \{K_i, \, i = 1, .., v{-}1\}, \, \boldsymbol{u}_{1f}, \, \boldsymbol{y}_{1f}\right),$$

highlighted in Fig. 8.1 by the yellow box, is exploited by the proposed diagnosis system. In contrast to this, the dynamics of the remaining controlled subsystems F_i, $(i = v, .., N)$ are represented by the model

$$E_1^{\{v,..,N\}} = \text{con}\left(\left\{ F_i|_{\boldsymbol{w}_i=\boldsymbol{0}}, \, i = v, .., N \right\} \cup \{K_i, \, i = v{-}1, .., N\}, \, \boldsymbol{p}_1, \, \boldsymbol{q}_1\right),$$

which constitutes a conscious model uncertainty (grey box in Fig. 8.1) and is not used for diagnosis.

To accomplish the design of the local diagnosis system, design agent D_1 must know all relevant models $F_i|_{\boldsymbol{w}_i=\boldsymbol{0}}$, K_i, $(i = 1, ..., v{-}1)$. Based on the initially available model information D_1 has to decide whether controlled subsystem F_i is relevant for fault diagnosis and has to gather the respective model set $\mathcal{M}_i = \left\{ F_i|_{\boldsymbol{w}_i=\boldsymbol{0}}, K_i \right\}$ from D_i through the communication network as shown in Fig. 8.1 by double framed arrows. This issue is formalised in the following problem:

Problem 8.1 (Plug-and-play diagnosis)

Problem:	*To detect and isolate considered faults $f \in \mathcal{F}$, the diagnosis system has to exploit a model of S_1 interacting with all other controlled subsystems F_i, $(i = 2, ..., N)$. This model only has to be as accurate as necessary to guarantee fault detection and fault isolation. The desired model has to be set up by D_1 that initially stores the models S_1 and K_1 of its subsystem and its local couplings.*
Restrictions:	*Subsystems are locally interconnected*
Given:	• *D_1 stores the model set $\mathcal{M}_1 = \{\{S_{1f},\ f \in \mathcal{F}\}, S_1, C_1, K_1\}$* • *$D_i$, $(\forall i \in \mathcal{N} \setminus \{1\})$ store the corresponding model set $\mathcal{M}_i = \{F_i, K_i\}$*
Find:	*Local conditions so as D_1 can decide whether or not the model set $\mathcal{M}_i$ is relevant to guarantee the detection and the isolation of considered faults $f \in \mathcal{F}$.*
Approach:	*Evaluate the relevance by means of the detectability and the isolability of the considered faults $f \in \mathcal{F}$.*

A solution to Problem 8.1 enables design agent D_1 to accomplish the design of the local diagnosis system (i.e., Problem 1.2 is solved).

Section 8.2 outlines the models of plug-and-play diagnosis in detail and states assumptions. As the dynamics of subsystem S_1 together with the physical interaction are solely approximated by the model $\hat{S}_{1f}^{\{1,..,v-1\}}$, whereas the model $E_{1f}^{\{v,..,N\}}$ is considered to be unknown, a robust fault detectability and a robust fault isolability conditions are stated in Section 8.3. Based on these conditions the detection threshold and the isolation thresholds are proposed that guarantee fault detection and isolation. Section 8.4 presents a robust residual generator which only makes use of the approximate model but generates an asymptotically correct residual. Finally, the iterative design steps of the diagnosis system supervised by design agent D_1 are summarised in Section 8.5 and is applied to the multizone furnace in Section 8.6.

Classification of plug-and-play diagnosis. This paragraph outlines some existing methods for model-based fault diagnosis in networked control systems with the aim to classify plug-and-play diagnosis. Therefore, this brief survey is dedicated to the model that is used by the diagnosis systems.

The publications [54, 56, 79, 191] have proposed architectures, where each subsystem is equipped with a local diagnosis system that uses the local measurements and accesses the local model only. The network connection between the local diagnosis systems is exploited for the continuous exchange of the local measurements or estimated states among the diagnosis systems with the aim to reconstruct the coupling input signal. It has been shown in [143] that the communication effort can be reduced if the local diagnosis systems make use of an exact model of the overall system in order to estimate the coupling input signal locally.

Other approaches interpret the coupling input as an unknown input and elucidate unknown input observer [60, 160, 174] or fault detection filters [93, 125]. Hence, only the subsystem model is used by the local diagnosis systems.

The approach proposed in this chapter follows the idea that the estimation of the coupling input signal only needs to be as precise as necessary to detect and isolate the fault. Hence, dispensable model information is neglected in both, the design phase and the execution phase of the local diagnosis system.

8.2 Models of plug-and-play diagnosis

This section introduces the models of plug-and-play diagnosis, which are amongst others the approximate model of the faulty subsystem together with the physical interaction that is used by the diagnosis system. Furthermore, the structure of the local diagnosis system is presented.

8.2.1 Approximate model for the design of the diagnosis system

The overall closed-loop system consists of N controlled subsystems S_i that are represented by the state-space model (2.12). A local interconnection structure is considered as represented by the local interconnection model (2.8) or as visualised by the interconnection graph shown in Fig. 3.7 on page 60. Due to this interconnection structure, subsystem S_1 is influenced by all other controlled subsystems F_i, $(i = 2, ..., N)$ through the physical couplings as shown in Fig. 8.1. As presented in Section 3.2.2, the controlled subsystems F_i, $(i = 2, ..., N)$ are subdivided into two groups according to their relevance for the design. The *relevant part* essentially characterises the physical interaction, while the *irrelevant part* is neglected. Accordingly, the dynamics of subsystem S_1 together with the relevant part is represented by the model

$$S_1^{\mathcal{N}_{\mathrm{S}1}} : \begin{cases} \dot{\boldsymbol{x}}_1^{\mathcal{N}_{\mathrm{S}1}}(t) &= \boldsymbol{A}_1^{\mathcal{N}_{\mathrm{S}1}} \boldsymbol{x}_1^{\mathcal{N}_{\mathrm{S}1}}(t) + \boldsymbol{B}_1^{\mathcal{N}_{\mathrm{S}1}} \boldsymbol{u}_1(t) + \boldsymbol{E}_1^{\mathcal{N}_{\mathrm{S}1}} \boldsymbol{q}_1(t), \quad \boldsymbol{x}_1^{\mathcal{N}_{\mathrm{S}1}}(0) = \boldsymbol{0} \\ \boldsymbol{y}_1(t) &= \boldsymbol{C}_{1f}^{\mathcal{N}_{\mathrm{S}1}} \boldsymbol{x}_1^{\mathcal{N}_{\mathrm{S}1}}(t) \\ \boldsymbol{p}_1(t) &= \boldsymbol{C}_{\mathrm{z}1}^{\mathcal{N}_{\mathrm{S}1}} \boldsymbol{x}_1^{\mathcal{N}_{\mathrm{S}1}}(t), \end{cases} \tag{8.1}$$

whereas the irrelevant part is represented by

$$E_1^{\mathcal{N}_{W1}}\,:\; \boldsymbol{q}_1(t) = \boldsymbol{E}_1^{\mathcal{N}_{W1}}(t) * \boldsymbol{p}_1(t),$$

as highlighted in Fig. 8.1. The approximate model

$$\hat{S}_1^{\mathcal{N}_{S1}}\,:\;\begin{cases}\dot{\hat{\boldsymbol{x}}}_1^{\mathcal{N}_{S1}}(t) &= \boldsymbol{A}_1^{\mathcal{N}_{S1}}\hat{\boldsymbol{x}}_1^{\mathcal{N}_{S1}}(t) + \boldsymbol{B}_1^{\mathcal{N}_{S1}}\boldsymbol{u}_1(t), \quad \hat{\boldsymbol{x}}_1^{\mathcal{N}_{S1}}(0) = \boldsymbol{0}\\ \hat{\boldsymbol{y}}_1(t) &= \boldsymbol{C}_1^{\mathcal{N}_{S1}}\hat{\boldsymbol{x}}_1^{\mathcal{N}_{S1}}(t)\end{cases} \tag{8.2}$$

is used for fault diagnosis. The irrelevant part $\boldsymbol{E}_1^{\mathcal{N}_{W1}}$ is considered as model uncertainty, which is described by the element-by-element related comparison system

$$\bar{E}_1^{\mathcal{N}_{W1}}\,:\; \bar{\boldsymbol{q}}_1(t) = \bar{\boldsymbol{E}}_1^{\mathcal{N}_{W1}}(t) * |\boldsymbol{p}_1(t)|,$$

where the relation $\bar{\boldsymbol{E}}_1^{\mathcal{N}_{W1}}(t) \geq |\boldsymbol{E}_1^{\mathcal{N}_{W1}}(t)|$ holds for all $t \geq 0$ (cf. Definition 2.14). One possibility to construct this comparison system is constituted by

$$\bar{E}_1^{\mathcal{N}_{W1}} = \text{con}\left(\left\{\left.\bar{F}_i\right|_{\boldsymbol{w}_i=\boldsymbol{0}},\, i \in \mathcal{N}_{W1}\right\} \cup \left\{\bar{K}_i,\, i \in \mathcal{N}_{W1} \cup \{v-1\}\right\},\, |\boldsymbol{z}_1|,\, \bar{\boldsymbol{s}}_1\right), \tag{8.3}$$

where $\left.\bar{F}_i\right|_{\boldsymbol{w}_i=\boldsymbol{0}}$ and $\bar{K}_i$ represent comparison systems of the model $\left.F_i\right|_{\boldsymbol{w}_i=\boldsymbol{0}}$ and K_i, respectively. These comparison systems are provided by the respective design agent D_i (see. Section 3.3).

Model of the faulty subsystem. The faulty subsystem S_{1f} is represented in state space by the model (2.26), where $f \in \mathcal{F} = \{\mathrm{f}_1, ..., \mathrm{f}_q\}$ is the set of considered faults (see Section 2.4.1). The dynamics of the faulty subsystem S_{1f} together with the controlled subsystems F_i that are relevant are modelled by

$$S_{1f}^{\mathcal{N}_{S1}} = \text{con}\left(\{S_{1f}, K_1\} \cup \left\{\left.F_i\right|_{\boldsymbol{w}_i=\boldsymbol{0}}, K_i,\, i = 2, .., v{-}1\right\}, \begin{pmatrix}\boldsymbol{u}_{1f}\\ \boldsymbol{p}_{1f}\end{pmatrix}, \begin{pmatrix}\boldsymbol{y}_{1f}\\ \boldsymbol{q}_{1f}\end{pmatrix}\right).$$

This model $S_{1f}^{\mathcal{N}_{S1}}$ is represented by

$$S_{1f}^{\mathcal{N}_{S1}}\,:\;\begin{cases}\dot{\boldsymbol{x}}_{1f}^{\mathcal{N}_{S1}}(t) &= \boldsymbol{A}_{1f}^{\mathcal{N}_{S1}}\boldsymbol{x}_{1f}^{\mathcal{N}_{S1}}(t) + \boldsymbol{B}_{1f}^{\mathcal{N}_{S1}}\boldsymbol{u}_{1f}(t) + \boldsymbol{E}_1^{\mathcal{N}_{S1}}\boldsymbol{q}_{1f}(t), \quad \boldsymbol{x}_{1f}^{\mathcal{N}_{S1}}(t_{\mathrm{f}}) = \boldsymbol{0}\\ \boldsymbol{y}_{1f}(t) &= \boldsymbol{C}_{1f}^{\mathcal{N}_{S1}}\boldsymbol{x}_{1f}^{\mathcal{N}_{S1}}(t)\\ \boldsymbol{p}_{1f}(t) &= \boldsymbol{C}_{z1}^{\mathcal{N}_{S1}}\boldsymbol{x}_{1f}^{\mathcal{N}_{S1}}(t).\end{cases}$$

According to the approximate model (8.2), an approximate model of the faulty subsystem S_{1f} under consideration of the physical interaction is represented by

$$\hat{S}_{1f}^{\mathcal{N}_{\rm S1}}: \begin{cases} \dot{\hat{\boldsymbol{x}}}_{1f}^{\mathcal{N}_{\rm S1}}(t) &= \boldsymbol{A}_{1f}^{\mathcal{N}_{\rm S1}} \hat{\boldsymbol{x}}_{1f}^{\mathcal{N}_{\rm S1}}(t) + \boldsymbol{B}_{1f}^{\mathcal{N}_{\rm S1}} \boldsymbol{u}_{1f}(t), \quad \hat{\boldsymbol{x}}_{1f}(t_{\rm f}) = \boldsymbol{0} \\ \hat{\boldsymbol{y}}_{1f}(t) &= \boldsymbol{C}_{1f}^{\mathcal{N}_{\rm S1}} \hat{\boldsymbol{x}}_{1f}^{\mathcal{N}_{\rm S1}}(t). \end{cases}$$

Assumptions on the fault-free closed-loop system. For the remainder of this chapter two assumption on the fault-free overall closed-loop system are made.

Consider the connection of the fault-free system $S_1^{\mathcal{N}_{\rm S1}}$ (represented by (8.1)) and the control station C_1 (described by (2.11)) according to

$$F_1^{\mathcal{N}_{\rm S1}} = \operatorname{con}\left(S_1^{\mathcal{N}_{\rm S1}} \cup C_1, \begin{pmatrix} \boldsymbol{w}_1 \\ \boldsymbol{p}_1 \end{pmatrix}, \begin{pmatrix} \boldsymbol{y}_1 \\ \boldsymbol{q}_1 \end{pmatrix} \right).$$

The resulting model is described in I/O relation by

$$F_1^{\mathcal{N}_{\rm S1}}: \begin{cases} \boldsymbol{y}_1(t) &= \hat{\boldsymbol{F}}_1^{\mathcal{N}_{\rm S1}}(t) * \boldsymbol{w}_1(t) + \boldsymbol{F}_{\rm yq1}^{\mathcal{N}_{\rm S1}}(t) * \boldsymbol{q}_1(t) \\ \boldsymbol{p}_1(t) &= \boldsymbol{F}_{\rm pw1}^{\mathcal{N}_{\rm S1}}(t) * \boldsymbol{w}_1(t) + \boldsymbol{F}_{\rm pq1}^{\mathcal{N}_{\rm S1}}(t) * \boldsymbol{q}_1(t). \end{cases}$$

The first assumption concerns the nominal control stations.

A 8.1 *There exist N decentralised control stations C_i, $(\forall i \in \mathcal{N})$ such that the overall closed-loop system as well as the systems $F_1^{\mathcal{N}_{\rm S1}}$ and $E_1^{\mathcal{N}_{\rm W1}}$ are I/O stable.*

Note that the conditions from Proposition 7.2 are sufficient for the fulfilment of the claims of assumption A 8.1. The second assumption concerns the comparison system

$$\bar{F}_1^{\mathcal{N}} = \operatorname{con}\left(\{ |F_1^{\mathcal{N}_{\rm S1}}|, \bar{E}_1^{\mathcal{N}_{\rm W1}} \}, |\boldsymbol{w}_1|, \bar{\boldsymbol{y}}_1 \right)$$

that is represented by

$$\bar{F}_1^{\mathcal{N}}: \quad \bar{\boldsymbol{y}}_1(t) = \bar{\boldsymbol{F}}_1^{\mathcal{N}_{\rm S1}}(t) * |\boldsymbol{w}_1|,$$

where $\bar{\boldsymbol{F}}_1^{\mathcal{N}_{\rm S1}}(t) = |\hat{\boldsymbol{F}}_1^{\mathcal{N}_{\rm S1}}(t)| + |\boldsymbol{F}_{\rm yq1}^{\mathcal{N}_{\rm S1}}(t)| * \bar{\boldsymbol{\Psi}}(t) * |\boldsymbol{F}_{\rm pw1}^{\mathcal{N}_{\rm S1}}(t)|$ with

$$\bar{\boldsymbol{\Psi}}(t) = \bar{\boldsymbol{E}}_1^{\mathcal{N}_{\rm W1}}(t) + \bar{\boldsymbol{E}}_1^{\mathcal{N}_{\rm W1}}(t) * |\boldsymbol{F}_{\rm pq1}^{\mathcal{N}_{\rm S1}}(t)| * \bar{\boldsymbol{\Psi}}(t). \tag{8.4}$$

A 8.2 *The upper bound $\bar{\boldsymbol{F}}_1^{\mathcal{N}}(t)$ exists, i.e., $\bar{\boldsymbol{F}}_1^{\mathcal{N}}(t) < \infty$ for all $t \geq 0$.*

According to assumption A 8.1, the impulse response functions $|\hat{\boldsymbol{F}}_1^{\mathcal{N}_{S1}}(t)|$, $|\boldsymbol{F}_{\text{yq1}}^{\mathcal{N}_{S1}}(t)|$, $|\boldsymbol{F}_{\text{pw1}}^{\mathcal{N}_{S1}}(t)|$ and $|\boldsymbol{F}_{\text{pq1}}^{\mathcal{N}_{S1}}(t)|$ are I/O stable. Thus, assumption A 8.2 claims that $\bar{\boldsymbol{\Psi}}(t) \geq \text{O}$ for all $t \geq 0$. Therefore, conditions can be found amongst others in [1, 116].

8.2.2 Structure of the local diagnosis system

This section presents the structure of the local diagnosis system that shall be designed.

In contrast to the classic diagnosis system introduced in Section 2.4.3, the proposed local diagnosis system evaluates the local measurements $\boldsymbol{u}_1(t)$ and $\boldsymbol{y}_1(t)$ using the approximate models $\hat{S}_1^{\mathcal{N}_{S1}}$ and $\hat{S}_{1f}^{\mathcal{N}_{S1}}$, $(\forall f \in \mathcal{F})$ to detect and isolate considered faults $f \in \mathcal{F}$.

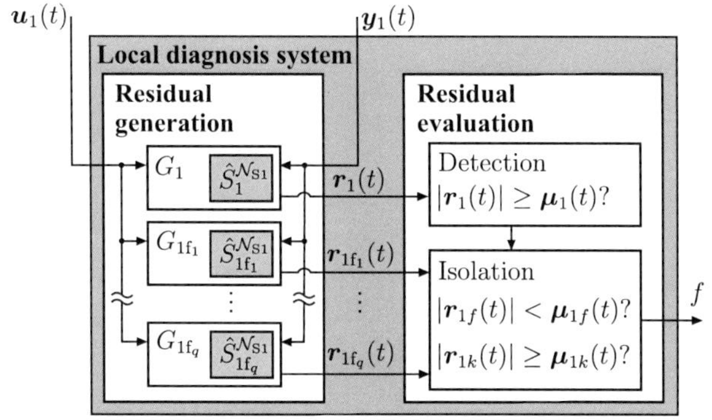

Figure 8.2: Structure of the local diagnosis unit: Residual generation and residual evaluation.

For fault detection, the local diagnosis system uses the fault-free approximate model $\hat{S}_1^{\mathcal{N}_{S1}}$. This *detection residual generator* is represented by the I/O-oriented model

$$G_1: \quad \boldsymbol{r}_1(t) = \boldsymbol{G}_{\text{ru1}}(t) * \boldsymbol{u}_1(t) + \boldsymbol{G}_{\text{ry1}}(t) * \boldsymbol{y}_1(t). \tag{8.5}$$

The moment, the residual $\boldsymbol{r}_1(t)$ crosses the given detection threshold $\boldsymbol{\mu}_1(t)$, a fault is detected.

For fault isolation, q additional *isolation residual generator* G_{1k}, $(k \in \mathcal{F} = \{f_1, ..., \text{f}_q\})$ are needed which exploit the respective approximate model $\hat{S}_{1k}^{\mathcal{N}_{S1}}$. Each isolation residual generator is represented by

$$G_{1k}: \quad \boldsymbol{r}_{1k}(t) = \boldsymbol{G}_{\text{ru1}k}(t) * \boldsymbol{u}_1(t) + \boldsymbol{G}_{\text{ry1}k}(t) * \boldsymbol{y}_1(t), \tag{8.6}$$

for $k \in \mathcal{F}$. Each isolation residual $\boldsymbol{r}_{1k}(t)$ is evaluated by means of an isolation threshold $\boldsymbol{\mu}_{1k}(t)$. After a fault has been detected, the fault isolation process starts. The fault f is called detected if all residuals $\boldsymbol{r}_{1k}(t)$, $(k \in \mathcal{F}, k \neq f)$ has exceeded their corresponding isolation threshold $\boldsymbol{\mu}_{1k}(t)$ and $\boldsymbol{r}_{1f}(t)$ has not crossed its threshold $\boldsymbol{\mu}_{1f}(t)$.

The design of the diagnosis system comprises the synthesis of the residual generation and the residual evaluation outlined as follows:

1. Design of the residual generation: The detection residual generator (8.5) and the isolation residual generators (8.6) have to be parametrised. This issue is considered in Section 8.4.
2. Design of the residual evaluation: The detection threshold $\boldsymbol{\mu}_1(t)$ and the isolation thresholds $\boldsymbol{\mu}_{1k}(t)$, $(\forall k \in \mathcal{F})$ have to be chosen. This is subject of discussion in Section 8.3.

These designs are processed by design agent D_1, which initially stores the model set

$$\mathcal{M}_1 = \{\{S_{1f},\, f \in \mathcal{F}\}\,,\, S_1,\, C_1,\, K_1\}\,.$$

Hence, to accomplish the design, D_1 has to set up the approximate models $\hat{S}_1^{\mathcal{N}_{\mathrm{S1}}}$ and $\hat{S}_{1f}^{\mathcal{N}_{\mathrm{S1}}}$, $(\forall f \in \mathcal{F})$ beforehand. A solution to this modelling problem is proposed in Section 8.5.

8.3 Design of the residual evaluation

This section, first, outlines the design issues, followed by the detailed analyses of the behaviour of the detection generated residual $\boldsymbol{r}_1(t)$. Finally, a robust detectability and robust isolability condition are stated that are the basis to chose the detection threshold and the isolation thresholds, respectively.

8.3.1 Design issues

The situation is considered in which residual generators represented by (8.5) and (8.6), respectively, exist and the generated residuals shall be analysed so that a detection and isolation threshold can be chosen. Due to the fact that this thesis considers faults that occur when the subsystem is in its operating point (cf. assumption A 2.6), the residual will differ from zero only if subsystem S_1 is stimulated. In particular, the generated detection residual $\boldsymbol{r}_1(t)$, thus, is analysed by means of the *generated residual model*

$$R_1^{\mathcal{N}} = \mathrm{con}\left(\{G_1,\, C_1,\, S_1^{\mathcal{N}_{\mathrm{S1}}}\} \cup \{E_1^{\mathcal{N}_{\mathrm{W1}}}\}\,,\, \boldsymbol{w}_1,\, \boldsymbol{r}_1\right)$$

which is represented by

$$R_1^{\mathcal{N}}: \quad \boldsymbol{r}_1(t) = \boldsymbol{R}_1^{\mathcal{N}}(t) * \boldsymbol{w}_1(t) \tag{8.7}$$

as highlighted in Fig. 8.3.

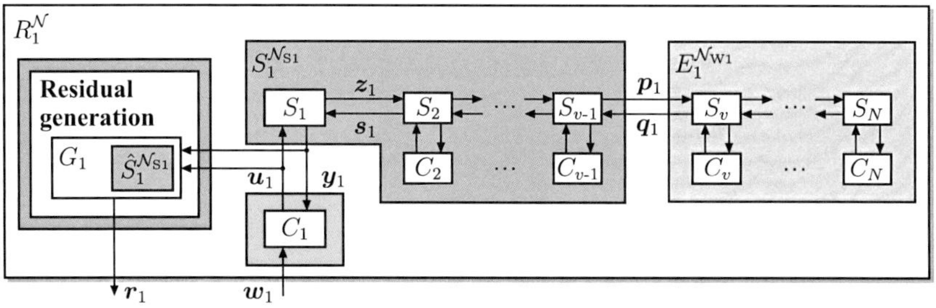

Figure 8.3: Structure of the generated residual model $R_1^{\mathcal{N}}$ that represents the behaviour of the generated detection residual $\boldsymbol{r}_1(t)$ stimulated through the reference signal $\boldsymbol{w}_1(t)$ in the fault-free case.

However, for the analysis of the residual, only the approximate models $\hat{S}_1^{\mathcal{N}_{S1}}$ and $\hat{S}_{1f}^{\mathcal{N}_{S1}}$, $(\forall f \in \mathcal{F})$, respectively, as well as the comparison system $\bar{E}_1^{\mathcal{N}_{W1}}$ are known by D_1. As a consequence, only a tolerance band can be derived in which the actual residual is located (Section 8.3.2). Due to the uncertainty of the residuals, the problem whether or not a fault can be detected or isolated is studied in Section 8.3.3. Moreover, it is shown Section 8.3.4 in that these tubes serve as detection threshold and isolation threshold.

8.3.2 Tolerance bands of the residuals

This section analyses the behaviour of the detection residual $\boldsymbol{r}_1(t)$ that is generated by the residual generator G_1 described by (8.5). As the overall closed-loop system model is not exactly known, only a tolerance band around the residual can be determined. The analysis is done for the fault-free as well as for the faulty case.

Fault-free case. Consider the healthy case, when no fault has been occurred. The behaviour from the input $\boldsymbol{w}_1(t)$ to the output $\boldsymbol{r}_1(t)$ is exactly represented by the generated residual model $R_1^{\mathcal{N}}$ modelled by (8.7).

As only the approximate model $\hat{S}_1^{\mathcal{N}_{S1}}$ is available for the analysis, the corresponding approximation $\hat{R}_1^{\mathcal{N}_{S1}}$ of the generated residual model $R_1^{\mathcal{N}}$ can be analysed. This approximation is represented by the approximate model

$$\hat{R}_1^{\mathcal{N}_{S1}} = \mathrm{con}\left(\left\{G_1,\, C_1,\, S_1^{\mathcal{N}_{S1}}\right\},\, \boldsymbol{w}_1,\, \hat{\boldsymbol{r}}_1\right)$$

depicted in Fig. 8.4. The approximate model is described by

$$\hat{R}_1^{\mathcal{N}_{S1}}: \quad \hat{\boldsymbol{r}}_1(t) = \hat{\boldsymbol{R}}_1^{\mathcal{N}_{S1}}(t) * \boldsymbol{w}_1(t) \tag{8.8}$$

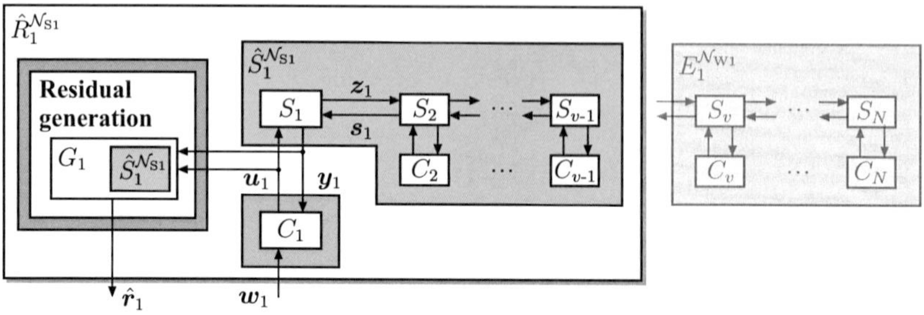

Figure 8.4: Structure of the approximate model $\hat{R}_1^{\mathcal{N}}$ that approximates the behaviour of the generated detection residual $\boldsymbol{r}_1(t)$ stimulated through the reference signal $\boldsymbol{w}_1(t)$. The irrelevant part $E_1^{\mathcal{N}_{\mathrm{W}1}}$ is ignored.

with

$$\hat{\boldsymbol{R}}_1^{\mathcal{N}_{\mathrm{S}1}}(t) = \boldsymbol{G}_{\mathrm{ru}1}(t) * \left(\boldsymbol{C}_1(t) - \boldsymbol{C}_1(t) * \boldsymbol{F}_1^{\mathcal{N}_{\mathrm{S}1}}(t)\right) + \boldsymbol{G}_{\mathrm{ry}1}(t) * \boldsymbol{F}_1^{\mathcal{N}_{\mathrm{S}1}}(t). \tag{8.9}$$

The behaviour of the error $\boldsymbol{r}_{\Delta 1}(t) = \boldsymbol{r}_1(t) - \hat{\boldsymbol{r}}_1(t)$ between the actual residual $\boldsymbol{r}_1(t)$ and its approximation $\hat{\boldsymbol{r}}_1(t)$ is represented by the comparison system

$$\bar{R}_{\Delta 1}^{\mathcal{N}_{\mathrm{S}1}}: \quad \bar{\boldsymbol{r}}_{\Delta 1}(t) = \bar{\boldsymbol{R}}_{\Delta 1}^{\mathcal{N}_{\mathrm{S}1}}(t) * |\boldsymbol{w}_1(t)|, \tag{8.10}$$

where the relation

$$\bar{\boldsymbol{r}}_{\Delta 1}(t) \geq |\boldsymbol{r}_1(t) - \hat{\boldsymbol{r}}_1(t)| \geq \boldsymbol{0} \tag{8.11}$$

holds for arbitrary bounded inputs $\boldsymbol{w}_1(t)$. The transfer function matrix $\bar{\boldsymbol{R}}_{\Delta 1}^{\mathcal{N}_{\mathrm{S}1}}(t)$ results to

$$\bar{\boldsymbol{R}}_{\Delta 1}^{\mathcal{N}_{\mathrm{S}1}}(t) = |\boldsymbol{G}_{\mathrm{ru}1}(t) * \left(\boldsymbol{C}_1(t) - \boldsymbol{C}_1(t) * \boldsymbol{F}_{\mathrm{yq}1}^{\mathcal{N}_{\mathrm{S}1}}(t)\right) + \boldsymbol{G}_{\mathrm{ry}1}(t) * \boldsymbol{F}_{\mathrm{yq}1}^{\mathcal{N}_{\mathrm{S}1}}(t)| * \bar{\boldsymbol{\Psi}}(t) * |\boldsymbol{F}_{\mathrm{pw}1}^{\mathcal{N}_{\mathrm{S}1}}(t)| \tag{8.12}$$

with $\bar{\boldsymbol{\Psi}}(t)$ from (8.4), where $\bar{\boldsymbol{\Psi}}(t) \geq \boldsymbol{\mathrm{O}}$ for all $t \geq 0$ due to assumptions A 8.1 and A 8.2.

As the residual generator G_1 uses the approximate model $\hat{S}_1^{\mathcal{N}_{\mathrm{S}1}}$, the equality

$$\hat{\boldsymbol{R}}_1^{\mathcal{N}_{\mathrm{S}1}}(t) = \boldsymbol{\mathrm{O}}, \quad \forall t \geq 0 \tag{8.13}$$

holds, i.e., $\hat{\boldsymbol{r}}_1(t) = \boldsymbol{0},\ \forall t \geq 0$. According to the relations (8.11) and (8.13) it can be concluded that the actual impulse response function $\boldsymbol{R}_1^{\mathcal{N}}(t)$ is enclosed by the tolerance band

$$\boldsymbol{\mathrm{O}} \leq |\boldsymbol{R}_1^{\mathcal{N}}(t)| \leq \bar{\boldsymbol{R}}_{\Delta 1}^{\mathcal{N}_{\mathrm{S}1}}(t), \quad \forall t \geq 0$$

which equivalently means that

$$|\boldsymbol{R}_1^{\mathcal{N}}(t)| \in \mathbb{T}_1^{\mathcal{N}_{\mathrm{S1}}}(t) := \left[\mathbf{O},\ \bar{\boldsymbol{R}}_{\Delta 1}^{\mathcal{N}_{\mathrm{S1}}}(t)\right]. \tag{8.14}$$

That is, the actual generated detection residual $\boldsymbol{r}_1(t)$ is located in a tolerance band according to

$$|\boldsymbol{r}_1(t)| \in \mathbb{T}_1^{\mathcal{N}_{\mathrm{S1}}}(t) * |\boldsymbol{w}_1(t)|, \quad \forall t \geq 0. \tag{8.15}$$

It is worth noting that the residual $\boldsymbol{r}_1(t)$ may differ from zero, even though no fault has occurred in subsystem S_1. This is due to the fact that the considered residual generators evaluate the measurements $\boldsymbol{u}_1(t)$ and $\boldsymbol{y}_1(t)$ by means of an approximate model. This issue is considered in Section 8.4.

Fig. 8.5a visualises an exemplary tube for the residual $\boldsymbol{r}_1(t)$. At $t = t_{\mathrm{S}}$ the controlled subsystem F_1 is stimulated through the step-wise reference input $\boldsymbol{w}_1(t) = \sigma(t - t_{\mathrm{S}})$.

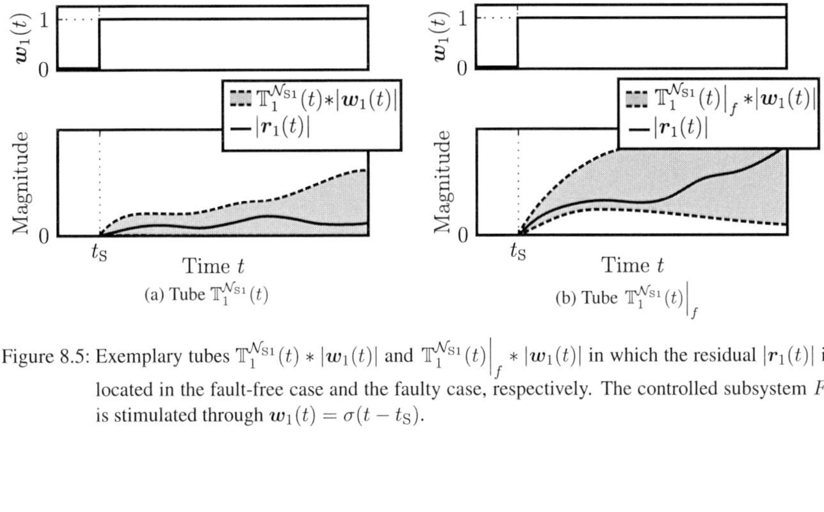

Figure 8.5: Exemplary tubes $\mathbb{T}_1^{\mathcal{N}_{\mathrm{S1}}}(t) * |\boldsymbol{w}_1(t)|$ and $\mathbb{T}_1^{\mathcal{N}_{\mathrm{S1}}}(t)\big|_f * |\boldsymbol{w}_1(t)|$ in which the residual $|\boldsymbol{r}_1(t)|$ is located in the fault-free case and the faulty case, respectively. The controlled subsystem F_1 is stimulated through $\boldsymbol{w}_1(t) = \sigma(t - t_{\mathrm{S}})$.

Faulty case. Under the consideration that a fault $f \in \mathcal{F}$ has occurred, the motion of the detection residual $\boldsymbol{r}_1(t)$ is reflected by the model

$$R_1^{\mathcal{N}}\big|_f = \mathrm{con}\left(\left\{G_1,\ C_1,\ S_{1f}^{\mathcal{N}_{\mathrm{S1}}}\right\} \cup \left\{E_1^{\mathcal{N}_{\mathrm{W1}}}\right\},\ \boldsymbol{w}_1,\ \boldsymbol{r}_1(t)\right).$$

described by

$$R_1^{\mathcal{N}}\big|_f: \quad \boldsymbol{r}_1(t) = \boldsymbol{R}_1^{\mathcal{N}}(t)\big|_f * \boldsymbol{w}_1(t).$$

In accordance with (8.8) and (8.10), the corresponding approximate model is represented by

$$\hat{R}_1^{\mathcal{N}_{S1}}\big|_f: \quad \hat{\boldsymbol{r}}_1(t) = \hat{\boldsymbol{R}}_1^{\mathcal{N}_{S1}}(t)\big|_f * \boldsymbol{w}_1(t)$$

and the error dynamics $\boldsymbol{r}_{\Delta 1}(t) = \boldsymbol{r}_1(t) - \hat{\boldsymbol{r}}_1(t)$ are described by the comparison system

$$\bar{R}_{\Delta 1}^{\mathcal{N}_{S1}}\big|_f: \quad \bar{\boldsymbol{r}}_{\Delta 1}(t) = \bar{\boldsymbol{R}}_{\Delta 1}^{\mathcal{N}_{S1}}(t)\big|_f * |\boldsymbol{w}_1(t)|,$$

where the relation (8.11) holds for arbitrary bounded inputs $\boldsymbol{w}_1(t)$. The transfer function matrices $\hat{\boldsymbol{R}}_1^{\mathcal{N}_{S1}}(t)\big|_f$ and $\bar{\boldsymbol{R}}_{\Delta 1}^{\mathcal{N}_{S1}}(t)\big|_f$ are constructed in accordance with (8.9) and (8.12), respectively, using the faulty controlled interacting subsystem $F_{1f}^{\mathcal{N}_{S1}}$.

From the relation (8.11), it can be concluded that the actual impulse response function $\boldsymbol{R}_1^{\mathcal{N}}(t)\big|_f$ is enclosed by the tolerance band

$$\left|\hat{\boldsymbol{R}}_1^{\mathcal{N}_{S1}}(t)\big|_f\right| - \bar{\boldsymbol{R}}_{\Delta 1}^{\mathcal{N}_{S1}}(t)\big|_f \leq \left|\boldsymbol{R}_1^{\mathcal{N}}(t)\big|_f\right| \leq \left|\hat{\boldsymbol{R}}_1^{\mathcal{N}_{S1}}(t)\big|_f\right| + \bar{\boldsymbol{R}}_{\Delta 1}^{\mathcal{N}_{S1}}(t)\big|_f,$$

where in contrast to the fault-free case (eqn (8.13)), the approximation $\hat{\boldsymbol{R}}_1^{\mathcal{N}_{S1}}(t)$ does not vanish. As a consequence, the actual detection residual $\boldsymbol{r}_1(t)$ that is generated if the fault f has occurred is enclosed in the tube

$$\mathbb{T}_1^{\mathcal{N}_{S1}}(t)\big|_f := \left[\max\left(\boldsymbol{0}, \left|\hat{\boldsymbol{R}}_1^{\mathcal{N}_{S1}}(t)\big|_f\right| - \bar{\boldsymbol{R}}_{\Delta 1}^{\mathcal{N}_{S1}}(t)\big|_f\right), \left|\hat{\boldsymbol{R}}_1^{\mathcal{N}_{S1}}(t)\big|_f\right| + \bar{\boldsymbol{R}}_{\Delta 1}^{\mathcal{N}_{S1}}(t)\big|_f\right] \tag{8.16}$$

according to

$$|\boldsymbol{r}_1(t)| \in \mathbb{T}_1^{\mathcal{N}_{S1}}(t)\big|_f * |\boldsymbol{w}_1(t)|, \quad \forall t \geq 0.$$

An exemplary tube $\mathbb{T}_1^{\mathcal{N}_{S1}}(t)\big|_f$ is highlighted in Fig. 8.5b.

8.3.3 Robust fault detectability and robust fault isolability

To be able to detect a fault or isolate a fault, the dynamics of fault-free system and the faulty systems must be distinguishable. Therefore, this section proposes a robust detectability condition and a robust isolability condition based on the tubes (8.14) and (8.16) to claim the separability of the fault-free and faulty uncertain system.

In the fault-free case, the dynamics $|\boldsymbol{R}_1^{\mathcal{N}}(t)|$ of the generated residual are known to be located within tube $\mathbb{T}_1^{\mathcal{N}_{S1}}(t)$ according to (8.14), whereas in the faulty case, the dynamic $\left|\boldsymbol{R}_1^{\mathcal{N}}(t)\big|_f\right|$ are enclosed by the tube $\mathbb{T}_1^{\mathcal{N}_{S1}}(t)\big|_f$ according to (8.16). In order to be able to distinguish between the fault-free case and the faulty case, there must exist a time instant in which both tubes do not intersect. This claim yields robust detectability as formalised in the following theorem.

Theorem 8.1 (Robust detectability) *Let the residual be generated by the detection residual generator G_1, let all controlled subsystems F_i, $(\forall i \in \mathcal{N})$ be in their operating points and let $\boldsymbol{w}_i = \boldsymbol{0}$, $(\forall i \in \mathcal{N} \setminus \{1\})$. The fault $f \in \mathcal{F}$ is detectable by the residual $\boldsymbol{r}_1(t)$ if*

$$\mathbb{T}_1^{\mathcal{N}_{S1}}(t) \cap \mathbb{T}_1^{\mathcal{N}_{S1}}(t)\big|_f = \emptyset, \quad \exists\, t \geq 0. \tag{8.17}$$

To clarify the detectability claim (8.17), an example is shown in Fig. 8.6. Figure 8.6a highlights the case in which the fault is not detectable. The motion of the residual $\boldsymbol{r}_1(t)$ can be represented by both models $R_1^{\mathcal{N}}$ or $R_1^{\mathcal{N}}\big|_f$. In contrast to this, Fig. 8.6b depicts the case in which the detectability claim (8.17) is satisfied such that it is clear which model represents the behaviour of $\boldsymbol{r}_1(t)$.

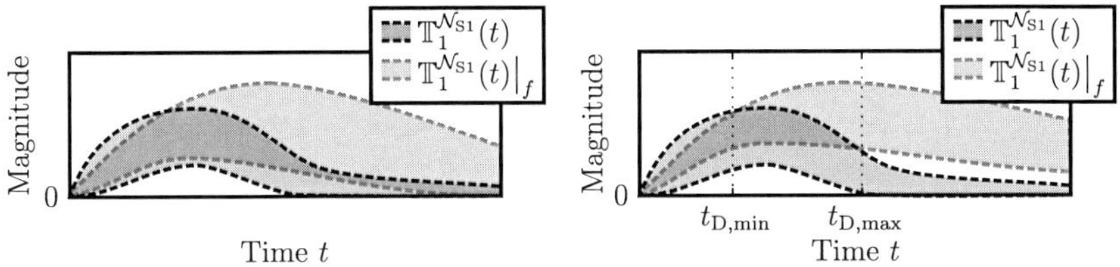

(a) The fault is not detectable since the tubes do intersect for all times $t \geq t_S$.

(b) The fault is detectabel since the tubes do not intersect for a certain time period.

Figure 8.6: Interpretation of the condition for robust detectability.

Generally, the more model information from other design agents is taken into account, the smaller the tubes will become. Based on this fact an algorithm is proposed in Section 8.5 in order to determine the necessary amount of model information to detect the faults.

The proposed robust detectability condition (as well as the robust detectability condition) depend upon the residual generator that is used for fault diagnosis. Accordingly, a fault may be detectable for a certain residual generator, while for another generator the fault is undetectable.

Robust fault isolability. In direct analogy, a robust isolability condition is derived in this paragraph.

As mentioned in Section 8.2.2, for fault isolation q further isolation residual generators G_{1k}, $(\forall k \in \mathcal{F})$ are required, where G_{1k} uses the approximate model $\hat{S}_{1k}^{\mathcal{N}_{S1}}$. Accordingly, the dynamics of the generated isolation residual using the residual generator G_{1k} are denoted by $\boldsymbol{R}_{1k}^{\mathcal{N}}(t)$ if the fault $k \in \mathcal{F}$ has been occurred or are denoted by $\boldsymbol{R}_{1k}^{\mathcal{N}}(t)\big|_f$ if the fault $f \neq k$ has been occurred. For these dynamics, tubes can be determined in direct analogy to the case of fault detection. In particular, $|\boldsymbol{R}_{1k}^{\mathcal{N}}(t)| \in \mathbb{T}_{1k}^{\mathcal{N}_{\mathrm{S}1}}(t)$ and $\left|\boldsymbol{R}_{1k}^{\mathcal{N}}(t)\big|_f\right| \in \mathbb{T}_{1k}^{\mathcal{N}_{\mathrm{S}1}}(t)\big|_f$.

To isolate a fault f, the dynamics $\boldsymbol{R}_{1k}^{\mathcal{N}}(t)$ and $\boldsymbol{R}_{1k}^{\mathcal{N}}(t)\big|_f$ of the generated isolation residual using the residual generator G_{1k}, $(k \neq f)$ have to be distinguishable so as to concluded that the fault k has not been occurred. Since it is only known that these dynamics are located within the tubes $\mathbb{T}_{1k}^{\mathcal{N}_{\mathrm{S}1}}(t)$ and $\mathbb{T}_{1k}^{\mathcal{N}_{\mathrm{S}1}}(t)\big|_f$, respectively, robust isolability have to be claimed.

Theorem 8.2 (Robust isolability) *Let the isolation residuals be generated by the isolation residual generator G_{1k}, $(\forall k \in \mathcal{F})$. Let $\boldsymbol{w}_i = \boldsymbol{0}$, $(\forall i \in \mathcal{N} \setminus \{1\})$ and let the fault f has been occurred and has been detected at time instant $t_{\mathrm{D}} \geq 0$. The fault f is isolable by the residuals $\boldsymbol{r}_{1k}$, $(\forall k \in \mathcal{F})$ if*

$$\mathbb{T}_{1k}^{\mathcal{N}_{\mathrm{S}1}}(t) \cap \mathbb{T}_{1k}^{\mathcal{N}_{\mathrm{S}1}}(t)\big|_f = \emptyset, \quad \exists\, t \geq t_{\mathrm{D}} \tag{8.18}$$

for all $k \in \mathcal{F} \setminus \{f\}$.

For instance, consider the set of faults $\mathcal{F} = \{\mathrm{f}_1, \mathrm{f}_2\}$. The fault $f = \mathrm{f}_1$ is isolable if there exists a time instant $t \geq t_{\mathrm{D}}$ in which the tube $\mathbb{T}_{1\mathrm{f}_2}^{\mathcal{N}_{\mathrm{S}1}}(t)\big|_{\mathrm{f}_1}$ does not intersect with the tube $\mathbb{T}_{1\mathrm{f}_2}^{\mathcal{N}_{\mathrm{S}1}}(t)$ (cf. Fig. 8.6b for the similar case of fault detection). Then, it can be guaranteed that motion of the isolation residual $\boldsymbol{r}_{\mathrm{f}_1}(t)$ is represented by the dynamics $\boldsymbol{R}_{1\mathrm{f}_1}^{\mathcal{N}}(t)$ and not by $\boldsymbol{R}_{1\mathrm{f}2}^{\mathcal{N}}(t)$.

Note that if the condition (8.18) is only satisfied for time instances $t < t_{\mathrm{D}}$, then the fault is not isolable. This is attributable to the fact that the isolation process, the moment a fault has been detected.

8.3.4 Thresholds for fault detection and fault isolation

This section proposes a detection threshold and isolation thresholds.

As it is known for the healthy case that the detection residual is located in the tube $\mathbb{T}_{1}^{\mathcal{N}_{\mathrm{S}1}}(t) * |\boldsymbol{w}_1(t)|$ according to (8.15), this tube serves as detection threshold, i.e.,

$$\boldsymbol{\mu}_1(t) := \mathbb{T}_{1}^{\mathcal{N}_{\mathrm{S}1}}(t) * |\boldsymbol{w}_1(t)|. \tag{8.19}$$

Accordingly, a fault f is detected if

$$|\boldsymbol{r}_1(t)| \notin \boldsymbol{\mu}_1(t), \quad \exists\, t \geq t_{\mathrm{f}} \geq 0.$$

The detectability condition (8.17) guarantees that there exists a reference signal $\boldsymbol{w}_1(t)$, for which the residual $\boldsymbol{r}_1(t)$ leaves the detection threshold $\boldsymbol{\mu}_1(t)$.

The isolation thresholds are chosen accordingly. That is, the isolation thresholds for the residuals $\boldsymbol{r}_{1k}(t)$, $(\forall k \in \mathcal{F})$ is

$$\boldsymbol{\mu}_{1k}(t) := \mathbb{T}_{1k}^{\mathcal{N}_{S1}}(t) * |\boldsymbol{w}_1(t)| \tag{8.20}$$

for all $k \in \mathcal{F}$. Then, the detected fault f is isolated, if

$$|\boldsymbol{r}_{1f}(t)| \in \boldsymbol{\mu}_{1f}(t), \quad \forall t \geq t_{\mathrm{D}} \tag{8.21}$$

and

$$|\boldsymbol{r}_{1k}(t)| \notin \boldsymbol{\mu}_{1k}(t), \quad \exists\, t \geq t_{\mathrm{D}}, \ \forall k \in \mathcal{F} \setminus \{f\}\,. \tag{8.22}$$

The isolability condition (8.18) guarantees that there exists a reference signal $\boldsymbol{w}_1(t)$ such that the fault f can be isolated according to (8.21) and (8.22). That means that only the residual $\boldsymbol{r}_{1f}(t)$ remains in its tube, whereas all other residuals $\boldsymbol{r}_{1k}(t)$, $(k \in \mathcal{F} \setminus \{f\})$ has left their tube for a (not necessarily same) time instant.

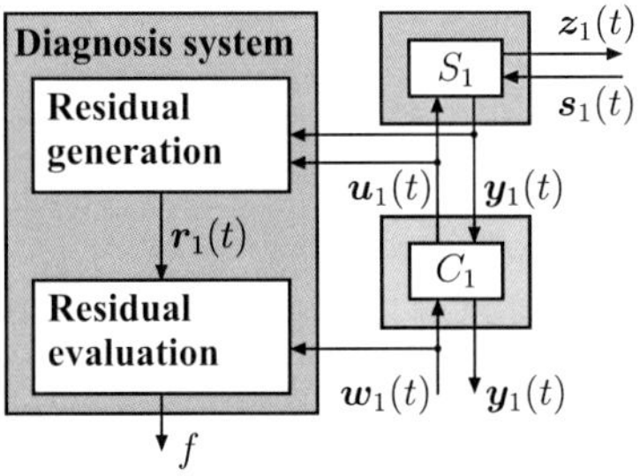

Figure 8.7: Required structure of the proposed diagnosis system: The measurements $\boldsymbol{u}_1(t)$ and $\boldsymbol{y}_1(t)$ are used for residual generation and the reference signal $\boldsymbol{w}_1(t)$ is required for residual evaluation.

In order to use the proposed thresholds (8.19) and (8.20), the diagnosis system must have access to the reference signal $\boldsymbol{w}_1(t)$. According to this claim, the required structure of the local diagnosis system is shown inFig. 8.7. Although a fault is detectable according to Theorem 8.1, it may cannot be detected by the detection threshold (8.19) and a specific reference signal. Accordingly, in view of the execution phase, the detectability should be verified using the type of reference signal (such as steps, sinus, ramps) that is used during the execution of the diagnosis system (if even a priori known). Then, a time interval $t \in [t_{\mathrm{D,min}} - t_{\mathrm{S}}, \quad t_{\mathrm{D,max}} - t_{\mathrm{S}}]$ can be extracted in which the fault will be detected after the faulty system has been stimulated at time instant $t_{\mathrm{S}} \geq t_{\mathrm{f}}$. An example is shown in Fig. 8.6b for an impulse-wise stimulation (i.e., $\boldsymbol{w}_1(t) = \delta(t)$). The time instant $t_{\mathrm{D,min}}$ denotes the earliest time instant when $\boldsymbol{r}_1(t)$ can leave the tube, whereas $t_{\mathrm{D,max}}$ denotes the latest. Note that these time depend upon the reference signal $\boldsymbol{w}_1(t)$.

8.4 Design of the residual generation

The considered residual generators only make use of an approximate model and, thus, ignore the dynamics of the irrelevant part $E_1^{\mathcal{N}_{\mathrm{W1}}}$. This section proposes a residual generator that generates a residual which is asymptotically independent of the influence from the unconsidered dynamics $E_1^{\mathcal{N}_{\mathrm{W1}}}$. It is shown that the width of the tube $\mathbb{T}_1^{\mathcal{N}_{\mathrm{S1}}}(t)$ is narrowed if the proposed residual generator is used.

8.4.1 Design issues

As mentioned in the previous sections, the considered residual generators make use of the approximate model $\hat{S}_1^{\mathcal{N}_{\mathrm{S1}}}$ and $\hat{S}_{1f}^{\mathcal{N}_{\mathrm{S1}}}$, $(\forall f \in \mathcal{F})$, respectively, and, thus, ignore the dynamics of the irrelevant part $E_1^{\mathcal{N}_{\mathrm{W1}}}$ completely. As a consequence, the measurements $\boldsymbol{u}_1(t)$ and $\boldsymbol{y}_1(t)$ that are evaluated by the residual generators are not consistent with the behaviour represented by the approximate model $\hat{S}_1^{\mathcal{N}_{\mathrm{S1}}}$ and $\hat{S}_{1f}^{\mathcal{N}_{\mathrm{S1}}}$, $(\forall f \in \mathcal{F})$, respectively. Due to this fact, the residual generators have to be designed in a way that the error between the supposed behaviour (reflected by $\hat{S}_1^{\mathcal{N}_{\mathrm{S1}}}$) and the actual behaviour (represented by $S_1^{\mathcal{N}}$) has barely an effect on the generated residual. In addition, the appropriate design of the residual generators will reduce the uncertainty of the generated residuals (i.e., the tubes introduced in Section 8.3 get tighter) so that less model information has to be gathered by design agent D_1 to design the diagnosis system.

8.4.2 A robust residual generator

This section proposes a PI-based residual generator that only uses the approximate model $\hat{S}_1^{\mathcal{N}_{\mathrm{S1}}}$. It will be shown by Corollary 8.1 that the generated residual vanishes asymptotically. The derivation of the residual generator G_1 for fault detection is presented which can easily be adapted to the design of identification residual generators G_{1f}.

The proposed residual generator essentially bases on the PI-based observer from [165], which uses a classical LUENBERGER observer that exploits the approximate model $\hat{S}_1^{\mathcal{N}_{\mathrm{S1}}}$ from (8.2). This LUENBERGER observer is represented by

$$
\begin{aligned}
\dot{\boldsymbol{x}}_{\mathrm{O1}}(t) &= \boldsymbol{A}_1^{\mathcal{N}_{\mathrm{S1}}}\boldsymbol{x}_{\mathrm{O1}}(t) + \boldsymbol{B}_1^{\mathcal{N}_{\mathrm{S1}}}\boldsymbol{u}_1 + \boldsymbol{N}_1\boldsymbol{r}_1(t) \\
\boldsymbol{r}_1(t) &= -\boldsymbol{C}_1^{\mathcal{N}_{\mathrm{S1}}}\boldsymbol{x}_{\mathrm{O1}}(t) + \boldsymbol{y}_1(t),
\end{aligned}
\tag{8.23}
$$

where $\boldsymbol{x}_{\mathrm{O}1}(t)$ is the estimation of the state $\boldsymbol{x}_1^{\mathcal{N}_{\mathrm{S}1}}(t)$ and the residual $\boldsymbol{r}_1(t)$ yields

$$\boldsymbol{r}_1(t) = \boldsymbol{y}_1(t) - \boldsymbol{y}_{\mathrm{O}1}(t) \tag{8.24}$$

with $\boldsymbol{y}_{\mathrm{O}1}(t) = \boldsymbol{C}_1^{\mathcal{N}_{\mathrm{S}1}}\boldsymbol{x}_{\mathrm{O}1}(t)$. The matrix $\boldsymbol{N}_1$ denotes gain matrix to be designed.

The main idea for residual generation is to treat the unmeasurable error input signal $\boldsymbol{q}_1(t)$ as an unknown input. The only information about the unknown input is that

$$\lim_{t\to\infty} \dot{\boldsymbol{q}}_1(t) = \boldsymbol{0} \tag{8.25}$$

for stepwise reference signals, i.e., $\boldsymbol{w}_1(t) = \sigma(t)\bar{\boldsymbol{w}}$ due to assumption A 8.1. Hence, the LUENBERGER observer (8.23) is extended by an integrator which aims at modelling the dynamics of the unknown input. The extended observer result to

$$\begin{aligned}
\begin{pmatrix} \dot{\boldsymbol{x}}_{\mathrm{O}1} \\ \dot{\boldsymbol{q}}_{\mathrm{O}1} \end{pmatrix} &= \begin{pmatrix} \boldsymbol{A}_1^{\mathcal{N}_{\mathrm{S}1}} & \boldsymbol{E}_1^{\mathcal{N}_{\mathrm{S}1}} \\ \mathrm{O} & \mathrm{O} \end{pmatrix} \begin{pmatrix} \boldsymbol{x}_{\mathrm{O}1}(t) \\ \boldsymbol{q}_{\mathrm{O}1}(t) \end{pmatrix} + \begin{pmatrix} \boldsymbol{B}_1^{\mathcal{N}_{\mathrm{S}1}} \\ \mathrm{O} \end{pmatrix} \boldsymbol{u}_1(t) + \begin{pmatrix} \boldsymbol{N}_1 \\ \boldsymbol{N}_2 \end{pmatrix} \boldsymbol{r}_1(t) \\
\boldsymbol{r}_1(t) &= \begin{pmatrix} -\boldsymbol{C}_1^{\mathcal{N}_{\mathrm{S}1}} & \mathrm{O} \end{pmatrix} \begin{pmatrix} \boldsymbol{x}_{\mathrm{O}1}(t) \\ \boldsymbol{q}_{\mathrm{O}1}(t) \end{pmatrix} + \boldsymbol{y}_1(t),
\end{aligned} \tag{8.26}$$

where the number of the additional integrator states matches the dimension of the unknown input, i.e., $\dim(\boldsymbol{q}_{\mathrm{O}1}(t)) = \dim(\boldsymbol{q}_1(t))$. The additional gain matrix $\boldsymbol{N}_2$ is introduced to stabilise the integrator dynamics. Inserting (8.24) into the state equation of (8.26) results finally to the residual generator

$$G_1: \left\{ \begin{aligned}
\begin{pmatrix} \dot{\boldsymbol{x}}_{\mathrm{O}1} \\ \dot{\boldsymbol{q}}_{\mathrm{O}1} \end{pmatrix} &= \begin{pmatrix} \boldsymbol{A}_1^{\mathcal{N}_{\mathrm{S}1}} - \boldsymbol{N}_1\boldsymbol{C}_1^{\mathcal{N}_{\mathrm{S}1}} & \boldsymbol{E}_1^{\mathcal{N}_{\mathrm{S}1}} \\ -\boldsymbol{N}_2\boldsymbol{C}_1^{\mathcal{N}_{\mathrm{S}1}} & 0 \end{pmatrix} \begin{pmatrix} \boldsymbol{x}_{\mathrm{O}1}(t) \\ \boldsymbol{q}_{\mathrm{O}1}(t) \end{pmatrix} + \begin{pmatrix} \boldsymbol{B}_1^{\mathcal{N}_{\mathrm{S}1}} \\ \mathrm{O} \end{pmatrix} \boldsymbol{u}_1(t) + \begin{pmatrix} \boldsymbol{N}_1 \\ \boldsymbol{N}_2 \end{pmatrix} \boldsymbol{y}_1(t) \\
\boldsymbol{r}_1(t) &= \begin{pmatrix} -\boldsymbol{C}_1^{\mathcal{N}_{\mathrm{S}1}} & \mathrm{O} \end{pmatrix} \begin{pmatrix} \boldsymbol{x}_{\mathrm{O}1}(t) \\ \boldsymbol{q}_{\mathrm{O}1}(t) \end{pmatrix} + \boldsymbol{y}_1(t).
\end{aligned} \right. \tag{8.27}$$

The next corollary describe the robustness property of the residual generator. Corollary 8.1 is a corollary of Theorem 2 in [165].

Corollary 8.1 (Robustness of the residual generator) *Consider the fault-free case, let all controlled subsystems F_i, ($\forall i \in \mathcal{N}$) be in their operating points and let $\boldsymbol{w}_1(t) = \sigma(t)\bar{\boldsymbol{w}}_1$, whereas $\boldsymbol{w}_i(t) = \boldsymbol{0}$, ($\forall i = \in \mathcal{N} \setminus \{1\}$). There exist the gain matrices $\boldsymbol{N}_1$ and $\boldsymbol{N}_2$ for the residual generator* (8.27) *such that*

$$\lim_{t\to\infty} \boldsymbol{r}_1(t) = \boldsymbol{0} \tag{8.28}$$

if

$$\left(\begin{pmatrix} \boldsymbol{A}_1^{\mathcal{N}_{\mathrm{S}1}} & \boldsymbol{E}_1^{\mathcal{N}_{\mathrm{S}1}} \\ \mathrm{O} & \mathrm{O} \end{pmatrix}, \begin{pmatrix} \boldsymbol{C}_1^{\mathcal{N}_{\mathrm{S}1}} & \mathrm{O} \end{pmatrix} \right) \text{ is detectable.} \tag{8.29}$$

Proof. In the first step a state transformation is applied to the residual generator (8.27). With $\boldsymbol{x}_{\delta 1}(t) = \boldsymbol{x}_1^{\mathcal{N}_{\mathrm{S}1}}(t) - \boldsymbol{x}_{\mathrm{O}1}(t)$ and $\boldsymbol{q}_{\delta 1}(t) = \boldsymbol{q}_1(t) - \boldsymbol{q}_{\mathrm{O}1}(t)$, the residual generator (8.27) reads

$$G_1 : \begin{cases} \begin{pmatrix} \dot{\boldsymbol{x}}_{\delta 1}(t) \\ \dot{\boldsymbol{q}}_{\delta 1}(t) \end{pmatrix} = \begin{pmatrix} \boldsymbol{A}_1^{\mathcal{N}_{\mathrm{S}1}} - \boldsymbol{N}_1 \boldsymbol{C}_1^{\mathcal{N}_{\mathrm{S}1}} & \boldsymbol{E}_1^{\mathcal{N}_{\mathrm{S}1}} \\ -\boldsymbol{N}_2 \boldsymbol{C}_1^{\mathcal{N}_{\mathrm{S}1}} & \mathrm{O} \end{pmatrix} \begin{pmatrix} \boldsymbol{x}_{\delta 1}(t) \\ \boldsymbol{q}_{\delta 1}(t) \end{pmatrix} + \begin{pmatrix} \mathrm{O} \\ \mathrm{I} \end{pmatrix} \dot{\boldsymbol{q}}_1(t) \\ \boldsymbol{r}_1(t) = \begin{pmatrix} \boldsymbol{C}_1^{\mathcal{N}_{\mathrm{S}1}} & \mathrm{O} \end{pmatrix} \begin{pmatrix} \boldsymbol{x}_{\delta 1}(t) \\ \boldsymbol{q}_{\delta 1}(t) \end{pmatrix}, \end{cases} \tag{8.30}$$

where the system matrix can be split into

$$\begin{pmatrix} \boldsymbol{A}_1^{\mathcal{N}_{\mathrm{S}1}} - \boldsymbol{N}_1 \boldsymbol{C}_1^{\mathcal{N}_{\mathrm{S}1}} & \boldsymbol{E}_1 \\ -\boldsymbol{N}_2 \boldsymbol{C}_1^{\mathcal{N}_{\mathrm{S}1}} & \mathrm{O} \end{pmatrix} = \underbrace{\begin{pmatrix} \boldsymbol{A}_1^{\mathcal{N}_{\mathrm{S}1}} & \boldsymbol{E}_1^{\mathcal{N}_{\mathrm{S}1}} \\ \mathrm{O} & \mathrm{O} \end{pmatrix}}_{\tilde{\boldsymbol{A}}} - \begin{pmatrix} \boldsymbol{N}_1 \\ \boldsymbol{N}_2 \end{pmatrix} \underbrace{\begin{pmatrix} \boldsymbol{C}_1^{\mathcal{N}_{\mathrm{S}1}} & \mathrm{O} \end{pmatrix}}_{\tilde{\boldsymbol{C}}}, \tag{8.31}$$

which is stabilisable, if and only if condition (8.29) is satisfied. Note that $\tilde{\boldsymbol{A}} \in \mathbb{R}^{\tilde{n}\times\tilde{n}}$ with $\tilde{n} = n_1 + \sum_{i=2}^{v-1} n_{\mathrm{F}i} + m_{\mathrm{q}1}$.

Now it is shown that (8.28) holds. It is well known from Hautus test of observability that condition (8.29) is satisfied, if and only if

$$\operatorname{rank} \begin{pmatrix} \lambda_i \mathbf{I} - \tilde{\boldsymbol{A}} \\ \tilde{\boldsymbol{C}} \end{pmatrix} = \tilde{n} \tag{8.32}$$

holds for all unstable eigenvalues λ_i of the matrix $\tilde{\boldsymbol{A}}$, where $\boldsymbol{q}_1 \in \mathbb{R}^{m_{\mathrm{q}1}}$. The system matrix $\tilde{\boldsymbol{A}}$

has at least the unstable eigenvalues $\lambda_i = 0$ for which condition (8.32) reads

$$\operatorname{rank}\begin{pmatrix} -\boldsymbol{A}_1^{\mathcal{N}_{S1}} & -\boldsymbol{E}_1^{\mathcal{N}_{S1}} \\ \boldsymbol{C}_1^{\mathcal{N}_{S1}} & \mathrm{O} \end{pmatrix} = \tilde{n}. \tag{8.33}$$

Consider the situation in which the feedback gains $\boldsymbol{N}_1$ and $\boldsymbol{N}_2$ exist such that the matrix on the left-hand side of eqn (8.31) is Hurwitz. Now, the steady state behaviour of the transformed residual generator (8.30) is analysed. For this situation, three conditions can be derived such that claim (8.28) holds. These conditions are the following are:

$$\lim_{t\to\infty}\left(\boldsymbol{A}_1^{\mathcal{N}_{S1}} - \boldsymbol{N}_1\boldsymbol{C}_1^{\mathcal{N}_{S1}}\right)\boldsymbol{x}_{\delta 1}(t) + \boldsymbol{E}_1^{\mathcal{N}_{S1}}\boldsymbol{q}_{\delta 1}(t) = \boldsymbol{0}, \tag{8.34}$$

$$\lim_{t\to\infty}\boldsymbol{N}_2\boldsymbol{C}_1^{\mathcal{N}_{S1}}\boldsymbol{x}_{\delta 1}(t) = \boldsymbol{0}, \tag{8.35}$$

$$\lim_{t\to\infty}\boldsymbol{C}_1^{\mathcal{N}_{S1}}\boldsymbol{x}_{\delta 1}(t) = \boldsymbol{0}. \tag{8.36}$$

The inserting of (8.36) into (8.34) and (8.35) leads to the system of linear equations

$$\lim_{t\to\infty}\begin{pmatrix} -\boldsymbol{A}_1^{\mathcal{N}_{S1}} & -\boldsymbol{E}_1^{\mathcal{N}_{S1}} \\ \boldsymbol{C}_1^{\mathcal{N}_{S1}} & \mathrm{O} \end{pmatrix}\begin{pmatrix} \boldsymbol{x}_{\delta 1}(t) \\ \boldsymbol{q}_{\delta 1}(t) \end{pmatrix} = \begin{pmatrix} 0 \\ 0 \end{pmatrix}, \tag{8.37}$$

which has a unique solution, if and only if condition (8.33) or condition (8.29), respectively, is satisfied. □

Condition (8.29) represents an existence condition for a residual generator (8.27). The gains matrices $\boldsymbol{N}_1$ and $\boldsymbol{N}_2$ has to be designed such that the residual generator (8.27) is asymptotically stable. To attenuate the influence of the unknown input high gains have to be chosen as proposed in [110]. Therefore, well known design methods such as pole placement [62, 119, 154] can be used. It has to be emphasised that an asymptotically stable residual generator is not necessary to satisfy the claim (8.28). It is only necessary that the residual generator is I/O stable. In case of an asymptotically stable residual generator, condition (8.33) reflects that the system has to provide at least as many linear independent measurements as unknown inputs.

The presented residual generator (8.27) is based on the classical LUENBERGER observer (8.23) that is extended by an exogenous integrator system in order to estimate unknown inputs, which have the property (8.25), asymptotically correct. In accordance with the inner model principle, the classical LUENBERGER observer (8.23) can be extended by arbitrary exogenous system that models the type of the unknown input signal (e.g., ramp, sinus) as it has been introduced by [98].

The following paragraph shows that since it is known that $\boldsymbol{r}_1(t)$ vanishes asymptotically, a tighter tube $\mathbb{T}_1^{\mathcal{N}_{S1}}(t)$ can be derived.

Analysis of the tube width. As mentioned in Section 8.3, only tubes can be determined which enclose the actual residual. Moreover, it has been shown that the tube width has effect on the detectability and isolability of the faults. In this paragraph, it is shown that tighter tubes result if the residual generator (8.27) is used. Essentially, as it is known that $\boldsymbol{r}_1(t)$ vanishes asymptotically, it can be concluded that the tube $\mathbb{T}_1^{\mathcal{N}_{S1}}(t)$ also vanishes asymptotically.

In the fault-free case, the detection residual is located in the interval

$$\mathbf{0} \leq |\boldsymbol{r}_1(t)| \leq \bar{\boldsymbol{R}}_{\Delta 1}^{\mathcal{N}_{S1}}(t) * |\boldsymbol{w}_1(t)|, \quad t \geq 0$$

according to (8.15) with $\bar{\boldsymbol{R}}_{\Delta 1}^{\mathcal{N}_{S1}}(t)$ described by (8.12). For step-wise reference signals (i.e., $\boldsymbol{w}_1(t) = \sigma(t)\bar{\boldsymbol{w}}_1$), the upper bound becomes

$$\bar{\boldsymbol{R}}_{\Delta 1}^{\mathcal{N}_{S1}}(t) * \sigma(t)\bar{\boldsymbol{w}}_1 = \int_0^t \bar{\boldsymbol{R}}_{\Delta 1}^{\mathcal{N}_{S1}}(\tau)\mathrm{d}\tau \cdot \bar{\boldsymbol{w}}_1. \tag{8.38}$$

Since $\bar{\boldsymbol{R}}_{\Delta 1}^{\mathcal{N}_{S1}}(t) \geq \mathbf{O},\ \forall t \geq 0$ the right hand side of (8.38) increases monotonically. However, due to the fact that the residual $\boldsymbol{r}_1(t)$ vanishes asymptotically if the residual generator (8.27) is used (cf. Corollary 8.1), the upper bound (8.14) can be determined to

$$\bar{\boldsymbol{R}}_{\Delta 1}^{\mathcal{N}_{S1}}(t) * \sigma(t)\bar{\boldsymbol{w}}_1 = \min\left(\int_0^t \bar{\boldsymbol{R}}_{\Delta 1}^{\mathcal{N}_{S1}}(\tau)\mathrm{d}\tau \cdot \bar{\boldsymbol{w}}_1,\ \left(\int_0^\infty \bar{\boldsymbol{R}}_{\Delta 1}^{\mathcal{N}_{S1}}(\tau)\mathrm{d}\tau - \int_0^t \bar{\boldsymbol{R}}_{\Delta 1}^{\mathcal{N}_{S1}}(\tau)\mathrm{d}\tau\right)\bar{\boldsymbol{w}}_1\right), \tag{8.39}$$

where the comparison is applied element-wise. The bound given in (8.39) is smaller than the bound given by (8.38) for large time t. Hence, the property of asymptotically convergence of the residual despite a limited amount of model information is reflected by a tighter tube $\mathbb{T}_1^{\mathcal{N}_{S1}}(t)$.

Figure 8.8 visualises an exemplary tighter tube $\mathbb{T}_1^{\mathcal{N}_{S1}}(t)$ for the residual $\boldsymbol{r}_1(t)$ in comparison with the tube shown in Fig. 8.5a.

However, under specific conditions on the faulty subsystem the tubes $\mathbb{T}_1^{\mathcal{N}_{S1}}(t)\big|_f * |\boldsymbol{w}_1(t)|$ in which the detection residual $|\boldsymbol{r}_1(t)|$ is located if the fault f has occurred will also vanish asymptotically. In the following, it is shown under which conditions

$$\lim_{t\to\infty} \boldsymbol{r}_1(t) = \mathbf{0} \tag{8.40}$$

is fulfilled in spite of the fault f has occurred. The conditions are stated for actuator faults (condition (8.41)), sensor faults (condition (8.42)) as well as process faults (condition (8.43)).

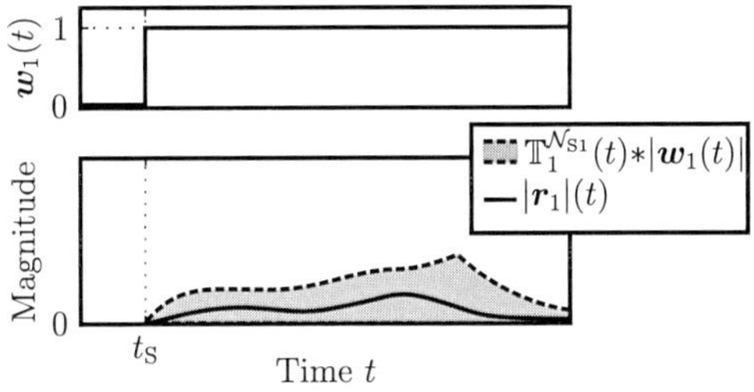

Figure 8.8: Tighter tube $\mathbb{T}_1^{\mathcal{N}_{S1}}(t)$ in comparison with the tube shown in Fig. 8.5a. The tube is narrowed if the PI-based residual generator is used and $\boldsymbol{w}_1(t) = \sigma(t - t_S)$.

For these faults, the estimation error $\boldsymbol{x}_{\delta 1}(t) = \boldsymbol{x}_1^{\mathcal{N}_{S1}}(t) - \boldsymbol{x}_{O1}(t)$ is the solution of

$$\begin{aligned} \dot{\boldsymbol{x}}_{\delta 1}(t) &= \left(\boldsymbol{A}_1^{\mathcal{N}_{S1}} - \boldsymbol{N}_1 \boldsymbol{C}_1^{\mathcal{N}_{S1}}\right) \boldsymbol{x}_{\delta 1}(t) + \boldsymbol{E}_1^{\mathcal{N}_{S1}} \boldsymbol{q}_{\delta 1}(t) + \boldsymbol{B}_{\delta 1} \boldsymbol{u}_{1f}(t) + \boldsymbol{N}_1 \boldsymbol{C}_{\delta 1} \boldsymbol{x}_{1f}(t) + \boldsymbol{A}_{\delta 1} \boldsymbol{x}_{1f}(t), \\ \boldsymbol{r}_1(t) &= \boldsymbol{C}_1^{\mathcal{N}_{S1}} \boldsymbol{x}_{\delta 1}(t) + \boldsymbol{C}_{\delta 1} \boldsymbol{x}_{1f}(t), \end{aligned}$$

where $\boldsymbol{B}_{\delta 1} = \boldsymbol{B}_{1f}^{\mathcal{N}_{S1}} - \boldsymbol{B}_1^{\mathcal{N}_{S1}}$, $\boldsymbol{C}_{\delta 1} = \boldsymbol{C}_{1f}^{\mathcal{N}_{S1}} - \boldsymbol{C}_1^{\mathcal{N}_{S1}}$ and $\boldsymbol{A}_{\delta 1} = \boldsymbol{A}_{1f}^{\mathcal{N}_{S1}} - \boldsymbol{A}_1^{\mathcal{N}_{S1}}$.

To ensure the existence of the gain matrices $\boldsymbol{N}_1$ and $\boldsymbol{N}_2$ in order that eqn (8.40) holds in case of an actuator fault, the equality

$$\lim_{t\to\infty} \begin{pmatrix} -\boldsymbol{A}_1^{\mathcal{N}_{S1}} & -\boldsymbol{E}_1^{\mathcal{N}_{S1}} \\ \boldsymbol{C}_1^{\mathcal{N}_{S1}} & \boldsymbol{O} \end{pmatrix} \begin{pmatrix} \boldsymbol{x}_{\delta 1}(t) \\ \boldsymbol{q}_{\delta 1}(t) \end{pmatrix} = \begin{pmatrix} \boldsymbol{B}_{\delta 1} \lim\limits_{t\to\infty} \boldsymbol{u}_{1f}(t) \\ \boldsymbol{0} \end{pmatrix}$$

has to be satisfied (cf. eqn (8.37)). If $\lim\limits_{t\to\infty} \boldsymbol{u}_{1f}(t) < \infty$ holds, a solution to the system of linear equations exits if

$$\operatorname{rank} \begin{pmatrix} -\boldsymbol{A}_1^{\mathcal{N}_{S1}} & -\boldsymbol{E}_1^{\mathcal{N}_{S1}} \\ \boldsymbol{C}_1^{\mathcal{N}_{S1}} & \boldsymbol{O} \end{pmatrix} = \operatorname{rank} \begin{pmatrix} -\boldsymbol{A}_1^{\mathcal{N}_{S1}} & -\boldsymbol{E}_1^{\mathcal{N}_{S1}} & \boldsymbol{B}_{\delta 1} \lim\limits_{t\to\infty} \boldsymbol{u}_{1f}(t) \\ \boldsymbol{C}_1^{\mathcal{N}_{S1}} & \boldsymbol{O} & \boldsymbol{O} \end{pmatrix}. \tag{8.41}$$

In direct analogy, there exist gain matrices $\boldsymbol{N}_1$ and $\boldsymbol{N}_2$ in order that the generated residual $\boldsymbol{r}_1(t)$ tends asymptotically to zeros in the case of a sensor fault if

$$\operatorname{rank} \begin{pmatrix} -\boldsymbol{A}_1^{\mathcal{N}_{S1}} & -\boldsymbol{E}_1^{\mathcal{N}_{S1}} \\ \boldsymbol{C}_1^{\mathcal{N}_{S1}} & \boldsymbol{O} \end{pmatrix} = \operatorname{rank} \begin{pmatrix} -\boldsymbol{A}_1^{\mathcal{N}_{S1}} & -\boldsymbol{E}_1^{\mathcal{N}_{S1}} & -\boldsymbol{N}_1 \boldsymbol{C}_{\delta 1} \lim\limits_{t\to\infty} \boldsymbol{x}_{1f}(t) \\ \boldsymbol{C}_1^{\mathcal{N}_{S1}} & \boldsymbol{O} & \boldsymbol{C}_{\delta 1} \lim\limits_{t\to\infty} \boldsymbol{x}_{1f}(t) \end{pmatrix} \tag{8.42}$$

providing that $\lim\limits_{t\to\infty} \boldsymbol{x}_1(t) < \infty$ holds. Finally, a process fault is considered. There exist gain matrices $\boldsymbol{N}_1$ and $\boldsymbol{N}_2$ in order that eqn. (8.40) is satisfied if

$$\operatorname{rank}\begin{pmatrix} -\boldsymbol{A}_1^{\mathcal{N}_{\mathrm{S1}}} & -\boldsymbol{E}_1^{\mathcal{N}_{\mathrm{S1}}} \\ \boldsymbol{C}_1^{\mathcal{N}_{\mathrm{S1}}} & \boldsymbol{O} \end{pmatrix} = \operatorname{rank}\begin{pmatrix} -\boldsymbol{A}_1^{\mathcal{N}_{\mathrm{S1}}} & -\boldsymbol{E}_1^{\mathcal{N}_{\mathrm{S1}}} & \boldsymbol{A}_{\delta 1} \lim\limits_{t\to\infty} \boldsymbol{x}_{1f}(t) \\ \boldsymbol{C}_1^{\mathcal{N}_{\mathrm{S1}}} & \boldsymbol{O} & \boldsymbol{O} \end{pmatrix} \tag{8.43}$$

under consideration that $\lim\limits_{t\to\infty} \boldsymbol{x}_{1f}(t) < \infty$ holds.

As the residual $\boldsymbol{r}_1(t)$ vanishes asymptotically in the faulty case if the conditions (8.41)–(8.43) hold, the fault has to be detected by its transient response.

Further residual generators using incomplete models. The presented residual generator (8.27) is one among many concepts which can be used to deal with model uncertainties. This paragraph gives a brief review about alternative residual generators of the form (8.5).

As mentioned before, the proposed residual generator (8.27) uses an exogenous system in order to estimate the unknown input. In contrast to this, the unknown-input observer aims at the complete decoupling of the unknown input on the estimated state (e.g. [60, 160, 174]). That means that in the fault-free case the upper bound $\boldsymbol{r}_1(t) = \boldsymbol{0}$ for all $t \geq 0$. For fault diagnosis, the system has to provide more linear independent measurements than linear unknown inputs. In particular, to cancel the unknown input, the system has to provide as many linear independent measurements as unknown inputs. To additionally estimate the system state and, thus, to generate a residual, further measurements are required. Hence, unknown-input observers cannot be used for single input single output as the multizone furnace (see Section 8.6).

Another approach aims at the explicitly usage of the comparison system $\bar{E}_1^{\mathcal{N}_{\mathrm{W1}}}$ for the residual generation. To take the upper bound $\bar{\boldsymbol{E}}_1^{\mathcal{N}_{\mathrm{W1}}}(t)$ of the irrelevant part into account, the residual generator (8.5) is extended by the input $\boldsymbol{q}_1(t)$ and output $\boldsymbol{p}_1(t)$. The resulting model reads

$$\begin{aligned} \boldsymbol{r}_1(t) &= \boldsymbol{G}_{\mathrm{ru1}}(t) * \boldsymbol{u}_1(t) + \boldsymbol{G}_{\mathrm{ry1}}(t) * \boldsymbol{y}_1(t) + \boldsymbol{G}_{\mathrm{rq1}}(t) * \boldsymbol{q}_1(t) \\ \boldsymbol{p}_1(t) &= \boldsymbol{G}_{\mathrm{pu1}}(t) * \boldsymbol{u}_1(t) + \boldsymbol{G}_{\mathrm{py1}}(t) * \boldsymbol{y}_1(t) + \boldsymbol{G}_{\mathrm{pq1}}(t) * \boldsymbol{q}_1(t). \end{aligned}$$

Accordingly, an upper bound on the residual results to

$$|\boldsymbol{r}_1(t)| \leq \bar{\boldsymbol{r}}_1(t) = \big(|\boldsymbol{G}_{\mathrm{ru1}}(t)| + \bar{\boldsymbol{G}}_{\Delta\mathrm{ru1}}(t)\big) * |\boldsymbol{u}_1(t)| + \big(|\boldsymbol{G}_{\mathrm{ry1}}(t)| + \bar{\boldsymbol{G}}_{\Delta\mathrm{ry1}}(t)\big) * |\boldsymbol{y}_1(t)|, \tag{8.44}$$

where $\bar{\boldsymbol{G}}_{\Delta\mathrm{ru1}}(t) = |\boldsymbol{G}_{\mathrm{rq1}}(t)| * \bar{\boldsymbol{\Psi}}(t) * |\boldsymbol{G}_{\mathrm{pu1}}(t)|$ and $\bar{\boldsymbol{G}}_{\Delta\mathrm{ry1}}(t) = |\boldsymbol{G}_{\mathrm{rq1}}(t)| * \bar{\boldsymbol{\Psi}}(t) * |\boldsymbol{G}_{\mathrm{py1}}(t)|$ with $\bar{\boldsymbol{\Psi}}(t) = \bar{\boldsymbol{E}}_1^{\mathcal{N}_{\mathrm{W1}}}(t) + \bar{\boldsymbol{E}}_1^{\mathcal{N}_{\mathrm{W1}}}(t) * |\boldsymbol{G}_{\mathrm{pq1}}(t)| * \bar{\boldsymbol{\Psi}}(t)$. However, the bound (8.44) will not be as tight as the bound (8.39) for step wise reference signals, as it can be concluded from the convolution of absolute value impulse response functions. Moreover, the asymptotic behaviour of the residual is not taken into account.

8.5 Plug-and-play diagnosis algorithm

This section proposes the plug-and-play diagnosis algorithm to design the diagnosis system in order to guarantee the detection of all faults $f \in \mathcal{F}$. The design encompasses the set up of the approximate model $\hat{S}_{1f}^{\mathcal{N}_{S1}}$, the parametrisation of the PI-based residual generator (8.27) and the set of the detection threshold $\boldsymbol{\mu}_1(t)$.

Initially, the design agents D_i only knows the model S_1 of the fault-free subsystem, the models S_{1f}, $(\forall f \in \mathcal{F})$ of the faulty subsystem, the model C_1 of the control station and the model K_1 of the local couplings. That is, the model set

$$\mathcal{M}_1 = \{S_1, C_1, K_1, \{S_{1f}, f \in \mathcal{F}\}\}$$

is a priori known by D_1. The design procedure, thus, requires that D_1 searches for the relevant models to set up the approximate model $S_{1f}^{\mathcal{N}_{S1}}$, designs the residual generator G_1 and sets the detection threshold $\boldsymbol{\mu}_1(t)$. The main idea of the design is that, initially, all controlled subsystems are assumed to be irrelevant for the design . Then, step by step further model information are taken into account until all faults $f \in \mathcal{F}$ can be detected according to Theorem 8.1. This procedure is summarised in Algorithm 8.1.

As initially all dynamics of other controlled subsystems are assumed to be irrelevant, the design agent starts Algorithm 4.1 to set up the comparison system $\bar{E}_1^{\{2,..,N\}}$ representing the aggregated effect of the these irrelevant dynamics (Step 1). Therefore, comparison systems $\bar{F}_i\big|_{w_i=0}$, $\bar{K}_i$, $(\forall i \in \mathcal{N})$ are transmitted among the design agents. During the loop (Steps 2–10) the existence of the residual generator is checked (Step 2), the residual generator is designed (Step 4), the tubes $\mathbb{T}_1^{\mathcal{N}_{S1}}(t)$ and $\mathbb{T}_1^{\mathcal{N}_{S1}}(t)\big|_f$ are calculated (Step 5) and the detectability condition is verified (Step 6). This loop is repeated until the approximate model is as precise as required to satisfy the detectability condition (Step 6). Model sets from other design agents are procured by D_1 in Step 9 if either the existence condition is violated (Step 2) or the detectability condition is violated (Step 6).

The number of exchanges of model information among the design agents is counted by the variable κ. If the design was successful, a diagnosis system is obtained that uses only as many models as necessary to guarantee fault detection. That is, the biggest admissible uncertainty of the interaction dynamics it can be dealt with for diagnosis purpose. Finally, the detection threshold is chosen in Step 11 and the diagnosis system is implemented and started up in Step 12. If D_1 has gathered all model sets $\mathcal{M}_i$, $(\forall i \in \mathcal{N})$ but either the residual generator does not exist (Step 3) or one of the faults is not detectable (Step 6), the design is aborted (Step 8).

To guarantee fault detection, the residual generator G_1 has to be asymptotically stable. Additionally, the shape of the tubes can be tuned by means of the feedback gains, as it can be seen by

Algorithm 8.1: Design of a local diagnosis system for S_1 to detect the faults $f \in \mathcal{F}$

Given: Control stations C_i, $(\forall i \in \mathcal{N})$ exist according to assumptions A 8.1 and A 8.2,
model set $\mathcal{M}_1 = (S_1, C_1, K_1, \{S_{1f},\ f \in \mathcal{F}\})$ available to D_1,
model set $\mathcal{M}_i = (S_i, C_i, K_i)$ available to D_i, $(\forall i \in \mathcal{N} \setminus \{1\})$

Initialise: $\kappa = 0$, $\mathcal{N}_{\mathrm{S1}} = \{1\}$, $\mathcal{N}_{\mathrm{W1}} = \{2, .., N\}$

Processing on D_1**:**

1. Procure the comparison systems $\bar{F}_i\big|_{\boldsymbol{w}_i=\mathbf{0}}$ and $\bar{K}_i$, $(\forall i \in \mathcal{N}_{\mathrm{W1}})$ using Algorithm 4.1 and set up the comparison system $\bar{E}_1^{\mathcal{N}_{\mathrm{W1}}}$ according to (8.3)
2. **while** $\mathcal{N}_{\mathrm{S1}} \subseteq \mathcal{N}$ **do**
3.   **if** condition (8.29) is satisfied **then** // Existence of the residual generator
4.   Design G_1 in order that (8.31) is Hurwitz
5.   Calculate $\mathbb{T}_1^{\mathcal{N}_{\mathrm{S1}}}(t)$ and $\mathbb{T}_1^{\mathcal{N}_{\mathrm{S1}}}(t)\big|_f$ according to (8.14) and (8.16)
6.   **if** condition (8.17) is satisfied for all $f \in \mathcal{F}$ **then** break // Detectability
7.   **end**
8.   **if** $\mathcal{N}_{\mathrm{S1}} = \mathcal{N}$ **then** STOP (no diagnosis system exists)
9.   Procure the model set $\mathcal{M}_i$ from D_i, $(i \in \mathcal{P}_1)$ and set $\kappa = \kappa + 1$,
    $\mathcal{N}_{\mathrm{S1}} = \mathcal{N}_{\mathrm{S1}} \cup \mathcal{P}_1$, $\mathcal{N}_{\mathrm{W1}} = \mathcal{N}_{\mathrm{W1}} \setminus \mathcal{P}_1$, $\mathcal{P}_1 = \mathcal{P}_i \setminus \mathcal{N}_{\mathrm{S1}}$
10. **end**
11. Set the detection threshold $\boldsymbol{\mu}_1(t) = \mathbb{T}_1^{\mathcal{N}_{\mathrm{S1}}}(t) * |\boldsymbol{w}_1(t)|$
12. Implement the residual generator G_1 and the detection threshold $\boldsymbol{\mu}_1(t)$ into hardware

Result: Local diagnosis system composed of residual generator G_1 and detection threshold $\boldsymbol{\mu}_1(t)$ to detect all faults $f \in \mathcal{F}$

eqn (8.12). That is, the influence of the irrelevant part $\bar{E}_1^{\mathcal{N}_{\mathrm{W1}}}$ can be attenuated by reducing the maximal magnitudes of the impulse response functions $\boldsymbol{G}_{\mathrm{ry1}}(t)$ and $\boldsymbol{G}_{\mathrm{ru1}}(t)$ (see also [110]).

Algorithm 8.1 can easily be extended to fault isolation by additionally checking the isolability condition (8.18) during the iteration and designing the isolation residual generators G_{1k}, $(\forall k \in \mathcal{F})$.

Remark 8.1 *The comparison systems* $\bar{F}_i\big|_{\boldsymbol{w}_i=\mathbf{0}}$ *and* $\bar{K}_i$ *are locally generated by the respective design agent* D_i *as mentioned in Section 3.3. Accordingly, the precision of the comparison systems* $\bar{F}_i\big|_{\boldsymbol{w}_i=\mathbf{0}}$ *and* $\bar{K}_i$ *and, thus, the accuracy of the comparison system* $\bar{E}_1^{\mathcal{N}_{\mathrm{W1}}}$ *depends upon the information the design agent is willing to provide.*

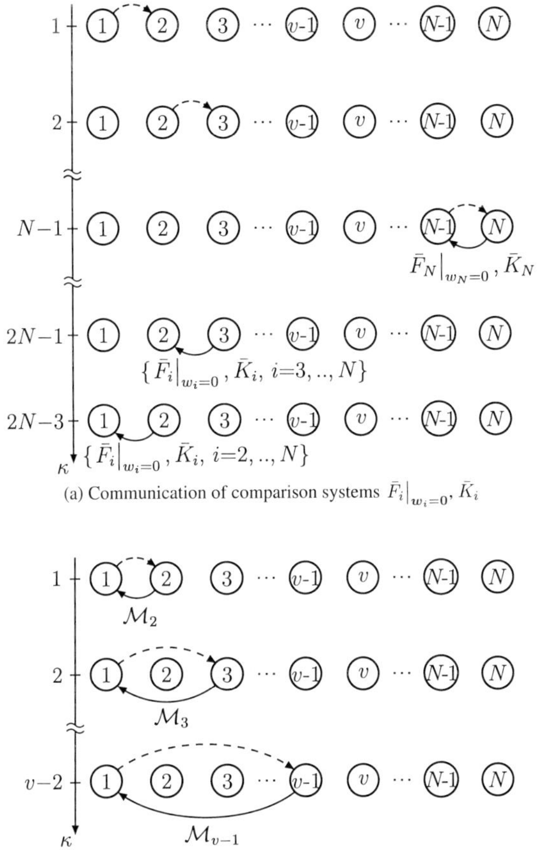

(a) Communication of comparison systems $\bar{F}_i\big|_{w_i=0}, \bar{K}_i$

(b) Communication of models and algorithms collected in $\mathcal{M}_i$

Figure 8.9: Communication graphs $\mathcal{G}_D(\kappa)$ during the processing of Algorithm 8.1: Dashed edges denote requests and solid edges denote response.

Communication among the design agents. During the processing of the Algorithm 8.1, the design agents communicate comparison systems as well as models and control algorithms among each other. The communication is highlighted by the communication graph $\mathcal{G}_D(\kappa)$ shown in Fig. 8.9. The interactions are subdivided into the request and response of comparison systems (Fig. 8.9a) with the aim to set up the comparison system $\bar{E}_1^{\{2,\ldots,N\}}$ and the transmission of models and algorithms (Fig. 8.9b) collected in the model sets $\mathcal{M}_i$, $(\forall i \in \mathcal{N}_{S1})$ to construct the approximate model $S_1^{\mathcal{N}_{S1}}$. The set up of the comparison system encompass $2N-3$

interactions (see Section 4.3), whereas the set up of the approximate model needs only $v - 2$ request-response messages.

Application to state estimation. Plug-and-play diagnosis can easily be adapted for the design of an observer for subsystem S_1. Essentially, the proposed residual generator estimates the states of the approximate model to generate the residual. Accordingly, this residual generator can be used to estimate the state of subsystem S_1 using an approximate model only. Based on Corollary 8.1, it can be shown that the estimation error will also vanish asymptotically (cf. conditions in [165]). Thus, the presented analysis of the generated residual can be adapted to analyse the estimation error. As a result, a tube can be derived which enclose the estimation error. The required model information can now be rated by means of the admissible tube width (i.e., uncertainty of the estimation error).

8.6 Example: Multizone furnace

The plug-and-play diagnosis algorithm is applied to design a diagnosis system to detect and isolate two possible faults that can occur in the multizone furnace (Section 2.5.2). It will be shown that an actuator fault can be detection based on the local model S_1 only, whereas the detection of a process fault requires additional model information from zone 2.

The zones S_i, $(i = 1, ..., 4)$ are modelled by (2.6) with (A.10a)–(A.10d) and are locally interconnected according to the model (2.8) with (A.11a)–(A.11d). The subsystems are initially controlled by proportional controllers which are represented by (2.11) with

$$\boldsymbol{A}_{\mathrm{C}i} = 0, \quad \boldsymbol{B}_{\mathrm{C}i} = 0, \quad \boldsymbol{C}_{\mathrm{C}i} = 0, \quad \boldsymbol{D}_{\mathrm{C}i} = 5, \qquad i = 1, ..., 4.$$

Considered faults. This paragraph introduces the actuator fault and the process fault to be detected. The model of the faulty subsystem is represented by (2.26).

The fault $f = \mathrm{f}_1$ describes a degraded heating power of the heater in zone 1. The reduced actuator power is modelled by the input matrix

$$\boldsymbol{B}_{1\mathrm{f}_1} = 10^{-2} \begin{pmatrix} 4.5 \\ 5 \end{pmatrix}.$$

The second fault $f = \mathrm{f}_2$ is a crack in the outer wall of the heater of zone 1. As a consequence, the temperature dynamics is slowed down. This fact is represented by the system matrix

$$A_{1f_2} = \begin{pmatrix} -0.23 & 0 \\ 0 & -0.2 \end{pmatrix}.$$

Design of the diagnosis system for fault detection. The design of the diagnosis system encompasses the assignment of relevant model information $\mathcal{M}_i$ which are required to parametrise the residual generator G_1 and to obtain the detection threshold $\mu_1(t)$. This design is supervised by design agent D_1 that processes Algorithm 8.1:

Step 1: Initially, the dynamics of all controlled zones F_i, $i = 2, 3, 4$ are considered to be irrelevant for fault detection. To set up the comparison system that describes the dynamics these irrelevant dynamics, design agent D_1 runs Algorithm 4.1. Thereafter, the available model set

$$\mathcal{M}_1 = \left\{ S_1,\, C_1,\, K_1,\, \{S_{1f},\, f = \mathrm{f}_1, \mathrm{f}_2\},\, \left\{ \bar{F}_i\big|_{w_i=0},\, \bar{K}_i,\, i = 2, 3, 4 \right\} \right\}.$$

is available to D_1 such that the comparison system $\bar{E}_1^{\{2,3,4\}}$ can be set up. The resulting upper bound $\bar{E}_1^{\{2,3,4\}}(t)$ together with the actual impulse response function $|E_1^{\{2,3,4\}}(t)|$ are shown in Fig. 8.10a.

(a) Upper bound $\bar{E}_1^{\{2,3,4\}}(t)$ and impulse response function $|E_1^{\{2,3,4\}}(t)|$

(b) Upper bound $\bar{E}_1^{\{3,4\}}(t)$ impulse response function $|E_1^{\{3,4\}}(t)|$

Figure 8.10: Application: Upper bound of the irrelevant dynamics in comparison with the actual impulse response function.

Step 3 $(\kappa{=}0)$: For $\mathcal{N}_{S1} = \{1\}$ the existence condition (8.29) is satisfied.

Step 4 $(\kappa{=}0)$: The gain matrices are designed by means of pole placement to result in an asymptotically stable residual generator G_1. In particular, with

$$N_1^\top = \begin{pmatrix} 4.44 & 0.35 \end{pmatrix}, \quad N_2 = 1.75$$

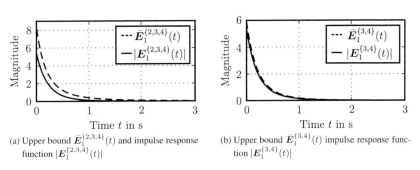

the eigenvalues of the system matrix

$$\begin{pmatrix} \boldsymbol{A}_1^{\{1\}} - \boldsymbol{N}_1 \boldsymbol{C}_1^{\{1\}} & \boldsymbol{E}_1^{\{1\}} \\ -\boldsymbol{N}_2 \boldsymbol{C}_1^{\{1\}} & \mathrm{O} \end{pmatrix}$$

result to

$$\lambda_1 = -2.56, \quad \lambda_2 = -0.23, \quad \lambda_3 = -0.1.$$

Step 6 (κ=0) : The detectability is verified by means of the tubes $\mathbb{T}_1^{\{1\}}(t) * \sigma(t)$ and $\mathbb{T}_1^{\{1\}}(t)\Big|_f * \sigma(t)$, $(f = \mathrm{f}_1, \mathrm{f}_2)$ as shown in Fig. 8.11a. The fault f_1 can already be detected, whereas the fault f_2 can remain undetected. Since the equalities (8.41) and (8.43) hold for both considered faults, the tubes $\mathbb{T}_1^{\{1\}}(t)\Big|_f * \sigma(t)$, $(f = \mathrm{f}_1, \mathrm{f}_2)$ vanish asymptotically. Note that detectability is verified using a step-wise reference signal in view of the application, where zone 1 is stimulated by a step-wise reference signal (see Section 8.3.3).

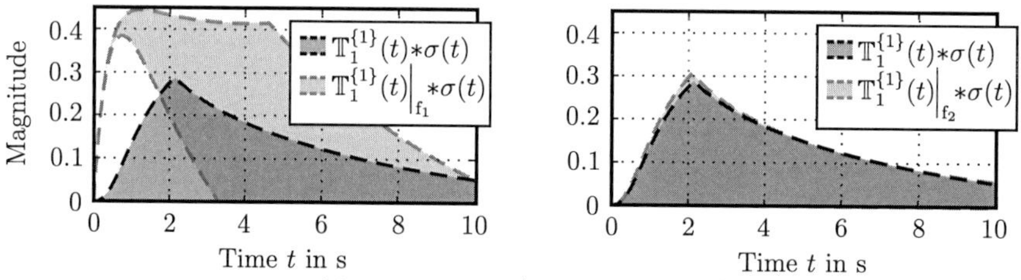

(a) Tubes $\mathbb{T}_1^{\{1\}}(t) * \sigma(t)$, $\mathbb{T}_1^{\{1\}}(t)\Big|_{\mathrm{f1}} * \sigma(t)$ and $\mathbb{T}_1^{\{1\}}(t)\Big|_{\mathrm{f2}} * \sigma(t)$

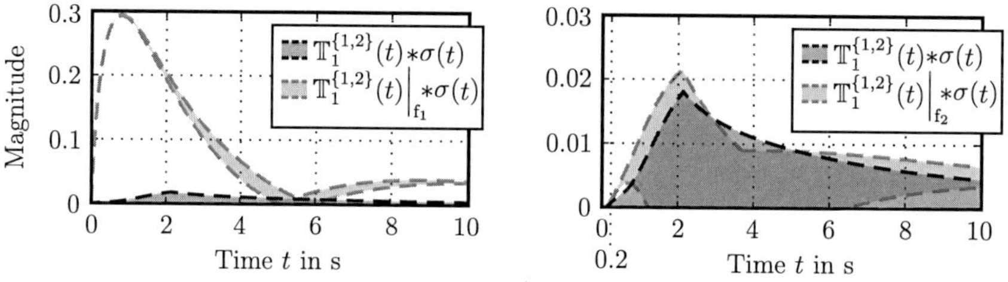

(b) Tubes $\mathbb{T}_1^{\{1,2\}}(t) * \sigma(t)$, $\mathbb{T}_1^{\{1,2\}}(t)\Big|_{\mathrm{f1}} * \sigma(t)$ and $\mathbb{T}_1^{\{1,2\}}(t)\Big|_{\mathrm{f2}} * \sigma(t)$

Figure 8.11: Application: Resulting tubes which enclose for the residual $\boldsymbol{r}_1(t)$ for the healthy case $\left(\mathbb{T}_1^{\mathcal{N}_{\mathrm{S1}}}(t) * |\boldsymbol{w}_1|\right)$, the faulty case $f = \mathrm{f}_1$ $\left(\mathbb{T}_1^{\mathcal{N}_{\mathrm{S1}}}(t)\Big|_{\mathrm{f1}} * |\boldsymbol{w}_1|\right)$ and the faulty case $f = \mathrm{f2}$ $\left(\mathbb{T}_1^{\mathcal{N}_{\mathrm{S1}}}(t)\Big|_{\mathrm{f2}} * |\boldsymbol{w}_1|\right)$ for the reference signal $\boldsymbol{w}_1(t) = \sigma(t)$.

Step 9 (κ=0) : Design agent D_1 requests the model set $\mathcal{M}_2$ from design agent D_2 and, thus, has available the model set

$$\mathcal{M}_1 = \left\{ \left\{ S_i, C_i, K_i,\ i = 1,2 \right\}, \left\{ S_{1f},\ f = \mathrm{f}_1, \mathrm{f}_2 \right\}, \left\{ \bar{F}_i \big|_{\boldsymbol{w}_i = \mathbf{0}}, \bar{K}_i,\ i = 3,4 \right\} \right\}.$$

Hence, the comparison system $\bar{E}_1^{\{3,4\}}$ is updated. Figure 8.10b shows that the updated upper bound $\bar{\boldsymbol{E}}_1^{\{3,4\}}(t)$ is very close to the original impulse response function $|\boldsymbol{E}_1^{\{3,4\}}(t)|$.

Step 3 (κ=1) : Still the claim (8.29) is satisfied for $\mathcal{N}_{\mathrm{S1}} = \{1,2\}$ in order that the residual generator can be designed.

Step 4 (κ=1) : The residual generator G_1 is redesigned using the approximate model $\hat{S}_1^{\{1,2\}}$. The feedback gains result to

$$\boldsymbol{N}_1^{\top} = \begin{pmatrix} 4.26 & 3.66 & 47.47 & -17.73 \end{pmatrix}, \quad \boldsymbol{N}_2 = 17.19$$

in order that

$$\begin{pmatrix} \boldsymbol{A}_1^{\{1,2\}} - \boldsymbol{N}_1 \boldsymbol{C}_1^{\{1,2\}} & \boldsymbol{E}_1^{\{1,2\}} \\ -\boldsymbol{N}_2 \boldsymbol{C}_1^{\{1,2\}} & \mathrm{O} \end{pmatrix}$$

is asymptotically stable. The particular eigenvalues are

$$\lambda_1 = -7.99, \quad \lambda_2 = -0.18, \quad \lambda_3 = -2.24, \quad \lambda_4 = -2.19, \quad \lambda_5 = -0.1.$$

Step 6 (κ=1) : Once again, the detectability is verified by means of the tubes $\mathbb{T}_1^{\{1,2\}}(t) * \sigma(t)$ and $\mathbb{T}_1^{\{1,2\}}(t)\Big|_f * \sigma(t)$, ($f = \mathrm{f}_1, \mathrm{f}_2$) as shown in Fig. 8.11b. Now, both faults are detectable by means of a step-wise reference signal.

Step 11: The detection threshold yields $\boldsymbol{\mu}_1(t) = \mathbb{T}_1^{\{1,2\}}(t) * |\boldsymbol{w}_1(t)|$. Therefore, the occurrence of the fault f_1 is immediately detected after stimulating the controlled subsystem through $\boldsymbol{w}_1(t) = \sigma(t)$, whereas the fault f_2 is detected approximately $200\,\mathrm{ms}$ after stimulation.

Step 12: The diagnosis system is started up. Figure 8.12 shows the result of a simulation, where the motion of the detection residual $\boldsymbol{r}_1(t)$ is depicted for the healthy as well as for the faulty cases. In the healthy case, the residual $|\boldsymbol{r}_1(t)|$ does not cross the detection threshold $\boldsymbol{\mu}_1(t)$. If the fault f_1 or f_2 occurs, the corresponding residual $|\boldsymbol{r}_1(t)|$ crosses the detection threshold. Thus, the respective fault is detected and the detection is stopped.

This example demonstrates the value of model information for diagnosis purpose. Figure 8.13 shows the impulse response functions $|\boldsymbol{S}_{1f}(t)|$ and $|\boldsymbol{S}_{1f}^{\mathcal{N}}(t)|$ of the isolated faulty subsystem $S_{1f}|_{\boldsymbol{s}_1=\mathbf{0}}$ and the faulty interacting subsystem $S_{1f}^{\mathcal{N}}$, respectively. Moreover, the impulse response functions $|\hat{\boldsymbol{S}}_{1f}^{\mathcal{N}_{\mathrm{S1}}}(t)|$ of the approximate system $\hat{S}_{1f}^{\mathcal{N}_{\mathrm{S1}}}$ are depicted. Although the consideration of the model from zone 2 does not significantly improve the approximation as shown

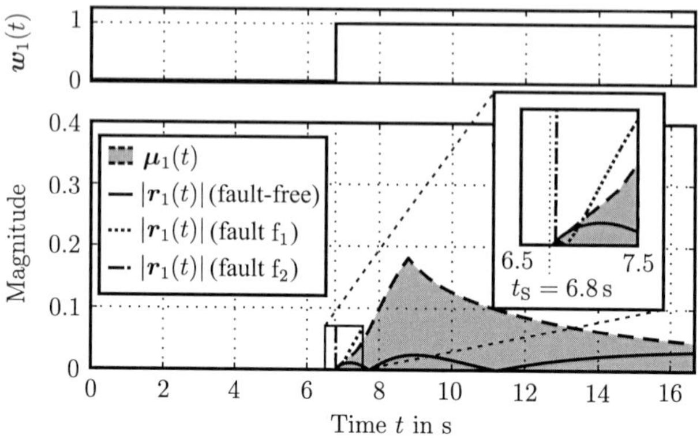

Figure 8.12: Application: Execution phase of the diagnosis system. In the healthy case, the residual $r_1(t)$ does not cross the detection threshold $\mu_1(t)$. In the faulty case, the residual crosses the detection threshold. After the fault has been detected the detection phase stops.

in Fig. 8.13, the simulation has shown that this improvement is necessary to guarantee detection of the fault f_2. Furthermore, Fig. 8.13 visualises that the additional including of the models from the design agents D_3 or D_4 do not significantly improve the approximation.

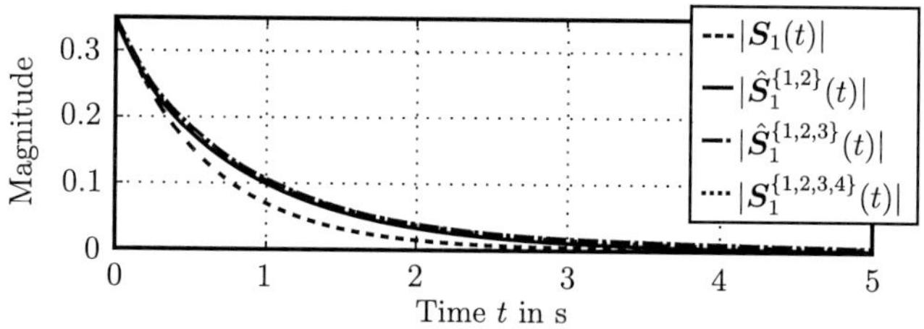

Figure 8.13: Application: Improvement of the approximation $S_1^{\mathcal{N}_{S1}}(t)$ by considering the models of neighbouring zones. The approximate model $\hat{S}_1^{\{1,2\}}$ is required for fault detection.

9 Conclusion

9.1 Summary

Plug-and-play control is devoted to the design of decentralised control stations at runtime by means of *design agents*. The situation is considered in which each subsystem is equipped with a design agent D_i that is responsible for the design of control station C_i. This local design must guarantee the adherence of a global control aim. As the design agents store only the model of its subsystem, the digital communication network is used to request further models from other design agents so as to *set up a model* of the subsystem under the influence of the physical interaction. The fundamental concern of this thesis is the study of the value of model information for the controller design. Thus, the higher the accuracy of this model in terms of its completeness is, the higher is the achievable overall closed-loop performance. This thesis has proposed three approaches to enable design agent D_i to accomplish the controller design. These approaches are characterised as follows:

1. A model has to be set up that *exactly* represents the dynamics of the physical interaction so that D_i is able to design C_i regarding the global control aim (Chapter 4). The substantial communication effort during the modelling phase and the high computational workload for the design is conditioned by a high closed-loop performance.

2. The global control objectives are decomposed into local design conditions which enable D_i to design C_i using the *local model* only (Chapters 5 and 6). The incompleteness of the model is handled by a worst-case design which inherents a high degree of conservatism but guarantee the adherence of the global control aim. The independence of the local design from models of other design agents makes it scalable, applicable to systems with privacy constraints and implementable on embedded systems with low processing capacity.

3. The *required model accuracy* for the controller design is evaluated for locally interconnected systems (Chapters 7 and 8) with the aim to reduce the conservatism within the local controller design. Consequently, the balance between the achievable closed-loop system performance and the admissible computational workload for the design becomes locally adjustable by design agent D_i.

The second approach aims at weakening the physical couplings among the controlled subsystem by design, which is achieved by the restriction of the operating set in spite of physical influences (Section 5.2) or by limiting the amplification of the coupling input signal (Section 6.2). As a result, the physical interaction is bounded, which enables the formulation of local design conditions that aim at boundedness (Proposition 5.1), stability (Theorem 6.1) and performance (Theorem 6.2 and Proposition 5.2) of the overall closed-loop system. It has been shown that this bound is known by design agent D_i so as it can accomplish the controller design with its initially available models. Nevertheless, the expected behaviour of the controlled subsystem is only known to be located within a tolerance band.

To narrow these tolerance bands the physical interaction with neighbouring subsystems are explicitly taken into account for the controller design in the third approach. The amount of model information to be considered is adjustable by D_i by a local decision threshold which reflects the acceptable model uncertainty for the design (Section 7.2) or which depends upon the design objectives (Section 8.3). It has been shown that the tolerance band width is halved if only the direct neighbours are taken into account, whereas the indirect neighbours do only have a minor effect on the tolerance band width.

To know the dynamics of the physical interaction exactly, the model available to D_i has to be without any uncertainty. The proposed Algorithm 4.1 to gather all models which are required to set up this exact model grounds on a depth-first search distributed among the design agents. It has been shown that the complexity of the algorithm scales linearly with number of subsystems and physical couplings.

When considering the transmission of model information over a communication network, the load of the network needs to be regarded. For all approaches, the communication among the design agents has been analysed by means of a *communication graph*. For the second approach at most a communication to the direct neighbours is established, whereas for the first and third approach the communication is extended to all those design agents which are associated to the subsystems that are strongly connected, respectively, strongly coupled to subsystem S_i (Sections 4.3, 5.2, 6.2, 7.2.4 and 8.5).

The proposed plug-and-play control methods has been applied to achieve *fault-tolerance* (Algorithms 4.2, 5.3, 7.3 and 8.1) and to *integrate new subsystems* (Algorithm 6.1). These application scenarios have been tested and evaluated on a thermofluid process (Sections 4.4.3, 5.5.2 and 6.5.2) by simulations and experiments and on a multizone furnace (Sections 7.4.3 and 8.6) by simulations. The experiments have indicated that the second approach is applicable to a real world processes despite the conservatism of the local controller design. Moreover, the investigations have shown that the physical interaction is appropriately be approximated by considering the dynamics of just a few controlled subsystems.

9.2 Outlook

The proposed methods for plug-and-play control could be extended into different directions. Three of these directions are pointed out in the following.

Networked control. This thesis considers a complete decentralised solution to the control of interconnected systems. If a communication among the control stations is introduced which is adapted to the interconnection structure of the subsystems, the coupling influence from other subsystems can be reduced [66, 145] or can even be cancelled completely [157]. As a consequence, the performance of the overall closed-loop system can be increased significantly. However, in networked control systems the exchange of signals among the control stations happens over an unreliable real-time communication network. In such networks the communication is subject to packet losses or delays which has considerable effect on the performance of the overall closed-loop system [37].

With the focus on plug-and-play control, the extension of the concepts towards the (re)design of networked controllers has the consequence that, on the one hand, less model information is required to achieve an adequate closed-loop performance but, on the other hand, the model that is used for the design of a control station must additionally represent the behaviour of the communication network.

Switching issues. Throughout this thesis, it is assumed that the design is performed when the subsystem is in its operating point in order to avoid phenomenons caused by switching between different control algorithms. However, the application of plug-and-play control to fault-tolerant control and the integration of new subsystems require the online design and the online exchange of the control algorithms, even if the system is not in its operating point. The switching between controlled systems will trigger undesired bumps in the transient behaviour or can cause in instability of the switched system [107]. The first issue has been considered in [135], where a smooth switching between two control algorithm has been investigated. Moreover, switching in the context of adding and removing subsystem has been investigated in [188], where a minimum dwell-time between these operations has been proposed to guarantee stability of the overall closed-loop system. Inspired by these results, local design conditions which depend upon the current situation of the process can developed to explicitly take the mentioned switching issues into account.

Clustering of controlled subsystems. The issue of finding groups of subsystems (so called clusters) which belong together in some sense is of grave interest. For instance, power networks are clustered to reduce the computational and administrative complexity [64] or wireless sensor networks are grouped to save energy [30]. There have been little research on the topic of finding an appropriate cluster of subsystems for the purpose of controller design. The approaches that have been presented in this thesis are restricted to chain-like interconnection structures. Hence, it seems promising to extend these approaches towards the clustering of controlled subsystems with an arbitrary interconnection structure.

Bibliography

Contributions of the author

[1] S. Bodenburg. *Local conditions to model the global system dynamics with local model information*. Tech. rep. Ruhr-Universität Bochum, Lehrstuhl für Automatisierungstechnik und Prozessinformatik, 2016.

[2] S. Bodenburg, V. Kraus, and J. Lunze. A design method for plug-and-play control of large-scale systems with a varying number of subsystems. In *Proc. American Control Conference*. 2016, pp. 5314–5321.

[3] S. Bodenburg and J. Lunze. Plug-and-play control - definition and realisation in MATLAB. *Automatisierungstechnik* 61 (2013), pp. 487–494.

[4] S. Bodenburg and J. Lunze. Plug-and-play control - theory and implementation. In *Proc. IEEE Conference on Industrial Informatics*. 2013, pp. 165–170.

[5] S. Bodenburg and J. Lunze. Plug-and-play control of interconnected systems with a changing number of subsystems. In *Proc. European Control Conference*. 2015, pp. 3525–3532.

[6] S. Bodenburg and J. Lunze. Plug-and-play reconfiguration of locally interconnected systems with limited model information. In *Proc. IFAC Workshop on Distributed Estimation and Control in Networked Systems*. 2015, pp. 20–27.

[7] S. Bodenburg and J. Lunze. A local search algorithm for plug-and-play reconfiguration of interconnected systems. *Automatisierungstechnik* 64.12 (2016), pp. 977–989.

[8] S. Bodenburg and J. Lunze. Cooperative reconfiguration of locally interconnected systems with limited model information: A plug-and-play approach. In *Proc. European Control Conference*. 2016, pp. 249–256.

[9] S. Bodenburg and J. Lunze. Plug-and-play diagnosis of locally interconnected systems with limited model information. In *Proc. Conference on Control and Fault-Tolerant Systems*. 2016, pp. 715–722.

[10] S. Bodenburg, S. Niemann, and J. Lunze. Experimental evaluations of a fault-tolerant plug-and-play controller. In *Proc. European Control Conference*. 2014, pp. 1945–1950.

[11] S. Bodenburg, D. Vey, and J. Lunze. Plug-and-play reconfiguration of decentralised controllers of interconnected systems. In *Proc. IFAC Symposium on Fault Detection, Supervision and Safety of Technical Processes*. 2015, pp. 353–359.

[12] D. Vey, S. Hügging, S. Bodenburg, and J. Lunze. Controller reconfiguration of physically interconnected systems by decentralized virtual actuators. In *Proc. IFAC Symposium on Fault Detection, Supervision and Safety of Technical Processes*. 2015, pp. 360–367.

Supervised theses

[13] G. Akgün. Implementierung und Simulation zweier Aufschwingalgorithmen für ein invertiertes Pendel. Bachelor thesis. Ruhr Universität Bochum, Lehrstuhl für Automatisierungstechnik und Prozessinformatik, 2014.

[14] A. Brodovski. Plug-and-play-Regelung mit vollständigen Modellinformationen. Bachelor thesis. Ruhr Universität Bochum, Lehrstuhl für Automatisierungstechnik und Prozessinformatik, 2014.

[15] B. Emir. Modellbildung physikalisch gekoppelter Systeme für das Konzept der Plug-and-play-Rekonfiguration. Bachelor thesis. Ruhr Universität Bochum, Lehrstuhl für Automatisierungstechnik und Prozessinformatik, 2015.

[16] Y. Erkul. Erstellung einer grafischen Benutzeroberfläche zur automatisierten Synthese zweier Reglerstrukturen für einen linearen Prozess. Bachelor thesis. Ruhr Universität Bochum, Lehrstuhl für Automatisierungstechnik und Prozessinformatik, 2012.

[17] P. Geisler. Plug-and-play-Rekonfiguration mit lokalen Informationen unter Verwendung von Zeitbereichsmethoden. Masters thesis. Ruhr Universität Bochum, Lehrstuhl für Automatisierungstechnik und Prozessinformatik, 2015.

[18] B. Guelbitti. Aufschwingen eines invertierten Pendels. Bachelor thesis. Ruhr Universität Bochum, Lehrstuhl für Automatisierungstechnik und Prozessinformatik, 2015.

[19] N. Höckner. Implementierung eines automatisierten Reglerentwurfs und Integration in das Plug-and-play-Regelungs Konzept. Bachelor thesis. Ruhr Universität Bochum, Lehrstuhl für Automatisierungstechnik und Prozessinformatik, 2013.

[20] A. Knaub. Modellprädiktive Plug-and-play-Rekonfiguration. Masters thesis. Ruhr Universität Bochum, Lehrstuhl für Automatisierungstechnik und Prozessinformatik, 2016.

[21] V. Kraus. Implementierung einer Prozessleiteinrichtung für ein 3-Tank System. Bachelor thesis. Ruhr Universität Bochum, Lehrstuhl für Automatisierungstechnik und Prozessinformatik, 2012.

[22] V. Kraus. Plug-and-play-Regelung an einem thermofluiden Prozess. Masters thesis. Ruhr Universität Bochum, Lehrstuhl für Automatisierungstechnik und Prozessinformatik, 2014.

[23] S. Niemann. Anwendung fehlertoleranter Regelung in vernetzten Systemen: Ein Plug-and-play-Regelungs Ansatz. Diploma thesis. Ruhr Universität Bochum, Lehrstuhl für Automatisierungstechnik und Prozessinformatik, 2013.

[24] S. Ok. Erstellung eines automatisierten Reglerentwurfsalgorithmus für das Plug-and-Play Regelungskonzept. Bachelor thesis. Ruhr Universität Bochum, Lehrstuhl für Automatisierungstechnik und Prozessinformatik, 2013.

[25] E. Polat. Analyse eines physikalisch gekoppelten Prozesses für die Erprobung von Plug-and-Play Regelung. Bachelor thesis. Ruhr Universität Bochum, Lehrstuhl für Automatisierungstechnik und Prozessinformatik, 2015.

[26] C. Schlichting. Simulative Untersuchungen von Plug-and-play-Rekonfiguration an einem Multizonenofen. Bachelor thesis. Ruhr Universität Bochum, Lehrstuhl für Automatisierungstechnik und Prozessinformatik, 2014.

[27] M. Schwung. Optimale Regelung eines invertierten Pendels. Bachelor thesis. Ruhr Universität Bochum, Lehrstuhl für Automatisierungstechnik und Prozessinformatik, 2015.

Further literature

[28] D. Abel, U. Epple, and G.-U. Spohr. Integration von Advanced Control in der Prozessindustrie. Wiley, 2008.

[29] R. Abraham and J. Lunze. Modelling and decentralized control of a multizone crystal growth furnace. *International Journal of Robust and Nonlinear Control* 2.2 (1992), pp. 107–122.

[30] M.M. Afsar and M.H. Tayarini-Najaran. Clustering in sensor networks: a literature survey. *Journal of Networks and Computer Applications* 46 (2014), pp. 198–226.

[31] A.A. Alam, A. Gattami, and K.H. Johansson. Suboptimal decentralized controller design for chain structures: application to vehicle formations. In *Proc. IEEE Conference on Decision and Control and European Control Conference*. 2011, pp. 4864–4870.

[32] A.M. Amani, S. Poorjandaghi, A. Afshar, and M.B. Menhaj. Fault tolerant control of large-scale systems subject to actuator fault using a cooperative approach. *Proc. of the Institution of Mechanical Engineers, Part I: Journal of Systems and Control Engineering* 228.2 (2014), pp. 63–77.

[33] T. Arai, Y. Aiyama, Y. Maeda, M. Sugi, and J. Ota. Agile assembly system by "plug and produce". *CIRP Annals - Manufacturing Technology* 49.1 (2000), pp. 1–4.

[34] T. Arai, Y. Aiyama, M. Sugi, and J. Ota. Holonic assembly systems with plug and produce. *Computers in Industry* 46.3 (2001), pp. 289–299.

[35] M. Araki. Stability of large-scale nonlinear systems - quadratic-order theory of composite-system method using M-matrices. *IEEE Trans. on Automatic Control* 23.2 (1978), pp. 129–142.

[36] B. Awerbuch. A new distributed depth-first-search algorithm. *Information Processing Letters* 20.3 (1985), pp. 147–150.

[37] J. Baillieul and P.J. Antsaklis. Control and communication challanges in networked real-time systems. *Proceedings of the IEEE* 95.1 (2007), pp. 9–28.

[38] L. Bakule. Decentralized control: an overview. *Annual Reviews in Control* 32.1 (2008), pp. 87–98.

[39] L. Bakule and J. Lunze. Complete decentralized design of decentralized controllers for serially interconnected systems. In *Proc. IWK TH Ilmenau*. 1985.

[40] L. Bakule and J. Lunze. Decentralized design of feedback control for large scale systems. *Kybernetika* 24.1 (1988), pp. 1–100.

[41] R. Bellman. Vector lyapunov functions. *SIAM Journal of Control* (1962).

[42] A. Bemporad, M. Heemels, and M. Johansson. Networked control systems. Springer-Verlag London, 2010.

[43] A. Bemporad and M. Morari. Robustness in Identification and Control. In. Ed. by A. Garulli and A. Tesi. Springer, 1999. Chap. Robust model predictive control: A survey, pp. 207–226.

[44] J. Bendtsen, K. Trangbaek, and J. Stourstrup. Closed-loop system identification with new sensors. In *Proc. IEEE Conference on Decision and Control*. 2008, pp. 2631–2636.

[45] J. Bendtsen, K. Trangbaek, and J. Stoustrup. Plug-and-play control: Improving control performance through sensor addition and pre-filtering. In *Proc. IFAC World Congress*. Seoul, 2008, pp. 336–341.

[46] J. Bendtsen, K. Trangbaek, and J. Stoustrup. Distributed decision making and control. In. Ed. by M. Thoma, F. Allgöwer, and M. Morari. Springer, 2012. Chap. Hierarchical model predictive control for plug-and-play resource distribution, pp. 337–355.

[47] J. Bendtsen, K. Trangbaek, and J. Stoustrup. Plug-and-play control - modifying control systems online. *IEEE Trans. Control Systems Technology* 21 (2013), pp. 79–93.

[48] B. Besselink, H. Sandberg, and K.H. Johansson. Clustering-based model reduction of networked passive systems. *IEEE Trans. on Automatic Control* 61.10 (2016), pp. 2958–2973.

[49] F. Blanchini. Feedback control for linear time-invariant systems with state and control bounds in the presence of disturbances. *IEEE Trans. on Automatic Control* 35.11 (1990), pp. 1231–1234.

[50] F. Blanchini. Set invariance in control. *Automatica* 35.11 (1999), pp. 1747–1767.

[51] F. Blanchini and S. Miani. Set-Theoretic Methods in Control. Birkhäuser, 2015.

[52] M. Blanke, C.W. Frei, F. Kraus, R.J. Patton, and M. Staroswiecki. What is fault-tolerant control? In *Proc. IFAC Symposium on Fault Detection, Supervision and Safety of Technical Processes*. 2000, pp. 40–51.

[53] M. Blanke, M. Kinnaert, J. Lunze, and M. Staroswiecki. Diagnosis and Fault-Tolerant Control. 3rd edition. Springer, 2016.

[54] F. Boem, R.M.G. Ferrari, and T. Parasini. Distributed fault detection and isolation of continuous-time non-linear systems. *European Journal of Control* 17.5–6 (2011), pp. 603–620.

[55] F. Boem, S. Riverso, G. Ferrari-Trecate, and T. Parasini. A plug-and-play fault diagnosis approach for large-scale systems. In *Proc. IFAC Symposium on Fault Detection, Supervision and Safety for Technical Processes*. 2015, pp. 601–606.

[56] F. Boem, R.M.G. Ferrari, T. Parasini, and M.M. Polycarpou. A distributed networked approach for fault detection of large-scale systems. *IEEE Trans. on Automatic Control* 62.1 (2017), pp. 18–33.

[57] S. Boyd and L. Vandenberghe. Convex Optimization. Cambridge University Press, 2004.

[58] S. Boyd, E. Feron, V. Balakrishnan, and L. El Ghaoui. Linear Matrix Inequalities in System and Control Theory. Siam, 1994.

[59] A.A. Cárdenas, A. Amin, and S. Sastry. Research challenges for the security of control systems. In *Proc. Conference on Hot topics in security*. 2008.

[60] J. Chen and R.J. Patton. Robust Model-Based Fault Diagnosis for Dynamic Systems. Springer, 1999.

[61] T.-Y. Cheung. Graph traversal techniques and the maximum flow problem in distributed computation. *IEEE Trans. on Software Engineering* 9.4 (1983), pp. 504–512.

[62] M. Chilali and P. Gahinet. H_∞ design with pole placement constraints: an LMI approach. *IEEE Trans. on Automatic Control* 41.3 (1996), pp. 358–367.

[63] I. Cidon. Yet another distributed depth-first-search algorithm. *Information Processing Letters* 26.6 (1988), pp. 301–305.

[64] E. Cotilia-Sanchez, P.D.H. Hines, C. Barrows, S. Blumsack, and M. Patel. Multi-attribute partitioning of power networks based on electrical distance. *IEEE Trans. on Power Systems* 28.4 (2013), pp. 4979–4987.

[65] S. Dashkovskiy, B.S. Rüffer, and F.R. Wirth. An ISS small-gain theorem for general networks. *Mathematics of Control, Signals and Systems* 19.2 (2007), pp. 93–122.

[66] O. Demir. Networked control of interconnected systems with identical subsystems. Shaker Verlag, 2013.

[67] O. Demir and J. Lunze. A decomposition approach to decentralized and distributed control of spatially interconnected systems. In *Proc. IFAC World Congress*. 2011, pp. 9109–9114.

[68] F. Deroo, M. Ulbrich, B.D.O. Anderson, and S. Hirche. Acceleration iterative distributed controller synthesis with a Barzili-Borwein step size. In *Proc. IEEE Conference on Decision and Control*. 2012, pp. 4864–4870.

[69] F. Deroo, M. Meinel, M. Ulbrich, and S. Hirche. Distributed control design with local model information and guaranteed stability. In *Proc. IFAC World Congress*. 2014, pp. 4864–4870.

[70] S. Ding. Model-Based Fault Diagnosis for Dynamic Systems. Springer, 2013.

[71] F. Dörfler, J.W. Simpson-Proco, and F. Bullo. Plug-and-play control and optimisation in microgrids. In *Proc. IEEE Conference on Decision and Control*. 2014.

[72] G.E. Dullerud and F. Paganini. A Course in Robust Control Theory - A Convex Approach. Springer, 2000.

[73] W. B. Dunbar. Distributed receding horizon control of dynamically coupled nonlinear systems. *IEEE Trans. on Automatic Control* 52.7 (2007), pp. 1249–1263.

[74] M. Farina, P. Colaneri, and R. Scattolini. Block-wise discretization accounting for structural constraints. *Automatica* 49.11 (2013), pp. 3411–3417.

[75] F. Farokhi. Decentralized Control of Networked Systems. PhD thesis. KTH Roayal Institute of Technology, Automatic Control Laboratory, 2014.

[76] F. Farokhi, C. Langbort, and K.H. Johansson. Control design with limited model information. In *Proc. American Control Conference*. 2011, pp. 4697–4704.

[77] F. Farokhi, C. Langbort, and K.H. Johansson. Optimal structured static state-feedback control design with limited model information for fully-actuated systems. *Automatica* 49.2 (2013), pp. 326–337.

[78] D.G. Feingold and R.S. Varga. Block diagonally dominant matrices and generalization of the Gerschgorin circle theorem. *Pacific Journal of Mathematics* 12.4 (1962), pp. 1241–1500.

[79] R.M.G. Ferrari, T. Parasini, and M.M. Polycarpou. Distributed fault detection and isolation of large-scale discrete-time nonlinear systems: an adaptive approximation approache. *IEEE Trans. on Automatic Control* 57.2 (2012), pp. 275–290.

[80] Władysław Findeisen. Decentralized and hierarchical control under consistency and disagreement of interests. *Automatica* 18.6 (1982), pp. 647–664.

[81] P.M. Gahinet. Explicit controller formulas for LMI-based H_∞ synthesis. *Automatica* 32.7 (1996), pp. 1007–1014.

[82] P.M. Gahinet and P. Apkarian. A linear matrix inequality approach to H_∞ control. *International Journal of Robust and Nonlinear Control* 4.4 (1994), pp. 421–448.

[83] M.R. Genesereth and S.P. Ketchpel. Software agents. Tech. rep. Stanford University Center for Integrated Facility Engineering, 1994.

[84] P. Göhner. Agentensysteme in der Automatisierungstechnik. Springer Vieweg, 2012.

[85] G.C. Goodwin, J.A. De Doná, and M.M. Seron. Constrained Control and Estimation: An Optimisation Approach. Springer, 2005.

[86] P. Grosdidier and M. Morari. Interaction measure for systems under decentralized control. *Automatica* 22.3 (1986), pp. 309–319.

[87] P.O. Gutman and M. Cwikel. Admissible sets and feedback control for discrete-time linear dynamical systems with bounded controls and states. *IEEE Trans. on Automatic Control* 31.4 (1986), pp. 373–376.

[88] D. Harel and Y. Feldman. Algorithmics: The Spirit of Computing. 3rd edition. Addison Wesley, 2004.

[89] Y.-C. Ho. Team decision theory and information structures. In *Proc. IEEE*. 1980, pp. 644–654.

[90] Y.-C. Ho and K.-C. Chu. Team decision theory and information structures in optimal control problems. *IEEE Trans. on Automatic Control* 17.1 (1972), pp. 15–28.

[91] S. Hodek, M. Loskyll, and J. Schlick. Feldgeräte und sematische Informationsmodelle. *Automatisierungstechnische Praxis* 12 (2012), pp. 44–51.

[92] S. Hodek and J. Schlick. Plug & Play Feldgeräteintegration - Methoden, Softwarekonzepte und technische Realisierung für eine ad hoc Feldgeräteintegration. In *VDI - Tagungsband Automation*. 2012.

[93] M. Hou and P.C. Müller. Fault detection and isolation observers. *International Journal of Control* 60.5 (1994), pp. 827–846.

[94] M. Hovd and S. Skogestad. Improved independent design of robust decentralized controllers. In *IFAC World Congress*. 1993, pp. 271–274.

[95] M. Hovd and S. Skogestad. Sequential design of decentralized controllers. *Automatica* 30.10 (1994), pp. 1601–1607.

[96] R. Izadi-Zamanabadi, K. Vinther, H. Mojallali, H. Rasmussen, and J. Stoustrup. Evaporator unit as a benchmark for plug and play and fault tolerant control. In *Proc. IFAC Symposium on Fault Detection, Supervision and Safety of Technical Processes*. 2012, pp. 701–706.

[97] T.N. Jensen. Plug & Play Control of Hydraulic Networks. PhD thesis. Aalborg University, Department of Electronic Systems Automation & Control, 2012.

[98] C.D. Johnson. On observer for systems with unknown and inaccessible inputs. *International Journal of Control* 21.5 (1975), pp. 825–831.

[99] C. Kambhampati, R.J. Patton, and F.J. Uppal. Reconfiguration in networked control systems: Fault tolerant control and plug-and-play. In *Proc. IFAC Symposium on Fault Detection, Supervision and Safety of Technical Processes*. 2006, pp. 126–131.

[100] E.C. Kerrigan. Robust Constraint Satisfaction: Invariant Sets and Predictive Control. PhD thesis. University of Cambridge, Department of Engineering, 2000.

[101] T. Knudsen, J. Bendtsen, and K. Trangbaek. Awareness and its use in incremental data driven modelling for plug and play process control. *European Journal of Control* 18.1 (2012), pp. 24–37.

[102] T. Knudsen, K. Trangbeak, and C.S. Kallesøe. Plug and play process control applied to a district heating system. In *Proc. IFAC World Congress*. 2008, pp. 325–330.

[103] I. Kolmanovsky and E.G. Gilbert. Theory and computation of disturbance invariant sets for discrete-time linear systems. *Mathematical Problems in Engineering* 4.4 (1998), pp. 317–367.

[104] S. Kundu and M. Anghel. Distributed coordinated control of large-scale nonlinear networks. In *Proc. IFAC Workshop on Distributed Estimation and Control in Networked Systems*. 2015, pp. 240–245.

[105] C. Langbort and J.C. Delvenne. Distributed design methods for linear quadratic control and their limitations. *IEEE Trans. on Automatic Control* 55.9 (2010), pp. 2085–2093.

[106] Z. Li, L. Sun, and Y.H. Li. Feedback control for linear systems with disturbances and input constraints. In *Proc. Conference on Intelligent Systems Design and Applications*. 2006, pp. 162–166.

[107] D. Liberzon. Switching in Systems and Control. Birkhauser, 2003.

[108] D. Limon, I. Alvarado, T. Alamo, and E.F. Camacho. Robust tube-based MPC for tracking of contrained linear systems with additive disturbances. *Journal of Process Control* 20.3 (2010), pp. 248–260.

[109] L. Litz. Dezentrale Regelung. Oldenburg Verlag, 1983.

[110] Y. Liu. Robust Nonlinear Control Design with Proportional-Integral-Observer Technique. PhD thesis. Universität Duisburg-Essen, Department of Mechanical and Process Engineering, 2011.

[111] L. Ljung. System Identification Toolbox - User's Guide. The Math Works, Inc., 2014.

[112] J. Löfberger. Yalmip: A toolbox for modeling and optimization in MATLAB. In *Proc. 2004 IEEE International Symposium on Computer Aided Control Systems Design*. 2004, pp. 284–289.

[113] M. Loskyll, I. Heck, J. Schlick, and M. Schwarz. Context-based orchestration for control of resource-efficient manufacturing processes. *Future Internet* 4.3 (2012), pp. 737–761.

[114] A. Luca, P. Rodriguez-Ayerbe, and D. Dumur. Invariant sets method for state-feedback control design. In *Proc. 17th Telecommunication forum*. 2009, pp. 681–684.

[115] J. Lunze. Dynamics of strongly coupled symmetric composite systems. *Interantional Journal of Control* 44.6 (1986), pp. 1617–1640.

[116] J. Lunze. Robust Multivariable Feedback Control. Akademie-Verlag Berlin, 1988.

[117] J. Lunze. Feedback Control of Large-Scale Systems. Prentice Hall, 1992.

[118] J. Lunze, ed.. Control Theory of Digitally Networked Dynamical Systems. Springer, 2014.

[119] J. Lunze. Regelungstechnik 1. 11. Springer, 2016.

[120] J. Lunze. Regelungstechnik 2. 9. Springer, 2016.

[121] J. Machowski, J. Bialek, and J. Bumby. Power Systems Dynamics: Stability and Control. John Wiley and Sons, 2008.

[122] M.S. Mahmoud. Decentralized Systems with Design Constraints. Springer, 2011.

[123] S.A.M. Makki and G. Havas. Distributed algorithms for depth-first search. *Information Processing Letters* 60.1 (1996), pp. 7–12.

[124] L. Márton, K. Schenk, and J. Lunze. Fault estimation and networked reconfiguration in large-scale control systems. In *Proc. IFAC World Congress*. 2017.

[125] M.A. Massoumnia. A geometric approach to the synthesis of failure detection filters. *IEEE Trans. on Automatic Control* 31.9 (1986), pp. 839–846.

[126] D.Q. Mayne and W. Langson. Robustifying model predictive control of constrained linear systems. *Electronic Letters* 37.23 (2001), pp. 1422–1423.

[127] D.Q. Mayne, M.M. Seron, and S.V. Raković. Robust model predictive control of constrained linear systems with bounded disturbances. *Automatica* 41.2 (2005), pp. 219–224.

[128] M. Metzger and G. Polaków. A survey on applications of agent technology in industrial process control. *IEEE Trans. on Industrial Informatics* 7.4 (2011), pp. 570–581.

[129] A. G. Michelsen and J. Stoustrup. High level model predictive control for plug-and-play process control with stability guaranty. In *Proc. IEEE Conference on Decision and Control*. 2010, pp. 2456–2461.

[130] A. G. Michelsen and K. Trangbaek. Local controller synthesis for plug and play process by decomposition. In *Proc. IEEE Conference on Control and Automation*. 2009, pp. 2223–2228.

[131] Microsoft. Plug and Play - Architecture and Driver Support. `www.web.archive.org/web/20040616023120/http://www.microsoft.com/whdc/system/pnppwr/pnp/default.mspx`. last view: 12/12/2017.

[132] B.E.A. Milani and C.E.T. Dórea. On invariant polyhedra of continous time systems subject to additive disturbances. *Automatica* 32.5 (1996), pp. 785–789.

[133] A. Mosebach, S. Roechner, and J. Lunze. Merging Control of Cooperative Vehicles. In *Proc. IFAC Symposium on Advances in Automotive Control*. 2016, pp. 168–174.

[134] M. Naumann, K. Wegener, and R.D. Schraft. Control architecture for robot cells to enable plug'n'produce. In *IEEE Conference on Robotics and Automation*. 2007, pp. 287–292.

[135] H. Niemann and J. Stoustrup. An architecture for implementation of multivariable controllers. In *Proc. American Control Conference*. 1999, pp. 4029–4033.

[136] Ü. Özgüner and W.R. Perkins. Optimal control of multilevel large-scale systems. *Interantional Journal of Control* 28.6 (1978), pp. 967–980.

[137] H.Van Dyke Parunak. Practical and Industrial Applications of Agent-Based Systems. Tech. rep. Environmental Research Institute of Michigan, 1998.

[138] R.J. Patton, C. Kambhampati, A. Casavola, P. Zhang, S. Ding, and D. Sauter. A generic strategy for fault-tolerance in control systems distributed over a network. *European Journal of Control* 13.2–3 (2007), pp. 280–296.

[139] S.V. Raković, E.C. Kerrigan, K.I. Kouramas, and D.Q. Mayne. *Invariant approximations of robustly positively invariant sets for constrained linear discrete-time systems subject to bounded disturbances*. Tech. rep. Department of Engineering, University of Cambridge, 2004.

[140] S.V. Raković, E.C. Kerrigan, K.I. Kouramas, and D.Q. Mayne. Invariant approximations of the minimal robust positively invariant set. *IEEE Trans. on Automatic Control* 50.3 (2005), pp. 406–410.

[141] S.V. Raković, E.C. Kerrigan, D.Q. Mayne, and K.I. Kouramas. Optimized robust control invariance for linear discrete-time systems: Theoretical foundations. *Automatica* 43.5 (2007), pp. 831–841.

[142] T. Reis and T. Stykel. Model Order Reduction: Theory, Research Aspects and Applications. In. Ed. by W.H.A. Schilders, H.A. van der Vorst, and J. Rommes. Springer, 2008. Chap. A survey on model reduction of coupled subsystems, pp. 133–155.

[143] V. Reppa, M.M. Polycarpou, and C.G. Panayiotou. Distributed sensor fault dignosis for a network of interconnected cyber-physical systems. *IEEE Trans. on Control of Network Systems* 2.1 (2015), pp. 11–23.

[144] J.H. Richter. Reconfigurable Control of Nonlinear Dynamical Systems - a Fault-Hiding Approach. Springer, 2011.

[145] S. Riverso. Distributed and Plug-and-Play Control for Constrained Systems. PhD thesis. University of Pavia, Control of Dynamic Systems Laboratory, 2014.

[146] S. Riverso, G. Ferrari-Trecate F. Boem, and T. Parisini. Fault diagnosis and control-reconfiguration in large-scale systems: a plug-and-play approach. In *Proc. IEEE Conference on Decision and Control*. 2014, pp. 4977–4982.

[147] S. Riverso, G. Ferrari-Trecate F. Boem, and T. Parisini. Plug-and-play fault diagnosis and control-reconfiguration for a class of nonlinear large-scale constrained systems. *IEEE Trans. on Automatic Control* 61.12 (2016), pp. 3963–3978.

[148] S. Riverso, M. Farina, and G. Ferrari-Trecate. Plug-and-play decentralized model predictive control for linear systems. *IEEE Trans. on Automatic Control* 58.10 (2013), pp. 2608–2614.

[149] S. Riverso and G. Ferrari-Trecate. Tube-based distributed control of linear constrained systems. *Automatica* 48.11 (2012), pp. 2860–2865.

[150] Hadi Saadat. Power System Analysis. WCB/McGraw-Hill, 1999.

[151] H. Sandberg and R.M. Murray. Model reduction of interconnected linear systems. *Optimal Control Applications and Methods* 30.3 (2009), pp. 225–245.

[152] K. Schenk and J. Lunze. Fault-tolerant control of networked systems with re-distribution of control tasks in case of faults. In *Proc. Conference on Control and Fault-Tolerant Systems*. 2016, pp. 723–729.

[153] C. Scherer, P.M. Gahinet, and M. Chilali. Multiobjective output-feedback control via LMI optimization. *IEEE Trans. on Automatic Control* 42.7 (1997), pp. 896–911.

[154] C. Scherer and S. Weiland. Linear Matrix Inequalities in Control. Delft Center for System and Control, 2005.

[155] R. Schneider. Convex Bodies: The Brunn-Minkowski Theory. 5th edition. Cambridge University Press, 2014.

[156] F. Schuette, D. Berneck, M. Eckmann, and S. Kakizaki. Advances in rapid control prototyping - results of a pilot project for engine control. *SAE Technical Paper 2005-01-1350* (2005).

[157] R. Schuh and J. Lunze. Self-organizing control of physically interconnected systems for disturbance attenuation. In *Proc. European Control Conference*. 2015, pp. 2191–2198.

[158] R. Schuh and J. Lunze. Self-organizing distributed control with situation-dependent communication. In *Proc. European Control Conference*. 2016, pp. 222–229.

[159] A. Sethi and S. Sethi. Flexibility in manufacturing: A survey. *International Journal of Flexible Manufacturing Systems* 2.4 (1990), pp. 289–328.

[160] I. Shames, A.M.H. Teixeira, H. Sandberg, and K.H. Johansson. Distributed fault detection for interconnected second-order systems. *Automatica* 47.12 (2011), pp. 2757–2764.

[161] M.B. Sharma, S.S. Iyengar, and N.K. Mandyam. An efficient distributed depth-first-search algorithm. *Information Processing Letters* 32.4 (1989), pp. 183–186.

[162] E.F. da Silva Neto and P. Berrie. Who's afraid of control in the field. Tech. rep. Endress+Hauser Process Solution AG, 2003.

[163] S. Skogestad and M. Morari. Robust performance of decentraized control systems by independent designs. *Automatica* 25.1 (1989), pp. 119–125.

[164] S. Skogestad and I. Postlethwaite. Multivariable Feedback Control. John Wiley and Sons Ltd, 2005.

[165] D. Söffker, T.-J. Yu, and P. C. Müller. State estimation of dynamical systems with non-linearities by using a proportional-integral observer. *International Journal of Systems Science* 29.9 (1995), pp. 1571–1582.

[166] T. Steffen. Control Reconfiguration of Dynamical Systems. Springer, 2005.

[167] F. Stoican and S. Olaru. Set-theoretic Fault-tolerant Control in Multisensor Systems. ISTE and Wiley, 2013.

[168] J. Stoustrup. Plug & play control: Control technology towards new challenges. *European Journal of Control* 15.3–4 (2009), pp. 311–330.

[169] J. Stoustrup, K. Trangbaek, and J.D. Bendtsen. A sensor fusion approach for exploiting new measurements in an existing controller. In *Proc. European Control Conference*. 2009, pp. 4096–4104.

[170] J. F. Sturm. Using Sedumi 1.02, A Matlab toolbox for optimization over symmetric cones. *Optimization Methods and Software* 11.1–4 (1999), pp. 625–653.

[171] S. Tarbouriech and C. Burgat. Positively invariant sets for constrained contonous time systems with cone properties. *IEEE Trans. on Automatic Control* 39.2 (1994), pp. 401–405.

[172] A. Teixeira, D. Pérez, H. Sandberg, and K.H. Johansson. Attack model and scenarios for networked control systems. In *Proc. Conference on High Confidence Networked Systems*. 2012, pp. 55–64.

[173] A. Teixeira, J. Araújo, H. Sandberg, and K.H. Johansson. Distributed actuator reconfiguration in networked control systems. In *Proc. IFAC Workhop on Distributed Estimation and Control in Networked Systems*. 2013, pp. 61–68.

[174] A. Teixeira, I. Shames, H. Sandberg, and K.H. Johansson. Distributed fault detection and isolation resilient to network model uncertainties. *IEEE Trans. on Cybernetics* 44.11 (2014), pp. 2024–2037.

[175] K.-C. Toh, M.J. Todd, and R.H. Tütüncü. Handbook on Semidefinite, Conic and Polynomial Optimization. In. Ed. by M.F. Anjos and J.B. Lasserre. Springer, 2012. Chap. On the implementation and usage of SDPT3 - A MATLAB Software package for semidefinite-quadratic-linear programming, version4.0, pp. 715–754.

[176] K. Trangbaek. Controller reconfiguration through terminal connections based on closed-loop system identification. In *Proc. IFAC Symposium on Robust Control Design*. 2009, pp. 25–30.

[177] K. Trangbaek and J. Bendtsen. Stable controller reconfiguration through terminal connections - A practical example. In *Proc. IEEE Conference on Control and Automation*. 2009, pp. 2037–2042.

[178] K. Trangbaek, J. Sourstrup, and J. Bendtsen. Stable Controller Reconfiguration through Terminal Connection. In *Proc. IFAC World Congress*. 2008, pp. 331–335.

[179] P. Trodden and A. Richards. Cooperative distributed MPC of linear systems with coupled constraints. *Automatica* 49.2 (2013), pp. 479–487.

[180] V. Turri, B. Besselink, J. Martensson, and K.H. Johansson. Fuel-efficient heavy-duty vehicle platooing by look-ahead control. In *Proc. Conference on Decision and Control*. 2014, pp. 654–660.

[181] VBN Aalborg University. VBN - Research Project: Plug and Play Process Control (P^3C). `www.vbn.aau.dk/en/projects/plug-and-play-process-control-p3c(8f9b8494-48ec-4b8f-897a-75b9e338c542).html`. last view: 12/12/2017.

[182] P.B. Usoro, F.C. Schweppe, L.A. Gould, and D.N. Wormley. A lagrange approach to set-theoretic control synthesis. *IEEE Trans. on Automatic Control* 27.2 (1982), pp. 393–399.

[183] R.S. Varga. Geršgorin and His Circles. Springer, 2004.

[184] D. D. Šiljak. Large-Scale Dynamical Systems. Dover Publication INC., 1978.

[185] D.D. Šiljak. Decentralized Control of Complex Systems. Academic Press INC., 1991.

[186] T. Wagner. Applying agents for engineering of industrial automation systems. *German Conference on Multiagent System Technologies* (2003), pp. 62–73.

[187] F. Wang and D. Liu, eds.. Networked Control Systems. Springer, 2008.

[188] H. Yang, B. Jiang, and M. Staroswiecki. Fault tolerant control for plug-and-play interconnected nonlinear systems. *Journal of the Franklin Institute* 353.10 (2016), pp. 2199–2217.

[189] M. N. Zeilinger, Y. Pu, S. Riverso, G. Ferrari-Trecate, and C. N. Jones. Plug and play distributed model predictive control base on distributed invariance and optimization. In *Proc. IEEE Conference on Decision and Control*. 2013, pp. 4193–4198.

[190] M.N. Zeilinger, C.N. Jones, D.M. Raimondo, and M. Morari. Real-time MPC - Stability through robust MPC design. In *Joint Proc. IEEE Conferece on Decision and Control and Chinese Control Conference*. 2009, pp. 3980–3986.

[191] Q. Zhang and X. Zhang. Distributed sensor fault dignosis in a class of interconnected nonlinear uncertain systems. *Annual Reviews in Control* 37.1 (2013), pp. 170–179.

[192] K. Zhou, J.C. Doyle, and K. Gover. Robust and Optimal Control. Prentice Hall, 1996.

[193] G.M. Ziegler. Lectures on Polytopes. 5th edition. Springer, 1995.

Appendices

A Models of the demonstration processes

A.1 Interconnected thermofluid process model

A.1.1 Nonlinear model

The continuous process is represented by the nonlinear differential equations

$$
\begin{aligned}
\dot{l}_1(t) &= \left(2\pi r_1 l_1(t) - \pi l_1^2(t)\right)^{-1} \Big(q_{\mathrm{F1}}(u_1(t)) - q_{\mathrm{1B}}(l_1(t), u_{\mathrm{1B}}) - q_{\mathrm{1S}}(l_1(t), u_{\mathrm{1S}})\Big) \\
\dot{l}_{\mathrm{B}}(t) &= A_{\mathrm{B}}^{-1}\Big(q_{\mathrm{2B}}(u_{\mathrm{2B}}) + q_{\mathrm{1B}}(l_1(t), u_{\mathrm{1B}}) + q_{\mathrm{SB}}(l_{\mathrm{S}}(t), u_{\mathrm{SB}}) - q_{\mathrm{BW}}(l_{\mathrm{B}}(t), u_{\mathrm{BW}}(t)) \\
&\qquad - q_{\mathrm{BS}}(l_{\mathrm{B}}(t), u_{\mathrm{BS}})\Big) \\
\dot{\vartheta}_{\mathrm{B}}(t) &= \left(A_{\mathrm{B}} l_{\mathrm{B}}(t)\right)^{-1} \Big(H_{\mathrm{B}} u_{\mathrm{HB}}(t) + q_{\mathrm{2B}}(u_{\mathrm{2B}})(\vartheta_2 - \vartheta_{\mathrm{B}}(t)) + q_{\mathrm{1B}}(l_{\mathrm{B}}(t), u_{\mathrm{1B}})(\vartheta_1 - \vartheta_{\mathrm{B}}(t)) \\
&\qquad + q_{\mathrm{SB}}(l_{\mathrm{S}}(t), u_{\mathrm{SB}})(\vartheta_{\mathrm{S}}(t) - \vartheta_{\mathrm{B}}(t))\Big) \\
\dot{l}_{\mathrm{S}}(t) &= A_{\mathrm{S}}^{-1}\Big(q_{\mathrm{3S}}(u_{\mathrm{3S}}) + q_{\mathrm{1S}}(l_1(t), u_{\mathrm{1S}}) + q_{\mathrm{BS}}(l_{\mathrm{B}}(t), u_{\mathrm{BS}}) - q_{\mathrm{SW}}(l_{\mathrm{S}}(t), u_{\mathrm{SW}}(t)) - \\
&\qquad q_{\mathrm{SB}}(l_{\mathrm{S}}(t), u_{\mathrm{SB}})\Big) \\
\dot{\vartheta}_{\mathrm{S}}(t) &= \left(A_{\mathrm{S}} l_{\mathrm{S}}(t)\right)^{-1} \Big(H_{\mathrm{S}} u_{\mathrm{HS}}(t) + q_{\mathrm{3S}}(u_{\mathrm{3S}})(\vartheta_3 - \vartheta_{\mathrm{S}}(t)) + q_{\mathrm{1S}}(l_1(t), u_{\mathrm{1S}})(\vartheta_1 - \vartheta_{\mathrm{S}}(t)) \\
&\qquad + q_{\mathrm{BS}}(l_{\mathrm{B}}(t), u_{\mathrm{BS}})(\vartheta_{\mathrm{B}}(t) - \vartheta_{\mathrm{S}}(t))\Big).
\end{aligned}
$$

The volume flows

$$
\begin{aligned}
q_{\mathrm{F1}}(u_1(t)) &= 10^{-4} \cdot 3.73 \cdot u_1(t) \\
q_{\mathrm{2B}}(u_{\mathrm{2B}}) &= 10^{-4} \cdot 2.28 \cdot u_{\mathrm{2B}} \\
q_{\mathrm{3S}}(u_{\mathrm{3S}}) &= 10^{-4} \cdot 1.42 \cdot u_{\mathrm{2B}}
\end{aligned}
$$

denote the flow from the fresh water supply to the storage tank T1 as well as from the storage tank T2 and T3 to the reactors TB and TS, respectively. Moreover,

$$
q_{\mathrm{1B}}(l_1(t), u_{\mathrm{1B}}) = u_{\mathrm{1B}} k_{\mathrm{1B}} \sqrt{2g\left(l_1(t) + \delta_1\right)}
$$

$$q_{1S}(l_1(t), u_{1S}) = u_{1S}k_{1S}\sqrt{2g\,(l_1(t) + \delta_1)}$$

represent the flow from the storage tank T1 to the reactors TB and TS, respectively, with the specific valve parameters k_{1B} and k_{1S} as well as with a correction constant δ_1.

$$\begin{aligned} q_{BS}(l_B(t), u_{BS}) &= u_{BS}k_{BS}\sqrt{2g\,(l_S(t) + \delta_S)} \\ q_{SB}(l_S(t), u_{SB}) &= u_{SB}k_{SB}\sqrt{2g\,(l_B(t) + \delta_B)} \end{aligned}$$

denote the flow from the reactor TB to the reactor TS and vice versa with the valve parameters k_{BS} and k_{SB} and the correction constants δ_B and δ_S. Finally,

$$\begin{aligned} q_{BW}(l_B(t), u_{BW}(t) &= u_{BW}(t)k_{BW}\sqrt{2g\,(l_S(t) + \delta_S)} \\ q_{SW}(l_S(t), u_{SW}(t) &= u_{SW}(t)k_{SW}\sqrt{2g\,(l_B(t) + \delta_B)} \end{aligned}$$

denote the outflows from the reactors TB and TS into the buffer tank TW with the valve parameters k_{BW} and k_{SW}. All parameters are listed in Table A.1.

Table A.1: Parameters of the nonlinear model of the pilot plant

Parameter	Value	Description
A_B	$0.07\,m^2$	Cross sectional area of reactor TB
A_S	$0.07\,m^2$	Cross sectional area of reactor TS
k_{1B}	$10^{-4} \cdot 1.09\,m^3/m$	Valve parameter of the valve between T1 and TB
k_{1S}	$10^{-5} \cdot 3.97\,m^3/m$	Valve parameter of the valve between T1 and TS
k_{BS}	$10^{-5} \cdot 4.72\,m^3/m$	Valve parameter of the valve between TB and TS
k_{BW}	$10^{-5} \cdot 4.92\,m^3/m$	Valve parameter of the valve between TB and TW
k_{SB}	$10^{-5} \cdot 1.96\,m^3/m$	Valve parameter of the valve between TS and TB
k_{SW}	$10^{-5} \cdot 1.94\,m^3/m$	Valve parameter of the valve between TS and TW
g	$9.81\,m/s^2$	Gravitation constant
H_B	$10^{-3} \cdot 3.9\,m^3K/s$	Heat coefficient of the heating in reactor TB
H_S	$10^{-3} \cdot 3.9\,m^3K/s$	Heat coefficient of the heating in reactor TS
r_1	$0.168\,m$	Radius of storage tank T1
δ_1	$0.22\,m$	Correction constant for tank T1
δ_B	$6.6\,m$	Correction constant for reactor TB
δ_S	$6.8\,m$	Correction constant for reactor TS
ϑ_1	$293.15\,K$	Temperature of the fluid in tank T1
ϑ_2	$293.15\,K$	Temperature of the fluid in tank T2
ϑ_3	$293.15\,K$	Temperature of the fluid in tank T3

The physical limitations of the pilot plant restrict the levels as well as the temperatures to

$$l_1 \in [0;\ 0.33]\,\mathrm{m}, \quad l_\mathrm{B} \in [0.26;\ 0.40]\,\mathrm{m}, \quad l_\mathrm{S} \in [0.26;\ 0.40]\,\mathrm{m},$$
$$\vartheta_\mathrm{B} \in [285.65;\ 323.15]\,\mathrm{K}, \qquad \vartheta_\mathrm{S} \in [285.65;\ 323.15]\,\mathrm{K}.$$

Furthermore, the control inputs are limited to

$$u_1 \in [0;\ 1], \quad u_\mathrm{B} \in [0;\ 1], \quad u_\mathrm{S} \in [0;\ 1], \quad u_\mathrm{HB} \in [0;\ 1], \quad u_\mathrm{HS} \in [0;\ 1]. \tag{A.1}$$

A.1.2 Linearised model

Since, the proposed methods for plug-and-play control rely on linear models, this section presents the linearised discrete-time model of the thermofluid process with three subsystems and the linearised continuous-time model with five subsystems.

Discrete-time model with three subsystems. The overall process is subdivided into three subsystems, representing the dynamics of the level in the tank T1 and respectively the level and temperature in the tanks TB and TS, respectively. The subsystem states are given by

$$\boldsymbol{x}_1(t) = \begin{pmatrix} l_\mathrm{B}(t) \\ \vartheta_\mathrm{B}(t) \end{pmatrix}, \qquad x_2(t) = l_1(t), \qquad \boldsymbol{x}_3(t) = \begin{pmatrix} l_\mathrm{S}(t) \\ \vartheta_\mathrm{S}(t) \end{pmatrix}$$

with the control actions

$$\boldsymbol{u}_1(t) = \begin{pmatrix} u_\mathrm{BW}(t) \\ u_\mathrm{HB}(t) \end{pmatrix}, \qquad u_2(t) = u_\mathrm{T1}(t), \qquad \boldsymbol{u}_3(t) = \begin{pmatrix} u_\mathrm{SW}(t) \\ u_\mathrm{HS}(t) \end{pmatrix}.$$

The nonlinear model is linearised around the operating point

$$\boldsymbol{x}_{\mathrm{OP},1} = \begin{pmatrix} 0.25\,\mathrm{m} \\ 301.15\,\mathrm{K} \end{pmatrix}, \qquad x_{\mathrm{OP},2} = 0.15\,\mathrm{m}, \qquad \boldsymbol{x}_{\mathrm{OP},3} = \begin{pmatrix} 0.25\,\mathrm{m} \\ 299.15\,\mathrm{K} \end{pmatrix} \tag{A.2}$$

with

$$\boldsymbol{u}_{\mathrm{OP},1} = \begin{pmatrix} 0.35 \\ 0.24 \end{pmatrix}, \qquad u_{\mathrm{OP},2} = 0.23, \qquad \boldsymbol{u}_{\mathrm{OP},3} = \begin{pmatrix} 0.18 \\ 0.1 \end{pmatrix}. \tag{A.3}$$

The other valves are opened with the constant angles

$$u_\mathrm{1B} = 0.3, \quad u_\mathrm{1S} = 0.3, \quad u_\mathrm{2B} = 0.5, \quad u_\mathrm{3S} = 0.3, \quad u_\mathrm{BS} = 0.2, \quad u_\mathrm{BS} = 0.3. \tag{A.4}$$

For this setup, the subsystems are represented by the linear models (2.6) for $i = 1, 2, 3$ with

$$\boldsymbol{A}_1 = 10^{-3}\begin{pmatrix} -0.28 & 0 \\ -3.68 & -15.32 \end{pmatrix}, \quad \boldsymbol{B}_1 = 10^{-2}\begin{pmatrix} -0.65 & 0 \\ 0 & 32.18 \end{pmatrix}, \quad \boldsymbol{C}_1 = \boldsymbol{E}_1 = \boldsymbol{C}_{z1} = \mathbf{I},$$

$$\boldsymbol{A}_2 = -10^{-3} \cdot 1.85, \qquad \boldsymbol{B}_2 = -10^{-3} \cdot 5.96, \qquad \boldsymbol{C}_2 = \boldsymbol{E}_2 = \boldsymbol{C}_{z2} = 1,$$

$$\boldsymbol{A}_3 = 10^{-3}\begin{pmatrix} -0.18 & 0 \\ -0.45 & 0.42 \end{pmatrix}, \quad \boldsymbol{B}_3 = 10^{-3}\begin{pmatrix} -8.81 & 0 \\ 0 & 46.32 \end{pmatrix}, \quad \boldsymbol{C}_3 = \boldsymbol{E}_3 = \boldsymbol{C}_{z3} = \mathbf{I}.$$

The physical couplings are modelled by (2.7) with

$$\boldsymbol{L}_{12} = 10^{-3}\begin{pmatrix} 1.67 \\ -40.24 \end{pmatrix}, \qquad \boldsymbol{L}_{13} = 10^{-4}\begin{pmatrix} 0.69 & 0 \\ -5.56 & 39.13 \end{pmatrix}, \tag{A.6a}$$

$$\boldsymbol{L}_{21} = \begin{pmatrix} 0 & 0 \end{pmatrix} \qquad \boldsymbol{L}_{23} = \begin{pmatrix} 0 & 0 \end{pmatrix} \tag{A.6b}$$

$$\boldsymbol{L}_{31} = 10^{-4}\begin{pmatrix} 1.13 & 0 \\ 9.04 & 61.93 \end{pmatrix}, \quad \boldsymbol{L}_{32} = 10^{-4}\begin{pmatrix} 6.14 \\ -98.16 \end{pmatrix} \tag{A.6c}$$

The parameters of the corresponding discrete-time subsystem models (5.12) are obtained from the continuous-time subsystem model using structure preserving discretisation schemes proposed in [74]. The sampling time yields $T = 0.2\,\mathrm{s}$. Hence, the subsystems are represented by the discrete-time subsystem model (5.12) for $i = 1, 2, 3$ with

$$\boldsymbol{A}_1 = \begin{pmatrix} 0.9999 & 0 \\ -0.0007 & 0.9969 \end{pmatrix}, \quad \boldsymbol{B}_1 = 10^{-3}\begin{pmatrix} -1.3 & 0 \\ 0 & 64.35 \end{pmatrix}, \quad \boldsymbol{E}_1 = 0.2 \cdot \mathbf{I}, \ \boldsymbol{C}_1 = \boldsymbol{C}_{z1} = \mathbf{I} \tag{A.7a}$$

$$\boldsymbol{A}_2 = 0.9996, \qquad \boldsymbol{B}_2 = -10^{-3} \cdot 1.19, \qquad \boldsymbol{E}_2 = 0.2, \qquad \boldsymbol{C}_2 = \boldsymbol{C}_{z2} = 1, \tag{A.7b}$$

$$\boldsymbol{A}_3 = \begin{pmatrix} 0.9999 & 0 \\ -0.0001 & 0.9979 \end{pmatrix}, \quad \boldsymbol{B}_3 = 10^{-3}\begin{pmatrix} -1.76 & 0 \\ 0 & 9.26 \end{pmatrix}, \quad \boldsymbol{E}_3 = 0.2\mathbf{I}, \quad \boldsymbol{C}_3 = \boldsymbol{C}_{z3} = \mathbf{I}. \tag{A.7c}$$

and the coupling gains from (A.6a)–(A.6c).

Continuous-time model with five subsystems. The overall process is subdivided into five scalar subsystems with the states

$$x_1(t) = l_{\mathrm{B}}(t), \quad x_2(t) = \vartheta_{\mathrm{B}}(t), \quad x_3(t) = l_1(t), \quad x_4(t) = l_{\mathrm{S}}(t), \quad x_5(t) = \vartheta_{\mathrm{S}}(t).$$

For the operating point (A.2)–(A.4), the parameters of the subsystem model (2.6) for $i = 1, ..., 5$ result to

$$\boldsymbol{A}_1 = -10^{-3} \cdot 2.79, \qquad \boldsymbol{B}_1 = -10^{-3} \cdot 6.51, \qquad \boldsymbol{C}_1 = \boldsymbol{E}_1 = \boldsymbol{C}_{\mathrm{z}1} = 1 \tag{A.8a}$$

$$\boldsymbol{A}_2 = -10^{-3} \cdot 15.32, \qquad \boldsymbol{B}_2 = 10^{-3} \cdot 32.18, \qquad \boldsymbol{C}_2 = \boldsymbol{E}_3 = \boldsymbol{C}_{\mathrm{z}3} = 1 \tag{A.8b}$$

$$\boldsymbol{A}_3 = -10^{-3} \cdot 1.85, \qquad \boldsymbol{B}_3 = -10^{-3} \cdot 5.96, \qquad \boldsymbol{C}_3 = \boldsymbol{E}_3 = \boldsymbol{C}_{\mathrm{z}3} = 1 \tag{A.8c}$$

$$\boldsymbol{A}_4 = -10^{-4} \cdot 1.82, \qquad \boldsymbol{B}_4 = -10^{-3} \cdot 8.81, \qquad \boldsymbol{C}_4 = \boldsymbol{E}_4 = \boldsymbol{C}_{\mathrm{z}4} = 1 \tag{A.8d}$$

$$\boldsymbol{A}_5 = -10^{-3} \cdot 10.42, \qquad \boldsymbol{B}_5 = 10^{-3} \cdot 46.32, \qquad \boldsymbol{C}_5 = \boldsymbol{E}_5 = \boldsymbol{C}_{\mathrm{z}5} = 1 \tag{A.8e}$$

and the physical couplings (2.7) yield

$$\boldsymbol{L}_{12} = 0, \quad \boldsymbol{L}_{13} = 10^{-3} \cdot 1.68, \quad \boldsymbol{L}_{14} = 10^{-4} \cdot 0.69 \quad \boldsymbol{L}_{15} = 0 \tag{A.9a}$$

$$\boldsymbol{L}_{21} = -10^{-3} \cdot 3.68, \; \boldsymbol{L}_{23} = -10^{-2} \cdot 4.24, \; \boldsymbol{L}_{24} = -10^{-4} \cdot 5.55, \; \boldsymbol{L}_{25} = 10^{-3} \cdot 3.91 \tag{A.9b}$$

$$\boldsymbol{L}_{31} = 0, \quad \boldsymbol{L}_{32} = 0, \quad \boldsymbol{L}_{34} = 0, \quad \boldsymbol{L}_{35} = 0 \tag{A.9c}$$

$$\boldsymbol{L}_{41} = 10^{-4} \cdot 1.13, \quad \boldsymbol{L}_{42} = 0, \quad \boldsymbol{L}_{43} = 10^{-5} \cdot 6.14, \quad \boldsymbol{L}_{45} = 0 \tag{A.9d}$$

$$\boldsymbol{L}_{51} = 10^{-4} \cdot 9.04, \quad \boldsymbol{L}_{52} = 10^{-3} \cdot 6.19, \quad \boldsymbol{L}_{53} = -10^{-3} \cdot 9.82, \; \boldsymbol{L}_{54} = -10^{-4} \cdot 4.49 \tag{A.9e}$$

A.2 Multizone furnace

A linear model for the mutizone furnace has been derived in [29] for 15 zones. In this thesis the model is adopted to a multizone furnace that consists of four zones. Each zone represents a subsystem which are modelled by (2.6) for $i = 1, 2, 3, 4$ with

$$\boldsymbol{A}_1 = \begin{pmatrix} -1.39 & -4.21 \\ -0.08 & -0.34 \end{pmatrix}, \quad \boldsymbol{B}_1 = \boldsymbol{E}_1 = \begin{pmatrix} 1.34 \\ 0.1 \end{pmatrix}, \quad \boldsymbol{C}_1 = \boldsymbol{C}_{\mathrm{z}1} = \begin{pmatrix} 0.2 & 0.8 \end{pmatrix}, \tag{A.10a}$$

$$\boldsymbol{A}_2 = \begin{pmatrix} -2.44 & -8.43 \\ -0.16 & -0.65 \end{pmatrix}, \quad \boldsymbol{B}_2 = \boldsymbol{E}_2 = \begin{pmatrix} 1.34 \\ 0.1 \end{pmatrix}, \quad \boldsymbol{C}_2 = \boldsymbol{C}_{z2} = \begin{pmatrix} 0.2 & 0.8 \end{pmatrix}, \tag{A.10b}$$

$$\boldsymbol{A}_3 = \begin{pmatrix} -2.44 & -8.43 \\ -0.16 & -0.65 \end{pmatrix}, \quad \boldsymbol{B}_3 = \boldsymbol{E}_3 = \begin{pmatrix} 1.34 \\ 0.1 \end{pmatrix}, \quad \boldsymbol{C}_3 = \boldsymbol{C}_{z3} = \begin{pmatrix} 0.2 & 0.8 \end{pmatrix}, \tag{A.10c}$$

$$\boldsymbol{A}_4 = \begin{pmatrix} -1.39 & -4.21 \\ -0.08 & -0.34 \end{pmatrix}, \quad \boldsymbol{B}_4 = \boldsymbol{E}_4 = \begin{pmatrix} 1.34 \\ 0.1 \end{pmatrix}, \quad \boldsymbol{C}_4 = \boldsymbol{C}_{z4} = \begin{pmatrix} 0.2 & 0.8 \end{pmatrix}, \tag{A.10d}$$

and the local couplings are modelled by (2.8a)–(2.8c) with

$$\boldsymbol{L}_{12} = 3.93, \tag{A.11a}$$

$$\boldsymbol{L}_{21} = 3.93, \qquad \boldsymbol{L}_{23} = 3.93, \tag{A.11b}$$

$$\boldsymbol{L}_{32} = 3.93, \qquad \boldsymbol{L}_{34} = 3.93, \tag{A.11c}$$

$$\boldsymbol{L}_{43} = 3.93. \tag{A.11d}$$

B Proofs

B.1 Proof of Theorem 4.1

Consider design agent D_i to be the vertex i of the interconnection graph $\mathcal{G}_{\mathrm{K}}$. To prove that the algorithm works totally correct for arbitrary directed graphs, it has to be shown

1. that the algorithm terminates and

2. that the algorithm works partially correct

according to [88]. Partial correctness means that if an answer is returned for a valid input it will be correct [88]. The algorithm can be separated into the forward-search phase, whenever a design agent is requested and the backtracking phase, whenever a design agent responses to the request. In the following, the reachability graph $\mathcal{G}_{\mathrm{R1}}(\mathcal{G}_{\mathrm{K}}^{\top})$ with vertex set $\mathcal{V}_{\mathrm{R1}}$ and edge set $\mathcal{E}_{\mathrm{R1}}$ is additionally utilised during this proof.

To prove the *termination*, the forward-search phase and the backtracking phase are investigated separately. During the forward-search phase all vertices i, $(i \in \mathcal{V}_{\mathrm{R1}})$ of the reachability graph are requested, because starting from vertex 1 all predecessors $j \in \mathcal{P}_1$ are requested within the loop (Line 4 - Line 16). Moving to vertex j (Line 5) all its predecessors $k \in \mathcal{P}_j$ are requested within the loop, etc. Thus, there exists a path $\mathrm{P}(1, \{j\}, k)$ in $\mathcal{G}_{\mathrm{R1}}(\mathcal{G}_{\mathrm{K}}^{\top})$. Moreover, the loop is started exactly once at each vertex i, $(i \in \mathcal{V}_{\mathrm{R1}})$ due to Line 3. As each edge (i, j), $((i, j) \in \mathcal{E}_{\mathrm{R1}}^{\top})$ is investigated exactly once (due to Line 4 and Line 5), it is concluded that the forward-search phase ends after $|\mathcal{E}_{\mathrm{R1}}|$ iterations. Furthermore, since the loop is only started once at each vertex i, $(i \in \mathcal{V}_{\mathrm{R1}})$, $|\mathcal{V}_{\mathrm{R1}}| - 1$ edges are traversed during the forward-search phase. During the backtracking phase, the $|\mathcal{V}_{\mathrm{R1}}| - 1$ edges are traversed a second time in the opposite direction and opposite order (RETURN in Line 18). That means, first, that the backtracking phase terminates in the start vertex 1 (Line 17) and, second, that the backtracking phase ends after $|\mathcal{V}_{\mathrm{R1}}| - 1$ iterations. Accordingly, the whole algorithm requires $|\mathcal{E}_{\mathrm{R1}}| + |\mathcal{V}_{\mathrm{R1}}| - 1$ iterations.

In the following the *partial correctness* of the algorithm is proved which is to find all vertices strongly connected to vertex 1 of a given graph $\mathcal{G}_{\mathrm{K}}$. Consider the vertex i, its predecessor vertex h and successor vertex j as illustrated in Fig. 2.1. It will be shown that in the forward-search phase and the backtracking phase the vertex i is sorted correctly. Therefore, the time instants where the algorithm is processed at design agent D_i (forward-search phase and backtracking

Figure 2.1: Classification of vertex i and edge (i,j) to prove the partial correctness of the search algorithm

phase) are considered. Consider that there exists a path $\mathrm{P}(1,i)$ within the graph $\mathcal{G}_{\mathrm{R1}}(\mathcal{G}_{\mathrm{K}}^{\top})$. Since $j \in \mathcal{P}_i$, is can be concluded that there exists a path $\mathrm{P}(1,\{i\},j)$ within the graph $\mathcal{G}_{\mathrm{R1}}(\mathcal{G}_{\mathrm{K}}^{\top})$. Thus, it will be shown in the following that if there exists a path $\mathrm{P}(j,1)$, then the vertex i is sorted as strongly connected. First, the case in which j has already been marked as strongly connected is considered. Then vertex i is classified as strongly connected during the forward-search phase (cf. Line 11). Second, if vertex j has not yet been classified as strongly connected but later, vertex i is marked as strongly connected when it is backtracked from vertex j to vertex i (cf. Line 9). The third case considers the situation in which node j is not marked as strongly connected and also will not be. Since, initially all vertices are categorised as not strongly connected (cf. Initiate), this status of vertex i will not be changed. As the loop is performed at all vertices i, $(\forall i \in \mathcal{V}_{\mathrm{R1}})$ (shown before), this result applies to all the vertices in $\mathcal{V}_{\mathrm{R1}}$. This proves the partial correctness.

Together with the proved termination, the total correctness of Algorithm 4.1 has been proved.

B.2 Proof of Corollary 4.1

The system matrix of the reconfigured overall closed-loop system

$$\mathring{F} = \mathrm{con}\left(\left\{S_{1f}, \mathring{C}_1, K_1\right\} \cup \{F_i, K_i,\ i \in \mathcal{N} \setminus \{1\}\},\ \boldsymbol{w},\ \begin{pmatrix} \mathring{\boldsymbol{y}}_1 \\ \boldsymbol{y}_2 \\ \vdots \\ \boldsymbol{y}_N \end{pmatrix}\right)$$

yields a triangular matrix of the form

$$\begin{pmatrix} \begin{pmatrix} \boldsymbol{A}_{\mathrm{F1}}^{\mathcal{N}_{\mathrm{S1}}} & \mathrm{O} \\ * & \boldsymbol{A}_{\delta 1} \end{pmatrix} & \mathrm{O} \\ * & \boldsymbol{A}_{\mathrm{F}}^{\mathcal{N} \setminus \mathcal{N}_{\mathrm{S1}}} \end{pmatrix},$$

where $\boldsymbol{A}_{\mathrm{F1}}^{\mathcal{N}_{\mathrm{S1}}}$ denotes the system matrix of the fault-free system $F_1^{\mathcal{N}_{\mathrm{S1}}} = \mathrm{con}\left(\left\{S_1^{\mathcal{N}_{\mathrm{S1}}}, C_1\right\}, \boldsymbol{w}_1, \boldsymbol{y}_1\right)$, $\boldsymbol{A}_{\mathrm{F}}^{\mathcal{N} \setminus \mathcal{N}_{\mathrm{S1}}}$ represents the accumulated system matrix of all controlled subsystems F_i, $(\forall i \in \mathcal{N} \setminus \mathcal{N}_{\mathrm{S1}})$ which are not strongly connected to S_1 and $*$ denotes an arbitrary entry. In accordance with assumption A 4.2, the overall closed-loop system is asymptotically stable if and only if the matrix $\boldsymbol{A}_{\delta 1}$ is Hurwitz for the reconfigured case (cf. Section 2.4.2). There exists a feedback gain $\boldsymbol{M}_1$ such that $\boldsymbol{A}_{\delta 1}$ is Hurwitz, if and only if the pair $\left(\boldsymbol{A}_1^{\mathcal{N}_{S1}}, \boldsymbol{B}_{1f}^{\mathcal{N}_{S1}}\right)$ is stabilisable. The design yields a classical state-feedback design which can be reformulated by the LMI (4.7) as shown by [58, 154].

B.3 Proof of Theorem 5.1

The LMI (5.10a) reflects claim that the controlled subsystem state remains in the ellipsoidal operating region $\mathcal{X}_i$ (cf. (5.8a)). See [182] for a detailed derivation.

Now, it is shown that if the LMIs (5.10b) are satisfied, the adherence of the desired control limitations is guaranteed (cf. claim (5.8b)). Each of the m_i LMIs reflects the respective claim

$$|\boldsymbol{u}_i| = |\boldsymbol{k}_{i,j}^\top \boldsymbol{x}_i| \leq \varrho_j,$$

where $\boldsymbol{k}_{i,j}^\top$ represents the j-th row of the static state-feedback gain $\boldsymbol{K}_i$. First, the congruence transformation with

$$\boldsymbol{T} = \begin{pmatrix} 1 & \boldsymbol{0}^\top \\ \boldsymbol{0} & \boldsymbol{X}^{-1} \end{pmatrix} \tag{B.1}$$

is simultaneously applied on the m_i LMIs (5.10b), followed by the usage of the Schur complement to gain the following equivalent relations:

$$\begin{pmatrix} \varrho_j^2 & \boldsymbol{w}_j^\top \\ \boldsymbol{w}_j & \boldsymbol{X} \end{pmatrix} \succeq 0 \quad \Leftrightarrow \quad \begin{pmatrix} \varrho_j^2 & \boldsymbol{k}_{i,j}^\top \\ \boldsymbol{k}_{i,j} & \boldsymbol{X}^{-1} \end{pmatrix} \succeq 0 \quad \Leftrightarrow \quad \sqrt{\boldsymbol{k}_{i,j}^\top \boldsymbol{X} \boldsymbol{k}_{i,j}} \leq \varrho_j,$$

where $\boldsymbol{k}_{i,j}^\top = \boldsymbol{w}_j^\top \boldsymbol{X}^{-1}$. In the next step, the implication

$$\sqrt{\boldsymbol{k}_{i,j}^\top \boldsymbol{X} \boldsymbol{k}_{i,j}} \leq \varrho_j \quad \Rightarrow \quad |\boldsymbol{k}_{i,j}^\top \boldsymbol{x}_i| \leq \varrho_j, \;\; \boldsymbol{x}_i \in \mathcal{X}_i \tag{B.2}$$

is proved. Therefore the relation

$$|\boldsymbol{k}_{i,j}^\top \boldsymbol{x}_i| \leq \sqrt{\boldsymbol{k}_{i,j}^\top \boldsymbol{X} \boldsymbol{k}_{i,j}}, \;\; \forall \boldsymbol{x}_i \in \mathcal{X}_i. \tag{B.3}$$

is shown. For all states $\boldsymbol{x}_i \in \mathcal{X}_i$ the relation

$$1 - \boldsymbol{x}_i^\top \boldsymbol{X}^{-1} \boldsymbol{x}_i \leq 0$$

or equivalently

$$\begin{pmatrix} \boldsymbol{X} & \boldsymbol{x}_i \\ \boldsymbol{x}_i^\top & 1 \end{pmatrix} \succeq 0$$

by applying the Schur complement with (B.1) is satisfied. The pre and post multiplication of the latter matrix with the vector

$$\boldsymbol{t} = \begin{pmatrix} \boldsymbol{k}_{i,j} \\ -\boldsymbol{k}_{i,j}^\top \boldsymbol{x}_i \end{pmatrix}$$

results in

$$\boldsymbol{t}^\top \begin{pmatrix} \boldsymbol{X} & \boldsymbol{x}_i \\ \boldsymbol{x}_i^\top & 1 \end{pmatrix} \boldsymbol{t} = \boldsymbol{k}_{i,j}^\top \boldsymbol{X} \boldsymbol{k}_{i,j} - (\boldsymbol{k}_{i,j}^\top \boldsymbol{x}_i)^2 \geq 0$$

which is equivalent to the relation (B.3) and, hence, proves the implication (B.2).

B.4 Proof of Corollary 5.1

As stated in Lemma 5.1, asymptotic stability of F_i and robustly positively invariance of $\mathcal{I}_i$ for F_i is reflected by the LMI (5.11b). In order to proof that the ellipsoid $\mathcal{I}_i$ is located in the polytopic operating region if the LMIs (5.11c) holds, it has to be shown that for all $\boldsymbol{x}_i \in \mathcal{I}_i$ the relation $|\boldsymbol{p}_j \boldsymbol{x}_i| \leq b_j$ is satisfied. That is to show that

$$|\boldsymbol{p}_j^\top \boldsymbol{x}_i| \leq \sqrt{\boldsymbol{p}_j^\top \boldsymbol{X} \boldsymbol{p}_j}, \quad \forall \boldsymbol{x}_i \in \mathcal{X}_i.$$

if the LMIs (5.11c) are satisfied. Therefore, it is referred to the proof of Theorem 5.1 or, alternatively, see [58].

To assure that the initial state $\boldsymbol{x}_{i,0}$ is contained in the ellipsoid $\mathcal{I}_i$, i.e., $\boldsymbol{x}_{0,i} \in \mathcal{I}_i$, the inequality

$$\boldsymbol{x}_{i,0}^\top \boldsymbol{X}^{-1} \boldsymbol{x}_{i,0} \leq 1 \tag{B.4}$$

has to be satisfied. Applying the Schur complement on (B.4), the LMI (5.11d) is obtained

As shown in [58], the minimisation of the cost function (5.11a) leads to the maximisation of the volume of the ellipsoid, whereas the constraints (5.11c) restrict the size by the polytope $\mathcal{X}_i$. The constraint (5.11b) ensures that $\mathcal{I}_i$ is an RPI set for F_i.

B.5 Proof of Corollary 5.2

The first requirement claims the set $\mathcal{Y}_{\min}$ to be a summand of $\mathcal{X}$ as it has been shown in [155].

The second requirement claims that the overlap of $\mathcal{Z}$ with $\mathcal{Y}_{\min}$ shall be the tightest overlap while regarding the first claim. Indeed, the tightest complete overlap of a polytope $\mathcal{Z}$ with another polytope $\mathcal{Y}_{\min} \supseteq \mathcal{Z}$ results when each edge of $\mathcal{Y}_{\min}$ is tangent to the boundary of $\mathcal{Z}$. That is $h_{\mathcal{Y}_{\min}}(\boldsymbol{a}) - h_{\mathcal{Z}}(\boldsymbol{a}) = 0$. Both claims may contradict. However, to simultaneously satisfy the first requirement, a particular edge may not tangent to $\mathcal{Z}$ but has to be as close as possible. In other words, the distance $h_{\mathcal{Y}_{\min}}(\boldsymbol{a}) - h_{\mathcal{Z}}(\boldsymbol{a})$ has to be minimal and non negative whenever $\mathcal{F}_{\mathcal{Y}_{\min}}(\boldsymbol{a})$ is an edge of $\mathcal{Y}_{\min}$.

Hence, both requirements together claim that the set is a summand of $\mathcal{X}$ and simultaneously is the tightest overlap of $\mathcal{Z}$ which completes the proof.

B.6 Proof of Lemma 6.1

The dynamics $\boldsymbol{P}_i^{\mathcal{N}}(s)$ of the physical interaction (6.1) are constructed according to (6.2). An upper bound on these dynamics is given by

$$\begin{aligned}\|\boldsymbol{P}_i^{\mathcal{N}}(\mathrm{j}\omega)\| \leq \|\boldsymbol{L}_{\{i\},\mathcal{N}\setminus\{i\}}\| \left(\mathbf{I} - \|\mathrm{diag}(\boldsymbol{F}_{\mathrm{zs}j}(\mathrm{j}\omega))_{j\in\mathcal{N}\setminus\{i\}} \cdot \boldsymbol{L}_{\mathcal{N}\setminus\{i\}}\|\right)^{-1} \\ \|\mathrm{diag}(\boldsymbol{F}_{\mathrm{zs}j}(\mathrm{j}\omega))_{j\in\mathcal{N}\setminus\{i\}} \cdot \boldsymbol{L}_{\mathcal{N}\setminus\{i\},\{i\}}\|,\end{aligned} \tag{B.5}$$

where the relation

$$\|\left(\mathbf{I} - \mathrm{diag}(\boldsymbol{F}_{\mathrm{zs}j}(\mathrm{j}\omega))_{j\in\mathcal{N}\setminus\{i\}} \cdot \boldsymbol{L}_{\mathcal{N}\setminus\{i\}}\right)^{-1}\| \leq \left(\mathbf{I} - \|\mathrm{diag}(\boldsymbol{F}_{\mathrm{zs}j}(\mathrm{j}\omega))_{j\in\mathcal{N}\setminus\{i\}} \cdot \boldsymbol{L}_{\mathcal{N}\setminus\{i\}}\|\right)^{-1}$$

holds due to the satisfaction of (6.3).

The condition (6.3) can equivalently be expressed by

$$\|\boldsymbol{F}_{\mathrm{zs}j}(\mathrm{j}\omega)\| \cdot \|\boldsymbol{L}_{ji}\| < \varphi(\omega) - \sum_{k\in\mathcal{P}_j, k\neq i} \|\boldsymbol{F}_{\mathrm{zs}j}(\mathrm{j}\omega)\| \cdot \|\boldsymbol{L}_{jk}\|, \quad j \in \mathcal{N}\setminus\{i\}.$$

Now, these $N-1$ inequalities are lumped together to

$$\|\mathrm{diag}(\boldsymbol{F}_{\mathrm{zs}j}(\mathrm{j}\omega))_{j\in\mathcal{N}\setminus\{i\}}\| \cdot \|\boldsymbol{L}_{\mathcal{N}\setminus\{i\},\{i\}}\| < \left(\varphi(\omega)\mathbf{I} - \|\mathrm{diag}(\boldsymbol{F}_{\mathrm{zs}j}(\mathrm{j}\omega))_{j\in\mathcal{N}\setminus\{i\}}\| \cdot \|\boldsymbol{L}_{\mathcal{N}\setminus\{i\}}\|\right) \cdot \mathbf{1}. \tag{B.6}$$

where the norm applies block element wise, e.g.,

$$\left\|\begin{pmatrix} \mathrm{O} & \boldsymbol{F}_{\mathrm{zs}1}(\mathrm{j}\omega)\boldsymbol{L}_{12} \\ \boldsymbol{F}_{\mathrm{zs}2}(\mathrm{j}\omega)\boldsymbol{L}_{21} & \mathrm{O} \end{pmatrix}\right\| = \begin{pmatrix} \mathrm{O} & \|\boldsymbol{F}_{\mathrm{zs}1}(\mathrm{j}\omega)\boldsymbol{L}_{12}\| \\ \|\boldsymbol{F}_{\mathrm{zs}2}(\mathrm{j}\omega)\boldsymbol{L}_{21}\| & \mathrm{O} \end{pmatrix}. \tag{B.7}$$

The satisfaction of (B.6) implies the relation

$$\|\mathrm{diag}(\boldsymbol{F}_{\mathrm{zs}j}(\mathrm{j}\omega))_{j\in\mathcal{N}\setminus\{i\}}\| \cdot \|\boldsymbol{L}_{\mathcal{N}\setminus\{i\},\{i\}}\| < \big(\mathbf{I} - \|\mathrm{diag}(\boldsymbol{F}_{\mathrm{zs}j}(\mathrm{j}\omega))_{j\in\mathcal{N}\setminus\{i\}}\| \cdot \|\boldsymbol{L}_{\mathcal{N}\setminus\{i\}}\|\big) \cdot \varphi(\omega)\mathbf{1} \tag{B.8}$$

since $0 < \varphi(\omega) \leq 1$. Inserting (B.8) into (B.5) results to

$$\|\boldsymbol{P}_i^{\mathcal{N}}(\mathrm{j}\omega)\| < \|\boldsymbol{L}_{\{i\},\mathcal{N}\setminus\{i\}}\| \cdot \varphi(\omega)\mathbf{1} = \sum_{j\in\mathcal{P}_i} \|\boldsymbol{L}_{ij}\| \cdot \varphi(\omega) =: \bar{P}_i^{\{i\}}(\omega),$$

which proofs that (6.4) is a comparison system of (6.1). Hence, the proof is completed.

B.7 Proof of Theorem 6.3

First of all, the controlled extended subsystem $F_{\mathrm{W}i}$ is introduced, which is constructed in accordance with

$$F_{\mathrm{W}i} = \mathrm{con}\left(\{S_{\mathrm{W}i} \cup C_i\}, \begin{pmatrix} \boldsymbol{w}_i \\ \boldsymbol{s}_i \end{pmatrix}, \begin{pmatrix} \boldsymbol{y}_i \\ z_i \\ v_{\mathrm{z}i} \\ \boldsymbol{v}_{\mathrm{y}i} \end{pmatrix}\right)$$

and is represented by the state space model

$$F_{\mathrm{W}i}: \begin{cases} \dot{\boldsymbol{x}}_{\mathrm{FW}i}(t) &= \boldsymbol{A}_{\mathrm{FW}i}\boldsymbol{x}_{\mathrm{FW}i}(t) + \boldsymbol{B}_{\mathrm{FW}i}\boldsymbol{w}_i(t) + \boldsymbol{e}_{\mathrm{FW}i}\boldsymbol{s}_i(t), \quad \boldsymbol{x}_{\mathrm{FW}i}(0) = \mathbf{0} \\ \boldsymbol{y}_i(t) &= \boldsymbol{C}_{\mathrm{FW}i}\boldsymbol{x}_{\mathrm{FW}i}(t) \\ z_i(t) &= \boldsymbol{c}_{\mathrm{zFW}i}^\top \boldsymbol{x}_{\mathrm{FW}i}(t) \\ v_{\mathrm{z}i}(t) &= \boldsymbol{c}_{\mathrm{FWz}i}^\top \boldsymbol{x}_{\mathrm{FW}i}(t) \\ \boldsymbol{v}_{\mathrm{y}i}(t) &= \boldsymbol{C}_{\mathrm{FWy}i}\boldsymbol{x}_{\mathrm{FW}i}(t). \end{cases}$$

The requirements (6.25) and (6.27) as well as (6.30)–(6.32) on the H_∞-norm are rewritten by means of Lemma 2.2 into the equivalent claim

$$\boldsymbol{P} \succ 0 \tag{B.9}$$

$$\begin{pmatrix} \boldsymbol{A}_{\mathrm{FW}i}^\top \boldsymbol{P} + \boldsymbol{P}\boldsymbol{A}_{\mathrm{FW}i} & \boldsymbol{P}\boldsymbol{e}_{\mathrm{FW}i} & \boldsymbol{c}_{\mathrm{FWz}i} \\ \boldsymbol{e}_{\mathrm{FW}i}^\top \boldsymbol{P} & -\gamma_{\mathrm{zs}i} & 0 \\ \boldsymbol{c}_{\mathrm{FWz}i}^\top & 0 & -\gamma_{\mathrm{zs}i} \end{pmatrix} \prec 0, \tag{B.10}$$

$$\begin{pmatrix} \boldsymbol{A}_{\mathrm{FW}i}^{\top}\boldsymbol{P}+\boldsymbol{P}\boldsymbol{A}_{\mathrm{FW}i} & \boldsymbol{P}\boldsymbol{B}_{\mathrm{FW}i} & \boldsymbol{C}_{\mathrm{FWy}i}^{\top} \\ \boldsymbol{B}_{\mathrm{FW}i}^{\top}\boldsymbol{P} & -\gamma_{\mathrm{yw}i}\mathbf{I} & \mathbf{O} \\ \boldsymbol{C}_{\mathrm{FWy}i} & \mathbf{O} & -\gamma_{\mathrm{yw}i}\mathbf{I} \end{pmatrix} \prec 0, \tag{B.11}$$

$$\begin{pmatrix} \boldsymbol{A}_{\mathrm{FW}i}^{\top}\boldsymbol{P}+\boldsymbol{P}\boldsymbol{A}_{\mathrm{FW}i} & \boldsymbol{P}\boldsymbol{e}_{\mathrm{FW}i} & \boldsymbol{C}_{\mathrm{FW}i}^{\top} \\ \boldsymbol{e}_{\mathrm{FW}i}^{\top}\boldsymbol{P} & -\gamma_{\mathrm{ys}i} & \mathbf{0}^{\top} \\ \boldsymbol{C}_{\mathrm{FW}i} & \mathbf{0} & -\gamma_{\mathrm{ys}i}\mathbf{I} \end{pmatrix} \prec 0, \tag{B.12}$$

$$\begin{pmatrix} \boldsymbol{A}_{\mathrm{FW}i}^{\top}\boldsymbol{P}+\boldsymbol{P}\boldsymbol{A}_{\mathrm{FW}i} & \boldsymbol{P}\boldsymbol{B}_{\mathrm{FW}i} & \boldsymbol{c}_{\mathrm{zFW}i} \\ \boldsymbol{B}_{\mathrm{FW}i}^{\top}\boldsymbol{P} & -\gamma_{\mathrm{zw}i}\mathbf{I} & \mathbf{0} \\ \boldsymbol{c}_{\mathrm{zFW}i}^{\top} & \mathbf{0}^{\top} & -\gamma_{\mathrm{zw}i} \end{pmatrix} \prec 0, \tag{B.13}$$

where the LMI (B.10) captures the claims (6.25) and (6.31) if $\left(\sum_{j\in\mathcal{P}_i}||\boldsymbol{L}_{ij}||\right)^{-1} \geq \gamma_{\mathrm{zs}i} > 0$. As (B.10)–(B.13) are non-linear matrix inequalities in the decision variables $\boldsymbol{P}$, $\boldsymbol{A}_{\mathrm{C}i}$, $\boldsymbol{B}_{\mathrm{C}i}$, $\boldsymbol{C}_{\mathrm{C}i}$, $\boldsymbol{D}_{\mathrm{C}i}$, the linearisation proposed in [82] is applied. Accordingly, the matrix $\boldsymbol{P}$ is partitioned into

$$\boldsymbol{P} = \begin{pmatrix} \boldsymbol{X} & \boldsymbol{M} \\ \boldsymbol{M}^{\top} & \boldsymbol{U} \end{pmatrix}, \ \boldsymbol{P}^{-1} = \begin{pmatrix} \boldsymbol{Y} & \boldsymbol{N} \\ \boldsymbol{N}^{\top} & \boldsymbol{V} \end{pmatrix}$$

with the matrices $\boldsymbol{U}_i, \boldsymbol{V}_i \in \mathbb{R}^{n_{\mathrm{C}i}\times n_{\mathrm{C}i}}$, $\boldsymbol{M}_i, \boldsymbol{N}_i \in \mathbb{R}^{(n_i+n_{\mathrm{Wz}i}+n_{\mathrm{Wy}i})\times n_{\mathrm{C}i}}$ and the symmetric matrices $\boldsymbol{X}_i = \boldsymbol{X}_i^{\top}, \boldsymbol{Y}_i = \boldsymbol{Y}_i^{\top} \in \mathbb{R}^{(n_i+n_{\mathrm{Wz}i}+n_{\mathrm{Wy}i})\times(n_i+n_{\mathrm{Wz}}+n_{\mathrm{Wy}i})}$. Thereafter, applying a congruence transformation with

$$\boldsymbol{T} = \begin{pmatrix} \mathbf{I} & \boldsymbol{Y} \\ \mathbf{O} & \boldsymbol{N}^{\top} \end{pmatrix}$$

on the partitioned matrix $\boldsymbol{P}$ results in (6.34a), which is equivalent to (B.9). A further congruence transformation with

$$\tilde{\boldsymbol{T}} = \mathrm{diag}\big(\boldsymbol{T}^{\top},\ \mathbf{I},\ \mathbf{I}\big)$$

is simultaneously performed on the matrix inequalities (B.10)–(B.13) to obtain the equivalent LMIs (6.34b)–(6.34e), where the matrices are

$$\boldsymbol{A}_{\mathrm{LMI}} = \begin{pmatrix} \boldsymbol{X}\boldsymbol{A}_{\mathrm{SW}i} - \widetilde{\boldsymbol{B}}\boldsymbol{C}_{\mathrm{SW}i} & \widetilde{\boldsymbol{A}} \\ \boldsymbol{A}_{\mathrm{SW}i} - \boldsymbol{B}_{\mathrm{SW}i}\widetilde{\boldsymbol{D}}\boldsymbol{C}_{\mathrm{SW}i} & \boldsymbol{A}_{\mathrm{SW}i}\boldsymbol{Y} + \boldsymbol{B}_{\mathrm{SW}i}\widetilde{\boldsymbol{C}} \end{pmatrix}, \ \boldsymbol{B}_{\mathrm{LMI}} = \begin{pmatrix} \widetilde{\boldsymbol{B}} \\ \boldsymbol{B}_{\mathrm{SW}i}\widetilde{\boldsymbol{D}} \end{pmatrix}, \tag{B.14}$$

$$\boldsymbol{C}_{\mathrm{LMI}} = \begin{pmatrix} \boldsymbol{C}_{\mathrm{SW}i} \\ \boldsymbol{Y}\boldsymbol{C}_{\mathrm{SW}i} \end{pmatrix}, \ \boldsymbol{e}_{\mathrm{LMI}} = \begin{pmatrix} \boldsymbol{X}\boldsymbol{e}_{\mathrm{SW}i} \\ \boldsymbol{e}_{\mathrm{SW}i} \end{pmatrix}, \ \boldsymbol{C}_{\mathrm{zLMI}} = \begin{pmatrix} \boldsymbol{c}_{\mathrm{zSW}i} \\ \boldsymbol{Y}\boldsymbol{c}_{\mathrm{zSW}i} \end{pmatrix} \tag{B.15}$$

$$\boldsymbol{C}_{\text{yLMI}}=\begin{pmatrix}\boldsymbol{C}_{\text{SWy}i}\\ \boldsymbol{Y}\boldsymbol{C}_{\text{SWy}i}\end{pmatrix},\ \boldsymbol{c}_{\text{z}\varphi\text{LMI}}=\begin{pmatrix}\boldsymbol{c}_{\text{SWz}i}\\ \boldsymbol{Y}\boldsymbol{c}_{\text{SWz}i}\end{pmatrix}\ \boldsymbol{D}_{\text{LMI}i}=\boldsymbol{D}_{\text{Wu}i}\widetilde{\boldsymbol{D}}_i, \tag{B.16}$$

From $\boldsymbol{P}\boldsymbol{P}^{-1} = \mathbf{I}$ follows the equality (6.37). Since $\mathbf{I}-\boldsymbol{X}\boldsymbol{Y}$ is invertible as stated in Lemma 4.2 of [62] and $n_{\text{C}i} = n_i + n_{\text{Wz}i} + n_{\text{Wu}i}$ there exists always a decomposition (6.37) for which $\boldsymbol{M}$ and $\boldsymbol{N}$ are invertible. Thus, the controller parameters can be calculated uniquely by (6.35) and (6.36).

In order to guarantee the existence of a solution to the LMIs (6.34) for the extended subsystem SW_i, it is necessary that the pair $(\boldsymbol{A}_{\text{SW}i}, \boldsymbol{B}_{\text{SW}i})$ is stabilisable and the pair $(\boldsymbol{A}_{\text{SW}i}, \boldsymbol{C}_{\text{SW}i})$ is detectable [82].

B.8 Proof of Lemma 7.1

I/O stability of the system $\hat{P}_i^{\mathcal{N}_{\text{S}i}}$ and the system $E_1^{\mathcal{N}_{\text{W}1}}$ is proved by showing the existence of upper bounds of the amplitude plots of the transfer functions $\hat{\boldsymbol{P}}_1^{\mathcal{N}_{\text{S}1}}(s)$, $\boldsymbol{P}_{\text{sq}1}^{\mathcal{N}_{\text{S}1}}(s)$, $\boldsymbol{P}_{\text{pz}1}^{\mathcal{N}_{\text{S}1}}(s)$ and $\boldsymbol{P}_{\text{pq}1}^{\mathcal{N}_{\text{S}1}}(s)$ as well as $\boldsymbol{E}_1^{\mathcal{N}_{\text{W}1}}(s)$ for an arbitrary number of subsystems. The transfer functions can recursively be determined by (C.6)–(C.9).

In the following it is shown that the dynamics $\boldsymbol{P}_{\text{pq}1}^{\mathcal{N}_{\text{S}1}\cup\{v\}}(s)$ are bounded. Therefore, it is shown that there exists an upper bound

$$\|\boldsymbol{P}_{\text{pq}1}^{\mathcal{N}_{\text{S}1}\cup\{v\}}(\text{j}\omega)\| \leq \bar{P}_{\text{pq}1}^{\mathcal{N}_{\text{S}1}}(\omega) = \|\boldsymbol{F}_{\text{zs}v}(\text{j}\omega)\| \cdot \bar{\Psi}^{\mathcal{N}_{\text{S}1}}(\omega)$$

with

$$\bar{\Psi}^{\mathcal{N}_{\text{S}1}}(\omega) = \left(1 - \|\boldsymbol{L}_{v-1v}\| \cdot \|\boldsymbol{F}_{\text{zs}v}(\text{j}\omega)\| \cdot \|\boldsymbol{L}_{vv-1}\| \cdot \bar{P}_{\text{pq}1}^{\mathcal{N}_{\text{S}1}}(\omega)\right)^{-1}. \tag{B.17}$$

As the transfer function matrix $\boldsymbol{F}_{\text{zs}v}(s)$ is I/O stable, it is to prove that

$$\bar{\Psi}^{\mathcal{N}_{\text{S}1}}(\omega) < \infty, \quad \forall\omega \in \mathbb{R}.$$

by means of the explicit calculation of the recursion that starts by $\mathcal{N}_{\text{S}1} = \{1\}$ and $v = 2$. Based on the fulfilled relations (7.4) (cf. assumption A 7.1) the upper bound $\bar{\Psi}^{\mathcal{N}_{\text{S}1}}(\omega)$ result to

$$\begin{aligned}
\bar{\Psi}^{\{1\}}(\omega) &= \left(1 - \|\boldsymbol{L}_{12}\| \cdot \|\boldsymbol{F}_{\text{zs}2}(\text{j}\omega)\| \cdot \|\boldsymbol{L}_{21}\| \cdot \bar{P}_{\text{pq}1}^{\{1\}}(\omega)\right)^{-1} = 1\\
\bar{\Psi}^{\{1,2\}}(\omega) &= \left(1 - \|\boldsymbol{L}_{23}\| \cdot \|\boldsymbol{F}_{\text{zs}3}(\text{j}\omega)\| \cdot \|\boldsymbol{L}_{32}\| \cdot \bar{P}_{\text{pq}1}^{\{1,2\}}(\omega)\right)^{-1}\\
&= \left(1 - \|\boldsymbol{L}_{23}\| \cdot \|\boldsymbol{F}_{\text{zs}3}(\text{j}\omega)\| \cdot \|\boldsymbol{L}_{32}\| \cdot \|\boldsymbol{F}_{\text{zs}2}(\text{j}\omega)\|\right)^{-1}\\
&< \left(1 - \tfrac{1}{4}\right)^{-1} = \tfrac{4}{3}
\end{aligned}$$

$$
\begin{aligned}
\bar{\Psi}^{\{1,2,3\}}(\omega) &= \left(1 - \|\boldsymbol{L}_{34}\| \cdot \|\boldsymbol{F}_{\mathrm{zs4}}(\mathrm{j}\omega)\| \cdot \|\boldsymbol{L}_{43}\| \cdot \bar{P}_{\mathrm{pq1}}^{\{1,2,3\}}(\omega)\right)^{-1} \\
&< \left(1 - \|\boldsymbol{L}_{34}\| \cdot \|\boldsymbol{F}_{\mathrm{zs4}}(\mathrm{j}\omega)\| \cdot \|\boldsymbol{L}_{43}\| \cdot \|\boldsymbol{F}_{\mathrm{zs3}}(\mathrm{j}\omega)\| \cdot \tfrac{4}{3}\right)^{-1} \\
&< \left(1 - \tfrac{1}{4} \cdot \tfrac{4}{3}\right)^{-1} = 1.5 \\
&\vdots
\end{aligned}
$$

The continuation of this calculation finally ends in the upper bound

$$
\bar{\Psi}^{\mathcal{N}_{\mathrm{S1}} \cup \{v\}}(\omega) < 2, \quad \forall v \geq 1. \tag{B.18}
$$

Hence, the transfer function $\boldsymbol{P}_{\mathrm{pq1}}^{\mathcal{N}_{\mathrm{S1}}}(s)$ is bounded from above by

$$
\|\boldsymbol{P}_{\mathrm{pq1}}^{\mathcal{N}_{\mathrm{S1}} \cup \{v\}}(\mathrm{j}\omega)\| < 2\|\boldsymbol{F}_{\mathrm{zs}v}(\mathrm{j}\omega)\|,
$$

for arbitrary $v \geq 1$. In direct analogy, the upper bounds of the amplitude plots of the other transfer functions $\hat{\boldsymbol{P}}_{1}^{\mathcal{N}_{\mathrm{S1}}}(s)$, $\boldsymbol{P}_{\mathrm{sq1}}^{\mathcal{N}_{\mathrm{S1}}}(s)$, $\boldsymbol{P}_{\mathrm{pz1}}^{\mathcal{N}_{\mathrm{S1}}}(s)$ as well as $\boldsymbol{E}_{1}^{\mathcal{N}_{\mathrm{W1}}}(s)$ can be derived.

B.9 Proof of Theorem 7.2

Based on eqn. (7.8) the particular upper bound $\Delta_1^{v+1}(\omega)$ is derived as follows

$$
\begin{aligned}
\Delta_1^{v+1}(\omega) = \|\boldsymbol{P}_{\mathrm{sq1}}^{\mathcal{N}_{\mathrm{S1}} \cup \{v\}}(\mathrm{j}\omega)\| \cdot \|\boldsymbol{L}_{vv+1}\| \cdot \|\boldsymbol{F}_{\mathrm{zs}v+1}(\mathrm{j}\omega)\boldsymbol{L}_{v+1v}\| \cdot \|\boldsymbol{P}_{\mathrm{pz1}}^{\mathcal{N}_{\mathrm{S1}} \cup \{v\}}(\mathrm{j}\omega)\| \\
\left(1 - \|\boldsymbol{L}_{vv+1}\| \cdot \|\boldsymbol{F}_{\mathrm{zs}v+1}(\mathrm{j}\omega)\boldsymbol{L}_{v+1v}\| \cdot \|\boldsymbol{P}_{\mathrm{pq1}}^{\mathcal{N}_{\mathrm{S1}} \cup \{v\}}(\mathrm{j}\omega)\|\right)^{-1}.
\end{aligned} \tag{B.19}
$$

Inserting the equations (C.6)–(C.8) into (B.19) results in

$$
\begin{aligned}
\Delta_1^{v+1}(\omega) =& \|\hat{\boldsymbol{P}}_{\mathrm{sq1}}^{\mathcal{N}_{\mathrm{S1}}} \boldsymbol{\Psi}_{\mathrm{s}}^{\mathcal{N}_{\mathrm{S1}}} \boldsymbol{L}_{v-1v} \boldsymbol{F}_{\mathrm{zs}v}\| \cdot \|\boldsymbol{L}_{vv+1}\| \cdot \|\boldsymbol{F}_{\mathrm{zs}v+1} \boldsymbol{L}_{v+1v}\| \cdot \|\boldsymbol{\Psi}_{\mathrm{p}}^{\mathcal{N}_{\mathrm{S1}}} \boldsymbol{F}_{\mathrm{zs}v} \boldsymbol{L}_{vv-1} \cdot \hat{\boldsymbol{P}}_{\mathrm{pz1}}^{\mathcal{N}_{\mathrm{S1}}}\| \\
&\left(1 - \|\boldsymbol{L}_{vv+1}\| \cdot \|\boldsymbol{F}_{\mathrm{zs}v+1} \boldsymbol{L}_{v+1v}\| \cdot \|\boldsymbol{\Psi}_{\mathrm{p}}^{\mathcal{N}_{\mathrm{S1}}} \boldsymbol{F}_{\mathrm{zs}v}\|\right)^{-1}.
\end{aligned} \tag{B.20}
$$

with $\boldsymbol{\Psi}_{\mathrm{s}}^{\mathcal{N}_{\mathrm{S1}}}(\mathrm{j}\omega)$ and $\boldsymbol{\Psi}_{\mathrm{p}}^{\mathcal{N}_{\mathrm{S1}}}(\mathrm{j}\omega)$ from (C.10) and (C.11). For the sake of clarity, the argument $\mathrm{j}\omega$ is omitted here. In the next step, the elements of eqn. (B.20) are rearranged in order to obtain the upper bound

$$
\begin{aligned}
\Delta_1^{v+1}(\omega) \leq& \|\boldsymbol{L}_{vv+1}\| \cdot \|\boldsymbol{F}_{\mathrm{zs}v+1} \boldsymbol{L}_{v+1v}\| \cdot \|\boldsymbol{F}_{\mathrm{zs}v}\| \bar{\Psi}^{\mathcal{N}_{\mathrm{S1}}} \left(1 - \|\boldsymbol{L}_{vv+1}\| \cdot \|\boldsymbol{F}_{\mathrm{zs}v+1} \boldsymbol{L}_{v+1v}\| \cdot \|\boldsymbol{F}_{\mathrm{zs}v}\| \bar{\Psi}^{\mathcal{N}_{\mathrm{S1}}}\right)^{-1} \\
&\underbrace{\|\hat{\boldsymbol{P}}_{\mathrm{sq1}}^{\mathcal{N}_{\mathrm{S1}}}\| \cdot \|\boldsymbol{L}_{v-1v}\| \cdot \|\boldsymbol{F}_{\mathrm{zs}v} \boldsymbol{L}_{vv-1}\| \cdot \|\hat{\boldsymbol{P}}_{\mathrm{pz1}}^{\mathcal{N}_{\mathrm{S1}}}\| \cdot \bar{\Psi}^{\mathcal{N}_{\mathrm{S1}}}}_{\geq \Delta_1^{v}(\omega)}. \\
\leq& \Gamma^{v+1}(\omega) \cdot \Delta_1^{v}(\omega)
\end{aligned}
$$

with $\bar{\Psi}^{\mathcal{N}_{\mathrm{S1}}}$ from (B.17) where $\bar{\Psi}^{\mathcal{N}_{\mathrm{S1}}}(\omega) \geq \|\boldsymbol{\Psi}_{\mathrm{s}}^{\mathcal{N}_{\mathrm{S1}}}(\mathrm{j}\omega)\|$ and $\bar{\Psi}^{\mathcal{N}_{\mathrm{S1}}}(\omega) \geq \|\boldsymbol{\Psi}_{\mathrm{p}}^{\mathcal{N}_{\mathrm{S1}}}(\mathrm{j}\omega)\|$ as well as the substitution

$$\Gamma^{v+1}(\omega) \leq \|\boldsymbol{L}_{vv+1}\| \cdot \|\boldsymbol{F}_{\mathrm{z}sv+1}\boldsymbol{L}_{v+1v}\| \cdot \|\boldsymbol{F}_{\mathrm{z}sv}\| \bar{\Psi}^{\mathcal{N}_{\mathrm{S1}}} \left(1 - \|\boldsymbol{L}_{vv+1}\| \cdot \|\boldsymbol{F}_{\mathrm{z}sv+1}\boldsymbol{L}_{v+1v}\| \cdot \|\boldsymbol{F}_{\mathrm{z}sv}\| \bar{\Psi}^{\mathcal{N}_{\mathrm{S1}}}\right)^{-1}$$

Hence, to show the relation (7.12) it is proved that

$$\Delta_1^{v+1}(\omega) \leq \Delta_1^{v}(\omega)\Gamma^{v+1}(\omega) < \Delta_1^{v}(\omega) \tag{B.21}$$

which is done by showing that

$$\Gamma^{v+1}(\omega) < 1 \tag{B.22}$$

for arbitrary $v \geq 2$. Since the relations (7.4) from Lemma 7.1 hold (cf. assumption A 7.1), the following relations hold

$$\begin{aligned}\Gamma^{v+1}(\omega) &< 1/4 \cdot \bar{\boldsymbol{\Psi}}^{\mathcal{N}_{\mathrm{S1}}} \left(1 - 1/4 \cdot \bar{\boldsymbol{\Psi}}^{\mathcal{N}_{\mathrm{S1}}}\right)^{-1}. \\ &< 1\end{aligned}$$

due to the fact that $\bar{\Psi}^{\mathcal{N}_{\mathrm{S1}}}(\omega) < 2$, for all $\omega \in \mathbb{R}$ (cf. eqn. (B.18)).

As the inequality (B.22) and, hence, (B.21) holds, it has been proved that (7.12) is satisfied for arbitrary $v \geq 2$, which completes the proof.

B.10 Proof of Lemma 7.2

To show that the system (6.4) represents a comparison system of the physical interaction (7.1), the same steps of the proof of Lemma 6.1 has to be performed with the little change of using $F^{\mathcal{N}_{\mathrm{S1}}}$ instead of F_i, $(\forall i \in \mathcal{N}_{\mathrm{S1}})$.

In the following, it is proved that (7.28) is a comparison system of the irrelevant part $E_i^{\mathcal{N}_{\mathrm{W1}}}$. Since, $\boldsymbol{E}_i^{\mathcal{N}_{\mathrm{W1}}}(s) = \boldsymbol{E}_1^{\mathcal{N}_{\mathrm{W1}}}(s)$ for all $i \in \mathcal{N}_{\mathrm{S1}}$, it is focused on the case, where $i = 1$. Thus, the proof aims at showing the relation

$$\|\boldsymbol{E}_1^{\mathcal{N}_{\mathrm{W1}}}(\mathrm{j}\omega)\| < \bar{E}_1^{\{1\}}(\omega) = \|\boldsymbol{L}_{v-1v}\| \cdot \varphi(\omega). \tag{B.23}$$

The dynamics $\boldsymbol{E}_1^{\mathcal{N}_{\mathrm{W1}}}(s)$ of the uncertainty model $E_1^{\mathcal{N}_{\mathrm{W1}}}$ are given by (C.12)

$$\begin{aligned}\boldsymbol{E}_1^{\mathcal{N}_{\mathrm{W1}}}(s) = &\boldsymbol{L}_{\{v-1\},\mathcal{N}_{\mathrm{W1}}} \left(\mathbf{I} - \mathrm{diag}(\boldsymbol{F}_{\mathrm{z}si}(s))_{i\in\mathcal{N}_{\mathrm{W1}}} \boldsymbol{L}_{\mathcal{N}_{\mathrm{W1}}}\right)^{-1} \\ &\mathrm{diag}(\boldsymbol{F}_{\mathrm{z}si}(s))_{i\in\mathcal{N}_{\mathrm{W1}}} \boldsymbol{L}_{\mathcal{N}_{\mathrm{W1}},\{v-1\}}\end{aligned}$$

with the upper bound

$$\|\boldsymbol{E}_1^{\mathcal{N}_{\mathrm{W1}}}(\mathrm{j}\omega)\| \leq \|\boldsymbol{L}_{\{v-1\},\mathcal{N}_{\mathrm{W1}}}\| \left(\mathbf{I} - \|\mathrm{diag}(\boldsymbol{F}_{\mathrm{zs}i}(\mathrm{j}\omega))_{i\in\mathcal{N}_{\mathrm{W1}}} \cdot \boldsymbol{L}_{\mathcal{N}_{\mathrm{W1}}}\|\right)^{-1} \\ \|\mathrm{diag}(\boldsymbol{F}_{\mathrm{zs}i}(\mathrm{j}\omega))_{i\in\mathcal{N}_{\mathrm{W1}}}\, \boldsymbol{L}_{\mathcal{N}_{\mathrm{W1}},\{v-1\}}\|. \tag{B.24}$$

Now, the relations (7.27) are rearranged and lumped together

$$\|\mathrm{diag}(\boldsymbol{F}_{\mathrm{zs}i}(\mathrm{j}\omega))_{i\in\mathcal{N}_{\mathrm{W1}}}\|\cdot\|\boldsymbol{L}_{\mathcal{N}_{\mathrm{W1}},\{v-1\}}\| < \Big(\varphi(\omega)\mathbf{I} - \|\mathrm{diag}(\boldsymbol{F}_{\mathrm{zs}i}(\mathrm{j}\omega))_{i\in\mathcal{N}_{\mathrm{W1}}}\|\cdot\|\boldsymbol{L}_{\mathcal{N}_{\mathrm{W1}}}\|\Big)\cdot\mathbf{1}, \tag{B.25}$$

where the norm $\|\mathrm{diag}(\boldsymbol{F}_{\mathrm{zs}i}(\mathrm{j}\omega))_{i\in\mathcal{N}_{\mathrm{W1}}} \cdot \boldsymbol{L}_{\mathcal{N}_{\mathrm{W1}}}\|$ applies block-element-wise according to (B.7). If relation (B.25) holds, then the relation

$$\|\mathrm{diag}(\boldsymbol{F}_{\mathrm{zs}i}(\mathrm{j}\omega))_{i\in\mathcal{N}_{\mathrm{W1}}}\| \cdot \|\boldsymbol{L}_{\mathcal{N}_{\mathrm{W1}},\{v-1\}}\| < \Big(\mathbf{I} - \|\mathrm{diag}(\boldsymbol{F}_{\mathrm{zs}i}(\mathrm{j}\omega))_{i\in\mathcal{N}_{\mathrm{W1}}}\| \cdot \|\boldsymbol{L}^{\mathcal{N}_{\mathrm{W1}}}\|\Big)\cdot\varphi(\omega)\mathbf{1} \tag{B.26}$$

is satisfied. Inserting the relation (B.26) into (B.24) yields

$$\|\boldsymbol{E}_1^{\mathcal{N}_{\mathrm{W1}}}(\mathrm{j}\omega)\| < \|\boldsymbol{L}_{\{v-1\},\mathcal{N}_{\mathrm{W1}}}\| \cdot \varphi(\omega)\mathbf{1} = \|\boldsymbol{L}_{v-1v}\| \cdot \varphi(\omega) = \bar{E}_1^{\mathcal{N}_{\mathrm{W1}}}(\omega).$$

which proof the relation (B.23).

B.11 Proof of Theorem 7.3

The proof of the relation is done in two steps. First, the upper bound $\bar{P}_i^{\mathcal{N}_{\mathrm{S1}}}(\omega)$ is reformulated. As the $\bar{P}_i^{\{i\}}(\omega)$ results if the local conditions (6.3) are satisfied for all $i \in \mathcal{N}_{\mathrm{S1}}$, in the second step, these local conditions are also reformulated and are inserted into $\bar{P}_i^{\mathcal{N}_{\mathrm{S1}}}(\omega)$.

First, $\bar{P}_i^{\mathcal{N}_{\mathrm{S1}}}(\omega)$ is reformulated by means of (C.1)–(C.4) from Appendix C.1 as follows:

$$\begin{aligned}
\bar{P}_i^{\mathcal{N}_{\mathrm{S1}}}(\omega) &= \|\hat{\boldsymbol{P}}_i^{\mathcal{N}_{\mathrm{S1}}}(\mathrm{j}\omega)\| + \bar{P}_{\Delta i}^{\mathcal{N}_{\mathrm{S1}}}(\omega) \\
&= \|\hat{\boldsymbol{P}}_i^{\mathcal{N}_{\mathrm{S1}}}(\mathrm{j}\omega)\| + \|\boldsymbol{P}_{\mathrm{sq}i}^{\mathcal{N}_{\mathrm{S1}}}(\mathrm{j}\omega)\| \cdot \bar{E}_1^{\mathcal{N}_{\mathrm{W1}}}(\omega) \\
&\quad \left(1 - \|\boldsymbol{P}_{\mathrm{pq}i}^{\mathcal{N}_{\mathrm{S1}}}(\mathrm{j}\omega)\|\cdot\bar{E}_1^{\mathcal{N}_{\mathrm{W1}}}(\omega)\right)^{-1} \|\boldsymbol{P}_{\mathrm{pz}i}^{\mathcal{N}_{\mathrm{S1}}}(\mathrm{j}\omega)\| \\
&\leq \|\boldsymbol{L}_{\{i\},\mathcal{N}_{\mathrm{S1}}\setminus\{i\}}\| \cdot \bar{\boldsymbol{\Gamma}}_i(\omega) \cdot \|\boldsymbol{L}_{\mathcal{N}_{\mathrm{S1}}\setminus\{i\},\{i\}}\| + \|\boldsymbol{L}_{\{i\},\mathcal{N}_{\mathrm{S1}}\setminus\{i\}}\| \cdot \bar{\boldsymbol{\Gamma}}_i(\omega) \cdot \boldsymbol{t} \cdot \|\boldsymbol{L}_{v-1v}\|\varphi(\omega) \\
&\quad \left(1 - \boldsymbol{t}^\top \bar{\boldsymbol{\Gamma}}_i(\omega)\, \boldsymbol{t}\cdot\|\boldsymbol{L}_{v-1v}\|\varphi(\omega)\right)^{-1} \boldsymbol{t}^\top \bar{\boldsymbol{\Gamma}}_i(\omega) \cdot \|\boldsymbol{L}_{\mathcal{N}_{\mathrm{S1}}\setminus\{i\},i}\| \\
&= \|\boldsymbol{L}_{\{i\},\mathcal{N}_{\mathrm{S1}}\setminus\{i\}}\| \cdot \bar{\boldsymbol{\Gamma}}_i(\omega) \cdot \|\boldsymbol{L}_{\mathcal{N}_{\mathrm{S1}}\setminus\{i\},\{i\}}\| + \|\boldsymbol{L}_{\{i\},\mathcal{N}_{\mathrm{S1}}\setminus\{i\}}\| \cdot \bar{\boldsymbol{\Gamma}}_i(\omega) \cdot \|\boldsymbol{L}_{\mathcal{N}_{\mathrm{S1}}\setminus\{i\},\{v\}}\|\varphi(\omega) \\
&\quad \left(1 - \boldsymbol{t}^\top \bar{\boldsymbol{\Gamma}}_i(\omega) \cdot \|\boldsymbol{L}_{\mathcal{N}_{\mathrm{S1}}\setminus\{i\},\{v\}}\|\varphi(\omega)\right)^{-1} \boldsymbol{t}^\top \bar{\boldsymbol{\Gamma}}_i(\omega) \cdot \|\boldsymbol{L}_{\mathcal{N}_{\mathrm{S1}}\setminus\{i\},i}\|,
\end{aligned} \tag{B.27}$$

where

$$\begin{aligned} \boldsymbol{t}^\top &= \begin{pmatrix} \mathbf{0}^\top & 1 \end{pmatrix} \\ \bar{\boldsymbol{\Gamma}}_i(\omega) &= \left(\mathbf{I} - \|\mathrm{diag}(\boldsymbol{F}_{\mathrm{zs}i}(\mathrm{j}\omega))_{i\in\mathcal{N}_{\mathrm{S1}}\setminus\{i\}}\| \cdot \|\boldsymbol{L}_{\mathcal{N}_{\mathrm{S1}}\setminus\{i\}}\|\right)^{-1} \mathrm{diag}(\|\boldsymbol{F}_{\mathrm{zs}i}(\mathrm{j}\omega)\|)_{i\in\mathcal{N}_{\mathrm{S1}}\setminus\{i\}} \end{aligned} \quad \text{(B.28)}$$

and $\bar{E}_1^{\mathcal{N}_{\mathrm{W1}}}(\omega) = \|\boldsymbol{L}_{v-1v}\|\varphi(\omega)$ from (7.28). The inverse of (B.27) exists(cf. proof of Proposition 7.2). Moreover, $\|\cdot\|$ applies block-element-wise for the matrices $\boldsymbol{L}_{\{i\},\mathcal{N}_{\mathrm{S1}}\setminus\{i\}}$, $\boldsymbol{L}_{\mathcal{N}_{\mathrm{S1}}\setminus\{i\},\{i\}}$, $\boldsymbol{L}_{\mathcal{N}_{\mathrm{S1}}\setminus\{i\},\{v\}}$, $\boldsymbol{L}_{\mathcal{N}_{\mathrm{S1}}\setminus\{i\}}$ and $\mathrm{diag}(\boldsymbol{F}_{\mathrm{zs}i}(\mathrm{j}\omega))_{i\in\mathcal{N}_{\mathrm{S1}}\setminus\{i\}}$ according to (B.7).

In the second step of the proof, it is focused on the situation in which the local conditions

$$\|\boldsymbol{F}_{\mathrm{zs}i}(\mathrm{j}\omega)\| \sum_{j\in\mathcal{P}_i} \|\boldsymbol{L}_{ij}\| < \varphi(\omega), \quad \forall\omega\in\mathbb{R}, \quad \forall i \in \mathcal{N}_{\mathrm{S1}} \quad \text{(B.29)}$$

are satisfied. Then $\bar{P}_i^{\{i\}}(\omega) = \sum_{j\in\mathcal{P}_i} \|\boldsymbol{L}_{ij}\| \cdot \varphi(\omega)$ results. Lumping these local conditions (B.29) for all $i \in \mathcal{N}_{\mathrm{S1}} \setminus \{1\}$ together, the equivalent expression

$$\begin{aligned} &\mathrm{diag}(\|\boldsymbol{F}_{\mathrm{zs}i}(\mathrm{j}\omega)\|)_{i\in\mathcal{N}_{\mathrm{S1}}\setminus\{i\}}\left(\|\boldsymbol{L}_{\mathcal{N}_{\mathrm{S1}}\setminus\{i\},\{v\}}\| + \|\boldsymbol{L}_{\mathcal{N}_{\mathrm{S1}}\setminus\{i\},\{i\}}\|\right) \\ &\qquad < \varphi(\omega)\mathbf{1} - \mathrm{diag}(\|\boldsymbol{F}_{\mathrm{zs}i}(\mathrm{j}\omega)\|)_{i\in\mathcal{N}_{\mathrm{S1}}\setminus\{i\}} \cdot \|\boldsymbol{L}_{\mathcal{N}_{\mathrm{S1}}\setminus\{i\}}\| \\ &\qquad = \left(\mathbf{I}\varphi(\omega) - \mathrm{diag}(\|\boldsymbol{F}_{\mathrm{zs}i}(\mathrm{j}\omega)\|)_{i\in\mathcal{N}_{\mathrm{S1}}\setminus\{i\}} \cdot \|\boldsymbol{L}_{\mathcal{N}_{\mathrm{S1}}\setminus\{i\}}\|\right) \cdot \mathbf{1} \end{aligned} \quad \text{(B.30)}$$

results. The satisfaction of the relation (B.30) implies the satisfaction of the relation

$$\begin{aligned} &\mathrm{diag}(\|\boldsymbol{F}_{\mathrm{zs}i}(\mathrm{j}\omega)\|)_{i\in\mathcal{N}_{\mathrm{S1}}\setminus\{i\}}\left(\|\boldsymbol{L}_{\mathcal{N}_{\mathrm{S1}}\setminus\{i\},\{v\}}\| + \|\boldsymbol{L}_{\mathcal{N}_{\mathrm{S1}}\setminus\{i\},\{i\}}\|\right) \\ &\qquad < \varphi(\omega)\mathbf{1} - \mathrm{diag}(\|\boldsymbol{F}_{\mathrm{zs}i}(\mathrm{j}\omega)\|)_{i\in\mathcal{N}_{\mathrm{S1}}\setminus\{i\}} \cdot \|\boldsymbol{L}_{\mathcal{N}_{\mathrm{S1}}\setminus\{i\}}\| \\ &\qquad = \left(\mathbf{I} - \mathrm{diag}(\|\boldsymbol{F}_{\mathrm{zs}i}(\mathrm{j}\omega)\|)_{i\in\mathcal{N}_{\mathrm{S1}}\setminus\{i\}} \cdot \|\boldsymbol{L}_{\mathcal{N}_{\mathrm{S1}}\setminus\{i\}}\|\right) \cdot \varphi(\omega)\mathbf{1} \end{aligned}$$

since $0 < \varphi(\omega) \leq 1$. This relation can be rewritten using (B.28) into the equivalent expression

$$\bar{\boldsymbol{\Gamma}}_i(\mathrm{j}\omega)\ \|\boldsymbol{L}_{\mathcal{N}_{\mathrm{S1}}\setminus\{i\},v}\| + \bar{\boldsymbol{\Gamma}}_i(\mathrm{j}\omega)\ \|\boldsymbol{L}_{\mathcal{N}_{\mathrm{S1}}\setminus\{i\},i}\| < \varphi(\omega)\mathbf{1}. \quad \text{(B.31)}$$

Similarly, the inequality (B.31) is rearranged once more under consideration of $0 < \varphi(\omega) \leq 1$ into

$$\boldsymbol{t}^\top \bar{\boldsymbol{\Gamma}}_i(\mathrm{j}\omega) \cdot \|\boldsymbol{L}_{\mathcal{N}_{\mathrm{S1}}\setminus\{i\},\{i\}}\| < \boldsymbol{t}^\top\left(\varphi(\omega)\mathbf{1} - \bar{\boldsymbol{\Gamma}}_i(\mathrm{j}\omega) \cdot \|\boldsymbol{L}_{\mathcal{N}_{\mathrm{S1}}\setminus\{i\},\{v\}}\|\right). \quad \text{(B.32)}$$

If (B.32) holds, then

$$\boldsymbol{t}^\top \bar{\boldsymbol{\Gamma}}_i(\mathrm{j}\omega) \cdot \|\boldsymbol{L}_{\mathcal{N}_{\mathrm{S1}}\setminus\{i\},\{i\}}\| < \boldsymbol{t}^\top\left(\mathbf{1} - \bar{\boldsymbol{\Gamma}}_i(\mathrm{j}\omega) \cdot \|\boldsymbol{L}_{\mathcal{N}_{\mathrm{S1}}\setminus\{i\},\{v\}}\| \cdot \varphi(\omega)\right)$$

$$= 1 - \boldsymbol{t}^\top \bar{\boldsymbol{\Gamma}}_i(\mathrm{j}\omega) \cdot \|\boldsymbol{L}_{\mathcal{N}_{\mathrm{S1}}\setminus\{i\},\{v\}}\| \cdot \varphi(\omega) \tag{B.33}$$

is satisfied too. The last step of the proof is to insert (B.33) in (B.27) and using (B.31) to conclude that

$$\begin{aligned}\bar{P}_i^{\mathcal{N}_{\mathrm{S1}}}(\omega) &< \|\boldsymbol{L}_{i,\mathcal{N}_{\mathrm{S1}}\setminus\{i\}}\| \left(\bar{\boldsymbol{\Gamma}}_i(\mathrm{j}\omega)\ \|\boldsymbol{L}_{\mathcal{N}_{\mathrm{S1}}\setminus\{i\},\{i\}}\| + \bar{\boldsymbol{\Gamma}}_i(\mathrm{j}\omega)\ \|\boldsymbol{L}_{\mathcal{N}_{\mathrm{S1}}\setminus\{i\},v}\|\right)\\ &< \|\boldsymbol{L}_{i,\mathcal{N}_{\mathrm{S1}}\setminus\{i\}}\| \cdot \varphi(\omega)\boldsymbol{1}\\ &= \sum_{j\in\mathcal{P}_i} \|\boldsymbol{L}_{ij}\|\varphi(\omega) = \bar{P}_i^{\{i\}}(\omega),\end{aligned}$$

which proves the relation (7.30). That means that if the local condition (6.3) is satisfied for all controlled subsystems F_i, $(\forall i \in \mathcal{N})$, then the resulting upper bound $\bar{P}_i^{\{i\}}(\omega)$ is strictly less than the upper bound $\bar{P}_i^{\mathcal{N}_{\mathrm{S1}}}(\omega)$ that results if the local condition (6.3) is satisfied for the irrelevant controlled subsystems F_i, $(\forall i \in \mathcal{N}_{\mathrm{W1}})$ only.

B.12 Proof of Proposition 7.2

The proof shows the existence of the upper bound $\bar{P}_1^{\mathcal{N}_{\mathrm{S1}}}$. In direct analogy, the existence of the upper bounds $\bar{P}_i^{\mathcal{N}_{\mathrm{S1}}}$, $(\forall i \in \mathcal{N}_{\mathrm{S1}} \setminus \{1\})$ can be shown Let the conditions (7.26) and (7.27) be satisfied. Then the upper bound $\bar{P}_1^{\mathcal{N}_{\mathrm{S1}}}(\omega)$ reads

$$\begin{aligned}\bar{P}_1^{\mathcal{N}_{\mathrm{S1}}}(\omega) = \|\hat{\boldsymbol{P}}_1^{\mathcal{N}_{\mathrm{S1}}}(\mathrm{j}\omega)\| + \|\boldsymbol{P}_{\mathrm{sq1}}^{\mathcal{N}_{\mathrm{S1}}}(\mathrm{j}\omega)\| \cdot \|\boldsymbol{L}_{v-1v}\|\varphi(\omega)\|\boldsymbol{P}_{\mathrm{pz1}}^{\mathcal{N}_{\mathrm{S1}}}(\mathrm{j}\omega)\|\\ \left(1 - \|\boldsymbol{P}_{\mathrm{pq1}}^{\mathcal{N}_{\mathrm{S1}}}(\mathrm{j}\omega)\| \cdot \|\boldsymbol{L}_{v-1v}\|\varphi(\omega)\right)^{-1}.\end{aligned} \tag{B.34}$$

From Lemma 7.1 follows that the amplitude plots $\|\hat{\boldsymbol{P}}_1^{\mathcal{N}_{\mathrm{S1}}}(\mathrm{j}\omega)\|$, $\|\boldsymbol{P}_{\mathrm{sq1}}^{\mathcal{N}_{\mathrm{S1}}}(\mathrm{j}\omega)\|$, $\|\boldsymbol{P}_{\mathrm{pz1}}^{\mathcal{N}_{\mathrm{S1}}}(\mathrm{j}\omega)\|$ and $\|\boldsymbol{P}_{\mathrm{pq1}}^{\mathcal{N}_{\mathrm{S1}}}(\mathrm{j}\omega)\|$ are bounded. Hence, the upper bound (B.34) exists if the inverse exists, i.e,

$$\|\boldsymbol{P}_{\mathrm{pq1}}^{\mathcal{N}_{\mathrm{S1}}}(\mathrm{j}\omega)\| \cdot \|\boldsymbol{L}_{v-1v}\|\varphi(\omega) < 1. \tag{B.35}$$

Due to assumption A 7.1, an upper bound on $\|\boldsymbol{P}_{\mathrm{pq}i}^{\mathcal{N}_{\mathrm{S1}}}(\mathrm{j}\omega)\|$ results to

$$\|\boldsymbol{P}_{\mathrm{pq1}}^{\mathcal{N}_{\mathrm{S1}}}(\mathrm{j}\omega)\| < \|\boldsymbol{F}_{\mathrm{pq}}^{\mathcal{N}_{\mathrm{S1}}}(\mathrm{j}\omega)\| \tag{B.36}$$

which can easily be comprehended by the comparison of the recursive calculations of $\boldsymbol{P}_{\mathrm{pq1}}^{\mathcal{N}_{\mathrm{S1}}}(\mathrm{j}\omega)$ according to (C.8) and $\boldsymbol{F}_{\mathrm{pq}}^{\mathcal{N}_{\mathrm{S1}}}(\mathrm{j}\omega)$ according to (C.17) under consideration of the relation (7.4). The multiplication of both sides of (B.36) with $\|\boldsymbol{L}_{v-1v}\|\varphi(\omega)$ results to

$$\|\boldsymbol{P}_{\mathrm{pq1}}^{\mathcal{N}_{\mathrm{S1}}}(\mathrm{j}\omega)\| \cdot \|\boldsymbol{L}_{v-1v}\|\varphi(\omega) < \|\boldsymbol{F}_{\mathrm{pq1}}^{\mathcal{N}_{\mathrm{S1}}}(\mathrm{j}\omega)\| \cdot \|\boldsymbol{L}_{v-1v}\|\varphi(\omega).$$

Finally, the relation $\|\boldsymbol{F}_{\mathrm{pq}}^{\mathcal{N}_{\mathrm{S1}}}(\mathrm{j}\omega)\| \cdot \|\boldsymbol{L}_{v-1v}\|\varphi(\omega) < 1$ holds in accordance with the claim (7.26) and due to the fact that $0 < \varphi(\omega) \leq 1$. Hence, the relation (B.35) has been proved which completes the proof. The recursive calculation of $\boldsymbol{P}_{\mathrm{pq}i}^{\mathcal{N}_{\mathrm{S1}}}(s)$, $(\forall i \in \mathcal{N}_{\mathrm{S1}} \setminus \{1\})$ is in accordance with (C.8), where the index 1 has to be exchanged by i.

B.13 Proof of Corollary 7.5

From Corollary 7.1 can be concluded that the reconfigured closed-loop system is I/O stable if $\mathring{F}^{\mathcal{N}_{\mathrm{S1}}}$ satisfies the cooperative stability condition $\mathcal{A}^{\mathcal{N}_{\mathrm{S1}}}$ in accordance with A 7.3. In the following, it is shown that the cooperative stability condition $\mathcal{A}^{\mathcal{N}_{\mathrm{S1}}}$ are fulfilled if the LMI (7.43) has a solution.

The system $\mathring{F}^{\mathcal{N}_{\mathrm{S1}}}$ results from the combination

$$\mathring{F}^{\mathcal{N}_{\mathrm{S1}}} = \mathrm{con}\left(\{S_{1f}, VA_1, C_1, K_1\} \cup \{F_i, K_i,\ i \in \mathcal{N}_{\mathrm{S1}} \setminus \{1\}\},\ \begin{pmatrix} \boldsymbol{w}^{\mathcal{N}_{\mathrm{S1}}} \\ \boldsymbol{q}_1 \end{pmatrix},\ \begin{pmatrix} \mathring{\boldsymbol{y}}^{\mathcal{N}_{\mathrm{S1}}} \\ \boldsymbol{p}_1 \end{pmatrix}\right),$$

where

$$\mathring{\boldsymbol{y}}^{\mathcal{N}_{\mathrm{S1}}} = \begin{pmatrix} \mathring{\boldsymbol{y}}_1^\top & \boldsymbol{y}_2^\top & \cdots & \boldsymbol{y}_{v-1}^\top \end{pmatrix}^\top$$

and is represented by the state space model

$$\mathring{F}^{\mathcal{N}_{\mathrm{S1}}}: \begin{cases} \begin{pmatrix} \dot{\tilde{\boldsymbol{x}}}_1(t) \\ \dot{\boldsymbol{x}}_{\delta 1}(t) \end{pmatrix} = \begin{pmatrix} \boldsymbol{A}_{\mathrm{F}}^{\mathcal{N}_{\mathrm{S1}}} & \mathrm{O} \\ \begin{pmatrix} -\boldsymbol{B}_1^{\mathcal{N}_{\mathrm{S1}}}\boldsymbol{D}_{\mathrm{C1}}\boldsymbol{C}_1^{\mathcal{N}_{\mathrm{S1}}} & \hat{\boldsymbol{B}}_1^{\mathcal{N}_{\mathrm{S1}}}\boldsymbol{C}_{\mathrm{C1}} \end{pmatrix} & \boldsymbol{A}_{\delta 1} \end{pmatrix} \begin{pmatrix} \tilde{\boldsymbol{x}}_1(t) \\ \boldsymbol{x}_{\delta 1}(t) \end{pmatrix} \\ \qquad + \begin{pmatrix} \boldsymbol{B}_{\mathrm{F}}^{\mathcal{N}_{\mathrm{S1}}} \\ \begin{pmatrix} \boldsymbol{B}_1^{\mathcal{N}_{\mathrm{S1}}}\boldsymbol{D}_{\mathrm{C1}} & \mathrm{O} \end{pmatrix} \end{pmatrix} \boldsymbol{w}^{\mathcal{N}_{\mathrm{S1}}}(t) + \begin{pmatrix} \boldsymbol{E}_{\mathrm{F1}}^{\mathcal{N}_{\mathrm{S1}}} \\ \mathrm{O} \end{pmatrix} \boldsymbol{q}_1(t), \\ \begin{pmatrix} \tilde{\boldsymbol{x}}_1(0) \\ \boldsymbol{x}_{\delta 1}(0) \end{pmatrix} = \begin{pmatrix} \boldsymbol{0} \\ \boldsymbol{0} \end{pmatrix} \\ \mathring{\boldsymbol{y}}^{\mathcal{N}_{\mathrm{S1}}} = \begin{pmatrix} \boldsymbol{C}_{\mathrm{F}}^{\mathcal{N}_{\mathrm{S1}}} & \mathrm{O} \end{pmatrix} \begin{pmatrix} \tilde{\boldsymbol{x}}_1(t) \\ \boldsymbol{x}_{\delta 1}(t) \end{pmatrix} \\ \boldsymbol{p}_1(t) = \begin{pmatrix} \boldsymbol{C}_{\mathrm{Fp1}}^{\mathcal{N}_{\mathrm{S1}}} & -\boldsymbol{C}_{\mathrm{p1}}^{\mathcal{N}_{\mathrm{S1}}} \end{pmatrix} \begin{pmatrix} \tilde{\boldsymbol{x}}_1(t) \\ \boldsymbol{x}_{\delta 1}(t) \end{pmatrix}, \end{cases}$$

where

$$\tilde{\boldsymbol{x}}_1(t) = \begin{pmatrix} \boldsymbol{x}_{1f}^{\mathcal{N}_{\mathrm{S1}}}(t) + \boldsymbol{x}_{\delta 1}(t) \\ \boldsymbol{x}_{\mathrm{C1}}(t) \end{pmatrix}.$$

Due to the triangular form of the system matrix, its eigenvalues are the union of the eigenvalues of the virtual actuator VA_1 from (7.42) and the eigenvalues of the fault-free cooperatively controlled subsystem $F^{\mathcal{N}_{S1}}$. Since, $F^{\mathcal{N}_{S1}}$ is I/O stable according to assumption A 7.3, the reconfigured controlled subsystem $\mathring{F}^{\mathcal{N}_{S1}}$ is I/O stable if and only if $\boldsymbol{A}_{\delta 1}$ is Hurwitz. There exists always a feedback gain $\boldsymbol{M}_1$, only if the pair $\left(\boldsymbol{A}_1^{\mathcal{N}_{S1}}, \boldsymbol{B}_{1f}^{\mathcal{N}_{S1}}\right)$ is stabilisable.

Now it is shown that the cooperative stability condition $\mathcal{A}^{\mathcal{N}_{S1}}$ are satisfied, if the LMI (7.43) is feasible. The transfer function matrix $\mathring{\boldsymbol{F}}_{\mathrm{pq}}^{\mathcal{N}_{S1}}(s)$ yields

$$\begin{aligned}\mathring{\boldsymbol{F}}_{\mathrm{pq}}^{\mathcal{N}_{S1}}(s) &= \begin{pmatrix}\boldsymbol{C}_{\mathrm{Fp1}}^{\mathcal{N}_{S1}} & -\boldsymbol{C}_{\mathrm{p1}}^{\mathcal{N}_{S1}}\end{pmatrix}\begin{pmatrix} s\mathbf{I}-\boldsymbol{A}_{\mathrm{F}}^{\mathcal{N}_{S1}} & \mathrm{O} \\ \begin{pmatrix}-\boldsymbol{B}_1^{\mathcal{N}_{S1}}\boldsymbol{D}_{\mathrm{C1}}\boldsymbol{C}_1^{\mathcal{N}_{S1}} & \hat{\boldsymbol{B}}_1^{\mathcal{N}_{S1}}\boldsymbol{C}_{\mathrm{C1}}\end{pmatrix} & s\mathbf{I}-\boldsymbol{A}_{\delta 1}\end{pmatrix}^{-1}\begin{pmatrix}\boldsymbol{E}_{\mathrm{F1}}^{\mathcal{N}_{S1}} \\ \mathrm{O}\end{pmatrix} \\ &= \begin{pmatrix}\boldsymbol{C}_{\mathrm{Fp1}}^{\mathcal{N}_{S1}} & -\boldsymbol{C}_{\mathrm{p1}}^{\mathcal{N}_{S1}}\end{pmatrix}\begin{pmatrix}\left(s\mathbf{I}-\boldsymbol{A}_{\mathrm{F1}}^{\mathcal{N}_{S1}}\right)^{-1} & \mathrm{O} \\ \boldsymbol{\Gamma}(s) & \left(s\mathbf{I}-\boldsymbol{A}_{\delta 1}\right)^{-1}\end{pmatrix}\begin{pmatrix}\boldsymbol{E}_{\mathrm{F1}}^{\mathcal{N}_{S1}} \\ \mathrm{O}\end{pmatrix} \\ &= \boldsymbol{C}_{\mathrm{Fp1}}^{\mathcal{N}_{S1}}\left(s\mathbf{I}-\boldsymbol{A}_{\mathrm{F1}}^{\mathcal{N}_{S1}}\right)^{-1}\boldsymbol{E}_{\mathrm{F1}}^{\mathcal{N}_{S1}} - \boldsymbol{C}_{\mathrm{p1}}^{\mathcal{N}_{S1}}\boldsymbol{\Gamma}(s),\end{aligned} \tag{B.37}$$

where

$$\boldsymbol{\Gamma}(s) = \left(s\mathbf{I}-\boldsymbol{A}_{\delta 1}\right)^{-1}\begin{pmatrix}-\boldsymbol{B}_1^{\mathcal{N}_{S1}}\boldsymbol{D}_{\mathrm{C1}}\boldsymbol{C}_1^{\mathcal{N}_{S1}} & \hat{\boldsymbol{B}}_1^{\mathcal{N}_{S1}}\boldsymbol{C}_{\mathrm{C1}}\end{pmatrix}\left(s\mathbf{I}-\boldsymbol{A}_{\mathrm{F}}^{\mathcal{N}_{S1}}\right)^{-1}.$$

Eqn. (B.37) can be rewritten into the equivalent form

$$\mathring{\boldsymbol{F}}_{\mathrm{pq}}^{\mathcal{N}_{S1}}(s) = \boldsymbol{F}_{\mathrm{pq}}^{\mathcal{N}_{S1}}(s) + \boldsymbol{\Theta}_{\Delta}(s)\boldsymbol{F}_{\mathrm{uq1}}^{\mathcal{N}_{S1}}(s),$$

where

$$\begin{aligned}\boldsymbol{\Theta}_{\Delta}(s) &= \boldsymbol{C}_{\mathrm{p1}}^{\mathcal{N}_{S1}}\left(s\mathbf{I}-\boldsymbol{A}_{\delta 1}\right)^{-1}\boldsymbol{B}_1^{\mathcal{N}_{S1}} \\ \boldsymbol{F}_{\mathrm{uq1}}^{\mathcal{N}_{S1}}(s) &= \boldsymbol{D}_{\mathrm{C1}}\boldsymbol{C}_1^{\mathcal{N}_{S1}}\left(s\mathbf{I}-\boldsymbol{A}_{\mathrm{F}}^{\mathcal{N}_{S1}}\right)^{-1}\boldsymbol{E}_{\mathrm{F1}}^{\mathcal{N}_{S1}} - \boldsymbol{C}_{\mathrm{C1}}\left(s\mathbf{I}-\boldsymbol{A}_{\mathrm{F}}^{\mathcal{N}_{S1}}\right)^{-1}\boldsymbol{E}_{\mathrm{F1}}^{\mathcal{N}_{S1}}.\end{aligned}$$

Based on the relation

$$\|\mathring{\boldsymbol{F}}_{\mathrm{pq}}^{\mathcal{N}_{S1}}(\mathrm{j}\omega)\| \leq \|\boldsymbol{F}_{\mathrm{pq}}^{\mathcal{N}_{S1}}(\mathrm{j}\omega)\| + \|\boldsymbol{\Theta}_{\Delta}(\mathrm{j}\omega)\| \cdot \|\boldsymbol{F}_{\mathrm{uq1}}^{\mathcal{N}_{S1}}(\mathrm{j}\omega)\|$$

the claim (7.34b) is satisfied if

$$\|\boldsymbol{\Theta}_{\Delta}(\mathrm{j}\omega)\| \leq \frac{1-\max_{\omega}\|\boldsymbol{F}_{\mathrm{pq}}^{\mathcal{N}_{S1}}(\mathrm{j}\omega)\| \cdot \|\boldsymbol{L}_{v-1v}\|}{\max_{\omega}\|\boldsymbol{F}_{\mathrm{uq1}}^{\mathcal{N}_{S1}}(\mathrm{j}\omega)\| \cdot \|\boldsymbol{L}_{v-1v}\|} =: \gamma_{\mathrm{VA}}. \tag{B.38}$$

Applying the relation of Lemma 2.2 leads to the nonlinear matrix inequality

$$\begin{pmatrix} \boldsymbol{P}\boldsymbol{A}_{\delta 1} + \boldsymbol{A}_{\delta 1}^{\top}\boldsymbol{P} & \boldsymbol{P}\boldsymbol{B}_1^{\mathcal{N}_{\mathrm{S1}}} & \boldsymbol{C}_1^{\mathcal{N}_{\mathrm{S1}}\,\top} \\ \star & -\gamma_{\mathrm{VA}}\mathbf{I} & \mathrm{O} \\ \star & \star & -\gamma_{\mathrm{VA}}\mathbf{I} \end{pmatrix} \prec 0$$

which is linearised by the congruence transformation with $\boldsymbol{T} = \mathrm{diag}\,(\boldsymbol{X}, \mathrm{I}, \mathbf{I})$ where $\boldsymbol{X} = \boldsymbol{P}^{-1}$. The result is the LMI (7.43).

The matrix $\boldsymbol{A}_{\delta 1}$ is Hurwitz and, thus, $F^{\mathcal{N}_{\mathrm{S1}}}$ is I/O stable (cf. claim (7.34a)), if and only if the LMI (7.43) is feasible. Moreover, the feasibility of the LMI (7.43) is sufficient for the claim (7.34b) to be satisfied. As $\mathcal{A}^{\mathcal{N}_{\mathrm{S1}}}$ is satisfied, the overall closed-loop stability is recovered in accordance with assumption A 7.3, which completes the proof.

C Construction rules of transfer functions

C.1 Physical interaction, relevant part and irrelevant part

This section presents, first, the construction rule of the transfer function of the physical interaction $P_i^{\mathcal{N}}$ followed by the determination of the transfer function of the relevant part $P_i^{\mathcal{N}_{S1}}$ and the irrelevant part $E_i^{\mathcal{N}_{W1}}$.

The transfer function of the physical interaction

$$P_i^{\mathcal{N}}: \quad \boldsymbol{s}_i(s) = \boldsymbol{P}_i^{\mathcal{N}}(s)\, \boldsymbol{z}_i(s)$$

reads

$$\boldsymbol{P}_i^{\mathcal{N}}(s) = \boldsymbol{L}_{\{i\},\mathcal{N}\setminus\{i\}} \left(\mathbf{I} - \operatorname{diag}(\boldsymbol{F}_{zsj}(s))_{j\in\mathcal{N}\setminus\{i\}}\, \boldsymbol{L}_{\mathcal{N}\setminus\{i\}}\right)^{-1} \operatorname{diag}(\boldsymbol{F}_{zsj}(s))_{j\in\mathcal{N}\setminus\{i\}}\, \boldsymbol{L}_{\mathcal{N}\setminus\{i\},\{i\}}.$$

The physical interaction $P_1^{\mathcal{N}}$ is subdivided into the relevant part $P_1^{\mathcal{N}_{S1}}$ and the irrelevant part $E_1^{\mathcal{N}_{W1}}$, where $\mathcal{N}_{S1} = \{1, 2, ..., v-1\}$ and $\mathcal{N}_{W1} = \{v, ..., N\}$. The relevant part of the physical interaction is modelled by

$$P_i^{\mathcal{N}_{S1}}: \begin{cases} \boldsymbol{s}_i(s) &= \hat{\boldsymbol{P}}_i^{\mathcal{N}_{S1}}(s)\, \boldsymbol{z}_i(s) + \boldsymbol{P}_{\mathrm{sq}i}^{\mathcal{N}_{S1}}(s)\, \boldsymbol{q}_1(s) \\ \boldsymbol{p}_1(s) &= \boldsymbol{P}_{\mathrm{pz}i}^{\mathcal{N}_{S1}}(s)\, \boldsymbol{z}_i(s) + \boldsymbol{P}_{\mathrm{pq}i}^{\mathcal{N}_{S1}}(s)\, \boldsymbol{q}_1(s) \end{cases}$$

for all $i \in \mathcal{N}_{S1}$. The transfer functions yield

$$\hat{\boldsymbol{P}}_i^{\mathcal{N}_{S1}}(s) = \boldsymbol{L}_{\{i\},\mathcal{N}_{S1}\setminus\{i\}}\, \boldsymbol{\Gamma}_i(s)\, \boldsymbol{L}_{\mathcal{N}_{S1}\setminus\{i\},\{i\}}, \tag{C.1}$$

$$\boldsymbol{P}_{\mathrm{sq}i}^{\mathcal{N}_{S1}}(s) = \boldsymbol{L}_{\{i\},\mathcal{N}_{S1}\setminus\{i\}}\, \boldsymbol{\Gamma}_i(s)\, \boldsymbol{T}, \tag{C.2}$$

$$\boldsymbol{P}_{\mathrm{pz}i}^{\mathcal{N}_{S1}}(s) = \boldsymbol{T}^\top \boldsymbol{\Gamma}_i(s)\, \boldsymbol{L}_{\mathcal{N}_{S1}\setminus\{i\},\{i\}}, \tag{C.3}$$

$$\boldsymbol{P}_{\mathrm{pq}i}^{\mathcal{N}_{S1}}(s) = \boldsymbol{T}^\top \boldsymbol{\Gamma}_i(s)\, \boldsymbol{T}, \tag{C.4}$$

where

$$\begin{aligned}\boldsymbol{\Gamma}_i(s) &= \left(\mathbf{I} - \operatorname{diag}(\boldsymbol{F}_{\mathrm{zs}j}(s))_{j\in\mathcal{N}_{\mathrm{S1}}\setminus\{i\}}\, \boldsymbol{L}_{\mathcal{N}_{\mathrm{S1}}\setminus\{i\}}\right)^{-1} \operatorname{diag}(\boldsymbol{F}_{\mathrm{zs}j}(s))_{j\in\mathcal{N}_{\mathrm{S1}}\setminus\{i\}},\\ \boldsymbol{T}^\top &= \begin{pmatrix}\mathbf{O} & \mathbf{I}\end{pmatrix}.\end{aligned} \tag{C.5}$$

For the special case $\mathcal{N}_{\mathrm{S1}} = \{i\}$ the transfer functions (C.1)–(C.4) become

$$\hat{\boldsymbol{P}}_i^{\{i\}}(s) = \boldsymbol{P}_{\mathrm{pq}i}^{\{i\}}(s) = \mathbf{O}, \quad \boldsymbol{P}_{\mathrm{sq}i}^{\{i\}}(s) = \boldsymbol{P}_{\mathrm{pz}i}^{\{i\}}(s) = \mathbf{I}.$$

Alternatively, the transfer functions (C.1)–(C.4) can be constructed recursively. Therefore, let $i = 1$. The transfer functions result to

$$\boldsymbol{P}_{\mathrm{sq1}}^{\mathcal{N}_{\mathrm{S1}}\cup\{v\}}(s) = \boldsymbol{P}_{\mathrm{sq1}}^{\mathcal{N}_{\mathrm{S1}}}(s)\,\boldsymbol{\Psi}_{\mathrm{s}}^{\mathcal{N}_{\mathrm{S1}}}(s)\,\boldsymbol{L}_{v-1v}\boldsymbol{F}_{\mathrm{zs}v}(s) \tag{C.6}$$

$$\boldsymbol{P}_{\mathrm{pz1}}^{\mathcal{N}_{\mathrm{S1}}\cup\{v\}}(s) = \boldsymbol{\Psi}_{\mathrm{p}}^{\mathcal{N}_{\mathrm{S1}}}(s)\,\boldsymbol{F}_{\mathrm{zs}v}(s)\,\boldsymbol{L}_{vv-1}\,\boldsymbol{P}_{\mathrm{pz1}}^{\mathcal{N}_{\mathrm{S1}}}(s) \tag{C.7}$$

$$\boldsymbol{P}_{\mathrm{pq1}}^{\mathcal{N}_{\mathrm{S1}}\cup\{v\}}(s) = \boldsymbol{\Psi}_{\mathrm{p}}^{\mathcal{N}_{\mathrm{S1}}}(s)\boldsymbol{F}_{\mathrm{zs}v}(s) \tag{C.8}$$

$$\hat{\boldsymbol{P}}_1^{\mathcal{N}_{\mathrm{S1}}\cup\{v\}}(s) = \hat{\boldsymbol{P}}_1^{\mathcal{N}_{\mathrm{S1}}}(s) + \boldsymbol{P}_{\mathrm{sq1}}^{\mathcal{N}_{\mathrm{S1}}}(s)\,\boldsymbol{\Psi}_{\mathrm{s}}^{\mathcal{N}_{\mathrm{S1}}}(s)\,\boldsymbol{L}_{v-1v}\,\boldsymbol{F}_{\mathrm{zs}v}(s)\,\boldsymbol{L}_{vv-1}\,\boldsymbol{P}_{\mathrm{pz1}}^{\mathcal{N}_{\mathrm{S1}}}(s), \tag{C.9}$$

where

$$\boldsymbol{\Psi}_{\mathrm{s}}^{\mathcal{N}_{\mathrm{S1}}}(s) = \left(\mathbf{I} - \boldsymbol{L}_{v-1v}\,\boldsymbol{F}_{\mathrm{zs}v}(s)\,\boldsymbol{L}_{vv-1}\,\boldsymbol{P}_{\mathrm{pq1}}^{\mathcal{N}_{\mathrm{S1}}}(s)\right)^{-1}, \tag{C.10}$$

$$\boldsymbol{\Psi}_{\mathrm{p}}^{\mathcal{N}_{\mathrm{S1}}}(s) = \left(\mathbf{I} - \boldsymbol{F}_{\mathrm{zs}v}(s)\,\boldsymbol{L}_{vv-1}\,\boldsymbol{P}_{\mathrm{pq1}}^{\mathcal{N}_{\mathrm{S1}}}(s)\,\boldsymbol{L}_{v-1v}\right)^{-1}. \tag{C.11}$$

Now, it is focused on the irrelevant part, which is modelled by

$$E_1^{\mathcal{N}_{\mathrm{W1}}}: \; \boldsymbol{q}_1(s) = \boldsymbol{E}_1^{\mathcal{N}_{\mathrm{W1}}}(s)\,\boldsymbol{p}_1(s).$$

The transfer function $\boldsymbol{E}_1^{\mathcal{N}_{\mathrm{W1}}}(s)$ is constructed according to

$$\begin{aligned}\boldsymbol{E}_1^{\mathcal{N}_{\mathrm{W1}}}(s) =& \boldsymbol{L}_{\{v-1\},\mathcal{N}_{\mathrm{W1}}}\left(\mathbf{I} - \operatorname{diag}(\boldsymbol{F}_{\mathrm{zs}i}(s))_{i\in\mathcal{N}_{\mathrm{W1}}}\,\boldsymbol{L}_{\mathcal{N}_{\mathrm{W1}}}\right)^{-1}\\ &\operatorname{diag}(\boldsymbol{F}_{\mathrm{zs}i}(s))_{i\in\mathcal{N}_{\mathrm{W1}}}\,\boldsymbol{L}_{\mathcal{N}_{\mathrm{W1}},\{v-1\}}.\end{aligned} \tag{C.12}$$

The recursive construction rule of the transfer function $\boldsymbol{E}_1^{\mathcal{N}_{\mathrm{W1}}}(s)$ is given by

$$\boldsymbol{E}_1^{\mathcal{N}_{\mathrm{W1}}\cup\{v-1\}}(s) = \boldsymbol{L}_{v-2v-1}\,\boldsymbol{\Psi}_{\mathrm{E}}^{\mathcal{N}_{\mathrm{W1}}}(s)\,\boldsymbol{F}_{\mathrm{zs}v-1}(s)\,\boldsymbol{L}_{v-1v-2},$$

where

$$\boldsymbol{\Psi}_{\mathrm{E}}^{\mathcal{N}_{\mathrm{W1}}}(s) = \left(\mathbf{I} - \boldsymbol{F}_{\mathrm{zs}v-1}(s)\,\boldsymbol{E}_1^{\mathcal{N}_{\mathrm{W1}}}(s)\right)^{-1}$$

and $\boldsymbol{E}_1^{\emptyset}(s) = \mathbf{O}$.

C.2 Cooperatively controlled subsystem

This section outlines the construction rule of the transfer functions of the cooperatively controlled subsystem

$$F^{\mathcal{N}_{\rm S1}}: \begin{cases} \boldsymbol{y}^{\mathcal{N}_{\rm S1}}(s) = \boldsymbol{F}_{\rm yw}^{\mathcal{N}_{\rm S1}}(s)\,\boldsymbol{w}^{\mathcal{N}_{\rm S1}}(s) + \boldsymbol{F}_{\rm yq}^{\mathcal{N}_{\rm S1}}(s)\,\boldsymbol{q}_1(s) \\ \boldsymbol{p}_1(s) = \boldsymbol{F}_{\rm pw}^{\mathcal{N}_{\rm S1}}(s)\,\boldsymbol{w}^{\mathcal{N}_{\rm S1}}(s) + \boldsymbol{F}_{\rm pq}^{\mathcal{N}_{\rm S1}}(s)\,\boldsymbol{q}_1(s), \end{cases}$$

The transfer functions are as follows:

$$\begin{aligned}\boldsymbol{F}_{\rm yw}^{\mathcal{N}_{\rm S1}}(s) =& \operatorname{diag}(\boldsymbol{F}_{{\rm yw}i}(s))_{i\in\mathcal{N}_{\rm S1}} + \operatorname{diag}(\boldsymbol{F}_{{\rm ys}i}(s))_{i\in\mathcal{N}_{\rm S1}}\,\boldsymbol{L}_{\mathcal{N}_{\rm S1}} \\ & \boldsymbol{\Gamma}_i(s)\operatorname{diag}(\boldsymbol{F}_{{\rm zw}i}(s))_{i\in\mathcal{N}_{\rm S1}}\end{aligned} \tag{C.13}$$

$$\begin{aligned}\boldsymbol{F}_{\rm yq}^{\mathcal{N}_{\rm S1}}(s) =& \Big(\operatorname{diag}(\boldsymbol{F}_{{\rm ys}i}(s))_{i\in\mathcal{N}_{\rm S1}} + \operatorname{diag}(\boldsymbol{F}_{{\rm ys}i}(s))_{i\in\mathcal{N}_{\rm S1}}\,\boldsymbol{L}_{\mathcal{N}_{\rm S1}} \\ & \boldsymbol{\Gamma}_i(s)\operatorname{diag}(\boldsymbol{F}_{{\rm zs}i}(s))_{i\in\mathcal{N}_{\rm S1}}\Big)\,\boldsymbol{T}\end{aligned} \tag{C.14}$$

$$\begin{aligned}\boldsymbol{F}_{\rm pw}^{\mathcal{N}_{\rm S1}}(s) =& \boldsymbol{T}^{\top}\Big(\operatorname{diag}(\boldsymbol{F}_{{\rm zw}i}(s))_{i\in\mathcal{N}_{\rm S1}} + \operatorname{diag}(\boldsymbol{F}_{{\rm zs}i}(s))_{i\in\mathcal{N}_{\rm S1}}\,\boldsymbol{L}_{\mathcal{N}_{\rm S1}} \\ & \boldsymbol{\Gamma}_i(s)\operatorname{diag}(\boldsymbol{F}_{{\rm zw}i}(s))_{i\in\mathcal{N}_{\rm S1}}\Big)\end{aligned} \tag{C.15}$$

$$\boldsymbol{F}_{\rm pq}^{\mathcal{N}_{\rm S1}}(s) = \boldsymbol{T}^{\top}\boldsymbol{\Gamma}_i(s)\operatorname{diag}(\boldsymbol{F}_{{\rm zs}i}(s))_{i\in\mathcal{N}_{\rm S1}}\,\boldsymbol{T}, \tag{C.16}$$

with the transformation matrix $\boldsymbol{T}$ from (C.5) and

$$\boldsymbol{\Gamma}_i(s) = \left(\mathbf{I} - \operatorname{diag}(\boldsymbol{F}_{{\rm zs}i}(s))_{i\in\mathcal{N}_{\rm S1}}\,\boldsymbol{L}_{\mathcal{N}_{\rm S1}}\right)^{-1}.$$

For the special case $\mathcal{N}_{\rm S1} = \{1\}$, the transfer functions result to

$$\boldsymbol{F}_{\rm yw}^{\{1\}}(s) = \boldsymbol{F}_{\rm yw1}(s), \quad \boldsymbol{F}_{\rm yq}^{\{1\}}(s) = \boldsymbol{F}_{\rm ys1}(s), \quad \boldsymbol{F}_{\rm pw}^{\{1\}}(s) = \boldsymbol{F}_{\rm zw1}(s), \quad \boldsymbol{F}_{\rm pq}^{\{1\}}(s) = \boldsymbol{F}_{\rm zs1}(s).$$

Moreover, the recursive construction rule of the transfer function $\boldsymbol{F}_{\rm pq}^{\mathcal{N}_{\rm S1}}(s)$ is given by

$$\boldsymbol{F}_{\rm pq}^{\mathcal{N}_{\rm S1}\cup\{v\}}(s) = \boldsymbol{\Psi}^{\mathcal{N}_{\rm S1}}(s)\,\boldsymbol{F}_{{\rm zs}v}(s), \tag{C.17}$$

where

$$\boldsymbol{\Psi}^{\mathcal{N}_{\rm S1}}(s) = \left(\mathbf{I} - \boldsymbol{F}_{{\rm zs}v}(s)\,\boldsymbol{L}_{vv-1}\,\boldsymbol{F}_{\rm pq}^{\mathcal{N}_{\rm S1}}(s)\,\boldsymbol{L}_{v-1v}\right)^{-1}.$$

D List of symbols

General convention

- **Scalars** are denoted by lower-case italic letters (x, u)
- **Vectors** are denoted by lower-case bold italic letters ($\boldsymbol{x}$, $\boldsymbol{u}$)
- **Matrices** are denoted by upper-case bold italic letters ($\boldsymbol{A}$, $\boldsymbol{B}$)
- **Models and Systems** are denoted by upper-case italic letters (S, C)
- **Transfer functions / impulse response functions** of a model bear the same letter: Matrix-valued functions and vector-valued functions are denoted by upper-case bold italic letters ($\boldsymbol{S}(s)$, $\boldsymbol{C}(t)$), scalar-valued functions are denoted by upper-case italic letters ($S(s)$, $C(t)$) and subscripts denote the I/O relation ($\boldsymbol{S}_{\mathrm{yu}}(s)$, $\boldsymbol{C}_{\mathrm{uw}}(t)$).

Indices and superscripts

$\hat{(\cdot)}$	Approximate model, dynamic or signal
$\bar{(\cdot)}$	Comparison system, upper bound of a transfer function, an impulse response function or a signal
$\mathring{(\cdot)}$	Reconfigured model, signal, parameter
$\dot{(\cdot)}$	Time derivative of a signal
$(\cdot)^\top$	Transpose of a matrix or a vector / Inversion of a graph
$(\cdot)^{\mathrm{H}}$	Conjugate transpose of a matrix
$(\cdot)^{-1}$	Inverse of a matrix
$(\cdot)^{\mathcal{A}}$	Model that is composed of models from D_i, ($\forall i \in \mathcal{A}$)
$(\cdot)_\Delta$	Error model, transfer function, impulse response function or signal
$(\cdot)_0$	Initial value of a signal at time $t = 0$
$(\cdot)_{ij}$	Entry of the i-th row and j-th column of a matrix
$(\cdot)_f$	Quantity in a faulty interconnected system

$(\cdot)_{\mathcal{A}}$	Accumulated entries of the columns $i \in \mathcal{A}$ and rows $j \in \mathcal{A}$ of a matrix
$(\cdot)_{\mathcal{A},\mathcal{B}}$	Accumulated entries of the columns $i \in \mathcal{A}$ and rows $j \in \mathcal{B}$ of a matrix
$(\cdot)\|_{cond}$	Model, signal under the condition $cond$

Systems and Models

D_i	i-th design agent
S	Overall system
S_i	i-th subsystem
$S_i\|_{\boldsymbol{s}_i=\mathbf{0}}$	i-th isolated subsystem, where $\boldsymbol{s}_i = \mathbf{0}$
$S_i^{\mathcal{N}}$	i-th interacting subsystem
K_i	i-th local coupling model
C	Controller
C_i	i-th control station
F	Overall closed-loop system
F_i	i-th controlled subsystem
$F_i\|_{\boldsymbol{w}_i=\mathbf{0}}$	i-th unforced controlled subsystem, where $\boldsymbol{w}_i = \mathbf{0}$
$F_i\|_{\boldsymbol{s}_i=\mathbf{0}}$	i-th isolated controlled subsystem, where $\boldsymbol{s}_i = \mathbf{0}$
$F_1^{\mathcal{N}}$	Controlled interacting subsystem from the perspective of subsystem S_1
$F^{\mathcal{N}_{\mathrm{S1}}}$	Cooperatively controlled subsystem
$P_i^{\mathcal{N}}$	Physical interaction with subsystem S_i
$P_i^{\mathcal{N}_{\mathrm{S1}}}$	Relevant part of the physical interaction $P_i^{\mathcal{N}}$
$E_i^{\mathcal{N}_{\mathrm{W1}}}$	Irrelevant part of the physical interaction $P_i^{\mathcal{N}}$
G_1	Detection residual generator of subsystem S_1
G_{1f}	Isolation residual generator for the fault f of subsystem S_1
$R_1^{\mathcal{N}}$	Model of the generated residual $\boldsymbol{r}_1(t)$
VA	Virtual actuator
W	Model representing a weight function for the H_∞-design
SW	Extended system
$\mathfrak{P}$	Optimisation problem of the tube-based MPC

Scalars and scalar-valued functions

n	Dimension of the state vector
m	Dimension of the input vector
r	Dimension of the output vector
m_{s}	Dimension of the coupling input vector
m_{q}	Dimension of the error input vector
r_{z}	Dimension of the coupling output vector
r_{p}	Dimension of the error output vector
s	Complex frequency
ω	Frequency
t	Continuous time
t_{f}	Time instant of fault occurrence
t_{D}	Time instant of fault detection
t_{I}	Time instant of fault isolation
t_{S}	Time instant of external stimulation
t	Continuous time
T	Sampling time
k	Discrete time
$\varphi(\omega)$	Function for the admissible amplification of the coupling signals
ε_i	Parameter for the admissible influence of the error system
h	Prediction horizon of the tube-based MPC
$stat_i$	Connectivity status of S_i to S_1 (Chapter 4)
req_i	Index of the design agent that requests D_i (Chapter 4)
f	fault
κ	Communication counter
$\xi(\omega)$	Decision threshold for a subsystem to be strongly coupled to S_1
$\xi_i(\omega)$	Local decision threshold for a subsystem to be strongly coupled to S_1
$\eta(\omega)$	Decision threshold for a subsystem to be strongly coupled to S_1: Admissible model uncertainty for the design
$h_{\mathcal{A}}(\boldsymbol{a})$	Support function of a convex set $\mathcal{A}$

$\Delta_1^v(\omega)$	Upper bound on the improvement of the approximation of the physical interaction $P_1^{\mathcal{N}}$ by considering the controlled subsystem F_v
$\delta(t)$	Dirac delta function
$\sigma(t)$	Unit step function

Vectors and vector-valued functions

$\boldsymbol{x}$	State vector
$\boldsymbol{u}$	Input vector
$\boldsymbol{y}$	Output vector
$\boldsymbol{s}$	Coupling input vector
$\boldsymbol{z}$	Coupling output vector
$\boldsymbol{q}$	Error input vector
$\boldsymbol{p}$	Error output vector
$\boldsymbol{v}$	Performance output
$\boldsymbol{r}$	Residual
$\hat{\boldsymbol{r}}$	Approximated residual
$\boldsymbol{r}_\Delta$	Error residual ($\boldsymbol{r}_\Delta = \boldsymbol{r} - \hat{\boldsymbol{r}}$)
$\boldsymbol{x}_{\mathfrak{P}}$	Prediction of the state $\boldsymbol{x}(k)$ determined by the tube-based MPC (Chapter 5)
$\boldsymbol{u}_{\mathfrak{P}}$	Prediction of the input $\boldsymbol{u}(k)$ determined by the tube-based MPC (Chapter 5)
$\boldsymbol{a}$	Auxiliary vector for the support function, support set and support halfspace
$\boldsymbol{\mu}$	Detection threshold
$\boldsymbol{\mu}_{\mathrm{f}}$	Isolation threshold
$\mathbf{0}$	Vector filled with 0 elements
$\mathbf{1}$	Vector filled with 1 elements

Matrices

$\boldsymbol{A}$	System matrix
$\boldsymbol{B}$	Input matrix
$\boldsymbol{C}$	Output matrix
$\boldsymbol{C}_{\mathrm{z}}$	Coupling output matrix

$\boldsymbol{D}$	Feedthrough matrix
$\boldsymbol{E}$	Coupling input matrix
$\boldsymbol{K}$	Static feedback gain of a controller
$\boldsymbol{L}$	Physical interconnection matrix
$\boldsymbol{M}$	Static feedback gain of a virtual actuator
$\boldsymbol{N}$	Gain matrix of a residual generator
$\boldsymbol{Q}, \boldsymbol{R}$	Weight matrix
$\boldsymbol{T}$	Transformation matrix
$\boldsymbol{X}, \boldsymbol{Y}$	Variable of a linear matrix inequality
O	Zero matrix
I	Identity matrix

Sets

$\mathbb{N}, \mathbb{N}_{+}, \mathbb{N}_{+0}$	Set of natural numbers, set of positive natural numbers and set of positive natural numbers including 0
$\mathbb{R}, \mathbb{R}_{+}, \mathbb{R}_{+0}$	Set of real numbers, set of positive real numbers and set of positive real numbers including 0
$\mathbb{C}, \mathbb{C}_{+}$	Set of complex numbers and set of complex numbers with positive real part
$\mathbb{R}^n$	Set of real vectors with dimension n
$\mathbb{R}^{n\times m}$	Set of real matrices with n rows and m columns
$\mathcal{N}$	Set of indices of all subsystems
$\mathcal{N}_{\mathrm{S1}}$	Set of indices of relevant subsystems
$\mathcal{N}_{\mathrm{W1}}$	Set of indices of irrelevant subsystems
$\mathcal{P}_i$	Set of predecessors of the i-th subsystem
$\mathcal{M}_i$	Model set available to design agent D_i
$\mathcal{F}$	Set of considered faults
$\mathcal{H}_{\mathcal{A}}(\boldsymbol{a})$	Support halfspace of a convex set $\mathcal{A}$
$\mathcal{F}_{\mathcal{A}}(\boldsymbol{a})$	Support set of a convex set $\mathcal{A}$
$\mathcal{A}$	Global stability conditions of the overall closed-loop system
$\mathcal{A}_i$	Local stability conditions of the controlled subsystem F_i
$\mathcal{A}^{\mathcal{N}_{\mathrm{S1}}}$	Cooperative stability conditions of the cooperative controlled subsystem $F^{\mathcal{N}_{\mathrm{S1}}}$

$\mathcal{A}_{\mathrm{D}}$	Global performance conditions of the overall closed-loop system
$\mathcal{A}_{\mathrm{D}i}$	Local performance conditions of the controlled subsystem F_i
$\mathcal{A}_{\mathrm{D}}^{\mathcal{N}_{\mathrm{S1}}}$	Cooperative performance conditions of the cooperative controlled subsystem$F^{\mathcal{N}_{\mathrm{S1}}}$
$\mathcal{B}_i$	Set of investigated edges (Chapter 4)
$\mathcal{X}$	Operating region of the state vector $\boldsymbol{x}$
$\mathcal{S}$	Operating region of the coupling input signal $\boldsymbol{s}$
$\mathcal{U}$	Operating region of control input $\boldsymbol{u}$
$\mathcal{Z}$	Operating region of the interconnection output signal $\boldsymbol{z}$
$\mathcal{I}$	Robustly positively invariant set
$\mathcal{I}_{\mathfrak{P}1}$	Robustly positively invariant set for subsystem S_i controlled by the tube-based MPC
$\mathcal{I}_{\min}$	Minimal robustly positively invariant set
$\mathcal{Y}$	Summand of a convex set
$\mathcal{Y}_{\min}$	Minimal summand of a convex set
$\mathcal{X}_{\mathfrak{P}}$	Tightened constraints of the tube-based MPC
$\mathcal{U}_{\mathfrak{P}}$	Tightened constraints of the tube-based MPC
$\mathscr{U}_{\mathfrak{P}}^{*}$	Optimal control sequence
$\mathscr{X}_{\mathfrak{P}1}^{*}$	Optimal state sequence
$\mathscr{U}_{\mathfrak{P}}$	Admissible control sequence
$\mathscr{U}_{1}$	Set of all admissible control sequence $\mathscr{U}_{\mathfrak{P}}$
$\mathcal{F}_{\mathfrak{P}1}$	Feasible set
$\mathbb{T}_i^{\mathcal{N}_{\mathrm{S1}}}$	Tube / tolerance band of a signal
$\mathcal{G}_{\mathrm{K}}$	Interconnection graph of the overall system
$\mathcal{G}_{\mathrm{K}i}$	Local interconnection graph
$\mathcal{G}_{\mathrm{R}i}(\mathcal{G})$	Reachability graph with respect to the vertex i of a given graph $\mathcal{G}$
$\mathcal{G}_{\mathrm{D}}(\kappa)$	Communication graph for time instant κ
$\mathcal{G}_{\mathrm{D}}$	Cumulated communication graph
$\mathcal{E}_{\mathrm{K}}$, $\mathcal{E}_{\mathrm{K}i}$	Edge set of the interconnection graph / local interconnection graph
$\mathcal{E}_{\mathrm{R}i}$	Edges set of the interconnection graph
$\mathcal{E}_{\mathrm{D}}$	Edge set of the communication graph
$\mathcal{V}_{\mathrm{K}}$, $\mathcal{V}_{\mathrm{K}i}$	Vertex set of the interconnection graph / local interconnection graph

$\mathcal{V}_{\mathrm{R}i}$	Vertex set reachability graph
$\mathcal{V}_{\mathrm{D}}$	Vertex set of the communication graph

Abbreviation

DLCRP	Linear constrained regulation problem in the presence of disturbances
FTC	Fault-tolerant control
I/O	input-output
LC	Level control
LMI	Linear matrix inequality
LQ	Linear-quadratic
MPC	Model predictive control
PnP	Plug-and-play
P^3C	Plug-and-play process control
PI	Proportional-integral
RPI	robustly positively invariant
TC	Temperature control

Persönliche Daten

Name:	Sven Mathias Bodenburg
Geburtsdatum:	31. August 1986
Geburtsort:	Braunschweig

Berufs- und Bildungsweg

Seit 12.2017	Lehrkraft für besondere Aufgaben, Technische Hochschule Georg Agricola, Bochum.
11.2011 - 05.2017	Wissenschaftlicher Mitarbeiter am Lehrstuhl für Automatisierungstechnik und Prozessinformatik (Prof. Dr.-Ing Jan Lunze), Ruhr-Universität Bochum.
10.2010 - 11.2012	Masterstudium der Elektrotechnik an der Ruhr-Universität Bochum mit Schwerpunkt Regelungstechnik, Titel der Masterarbeit: *Implementation of plug-and-play control in networked control systems.*
09.2006 - 08.2010	Diplomstudium der Elektrotechnik an der Ostfalia, Hochschule für angewandte Wissenschaften mit Schwerpunkt Automatisierungstechnik. Titel der Diplomarbeit: *Optimierung der Struktur eines Fahrprofilreglers zum automatisierten Nachfahren von Geschwindigkeitsverläufen und Integration eines Anfahrreglers für Handschaltgetriebe.*
02.2010 - 08.2010	Praktikum und Diplomarbeit bei IAV GmbH, Gifhorn.
08.2008 - 01.2009	Praktikum und Studienarbeit bei Vehico GmbH, Braunschweig.
09.2003 - 07.2006	Berufsausbildung zum Elektroniker für Automatisierungstechnik bei Volkswagen AG, Braunschweig.
2004 - 2006	Berufsbildende Schule 2 zum Erlangen der Fachhochschulreife in Abendschulform, Braunschweig.
1999 - 2003	Realschule Maschstraße, Braunschweig.